# Interacting Binaries

# NATO ASI Series

## Advanced Science Institutes Series

*A series presenting the results of activities sponsored by the NATO Science Committee, which aims at the dissemination of advanced scientific and technological knowledge, with a view to strengthening links between scientific communities.*

The series is published by an international board of publishers in conjunction with the NATO Scientific Affairs Division

| | | |
|---|---|---|
| A | Life Sciences | Plenum Publishing Corporation |
| B | Physics | London and New York |
| | | |
| C | Mathematical and Physical Sciences | D. Reidel Publishing Company Dordrecht, Boston and Lancaster |
| | | |
| D | Behavioural and Social Sciences | Martinus Nijhoff Publishers |
| E | Engineering and Materials Sciences | The Hague, Boston and Lancaster |
| | | |
| F | Computer and Systems Sciences | Springer-Verlag |
| G | Ecological Sciences | Berlin, Heidelberg, New York and Tokyo |

# Interacting Binaries

edited by

## P. P. Eggleton

and

## J. E. Pringle

Institute of Astronomy,
University of Cambridge,
Cambridge, U.K.

## D. Reidel Publishing Company

Dordrecht / Boston / Lancaster

Published in cooperation with NATO Scientific Affairs Division

Proceedings of the NATO Advanced Study Institute on
Interacting Binaries
Cambridge, U.K.
31 July-13 August, 1983

Library of Congress Cataloging in Publication Data

NATO Advanced Study Institute on Interacting Binaries (1983 : Cambridge,
    Cambridgeshire)
    Interaction binaries.

    (NATO ASI series, Series C, Mathematical and physical sciences; vol. 150)
    Includes index.
    1.  Stars, Double—Congresses.  I.  Eggleton, P.P. (Peter P.), 1940–
II.  Pringle, J. E. (James Edward), 1949–    .  III.  Title.  IV.  Series.
QB821.N38    1983            523.8′41            85-1897
 ISBN-13: 978-94-010-8856-5        e-ISBN-13: 978-94-009-5337-6
 DOI: 10.1007/978-94-009-5337-6

---

Published by D. Reidel Publishing Company
P.O. Box 17, 3300 AA  Dordrecht, Holland

Sold and distributed in the U.S.A. and Canada
by Kluwer Academic Publishers,
190 Old Derby Street, Hingham, MA 02043, U.S.A.

In all other countries, sold and distributed
by Kluwer Academic Publishers Group,
P.O. Box 322, 3300 AH  Dordrecht, Holland

D. Reidel Publishing Company is a member of the Kluwer Academic Publishers Group

---

# TABLE OF CONTENTS

# PREFACE

Peter P. Eggleton and James E. Pringle

Institute of Astronomy
Madingley Road
Cambridge
England

The 1970's can be described, in retrospect, as the "Decade of the Close Binary". Exciting observations with new technology, combined with classical work, both observational and theoretical, convinced the astronomical world that binary interaction of various kinds is not only interesting but common. Indeed, by 1975 almost anything unusual had a good chance of being interpreted as due to binary interaction. But astronomers are seldom overwhelmed by speculation, even their own, and solid observational work has confirmed or refuted such speculation, without regard to its plausibility. For instance, binarity has been found where it was perhaps least expected, in Barium stars, and refuted where it could most reasonably be expected, in Wolf-Rayets. Unfortunately, many other classes of potential binaries remain without the clearest evidence of binarity, for instance Be stars, symbiotics and blue stragglers.

This Advanced Study Institute was held to commemorate John Whelan (1945-1981), whose scientific career, sadly cut short in its prime, did much to further the careful study, theoretical and observational, of close binaries, as well as to encourage the spirit of international friendship and collaboration. His own interests covered a greater field, but "Interacting Binaries" seemed a reasonable restriction. We publish here 15 review talks, which still do not cover the whole topic, although they range widely. Even the luxury of two weeks, generously supported by the NATO Scientific Affairs Division, could not give full coverage; each topic can have (and has had) whole conferences devoted to it. In an age where more and more is known about less and less, we have attempted temporarily to reverse the trend.

# THE DISTRIBUTIONS OF PERIODS AND AMPLITUDES
# OF LATE-TYPE SPECTROSCOPIC BINARIES

R.F. Griffin

The Institute of Astronomy
Madingley Road
Cambridge    CB3 0HA
U.K.

ABSTRACT

Photoelectric radial-velocity measurements, made on an unhurried observing schedule, have permitted the discovery of groups of spectroscopic binaries in which the usual observational discrimination against long periods and small amplitudes is much reduced. The distributions of periods and amplitudes in these groups is quite different from those of binaries with published orbits. In particular, the $N - \log P$ distribution appears to be a monotonically rising function for at least the range $3 < P < 3000$ days.

A long time ago the organisers of this conference asked me if I would allow my name to appear on the conference literature. They put it in such a way that I felt obliged to agree, and <u>then</u> they told me that that would involve my giving a talk. When I objected that the binaries in which I am mostly interested were quite well separated and by no means interacting, they countered that <u>all</u> binary stars are interacting - there wouldn't even be any binary stars if it were not for <u>gravitational</u> interaction! But that still doesn't explain why I am down to speak first!

My interest in late-type binary stars dates from the late 1960s, when we acquired at the Cambridge Observatories here an instrument that seemed capable of providing a useful observational input to the study of binaries. That instrument was the first photoelectric radial-velocity spectrometer, which still lives in the dome over there and is in use almost every clear night. I have also been privileged to use the 200-inch

1

*P. P. Eggleton and J. E. Pringle (eds.), Interacting Binaries, 1–12.*
© *1985 by D. Reidel Publishing Company.*

telescope with a much more elegant radial-velocity spectrometer constructed by J.E. Gunn and myself; in addition, I have collaborated with M. Mayor in observations with his CORAVEL at Haute Provence, and have used the Victoria radial-velocity instrument on a guest-investigator basis. However, despite the attractions of overseas observing facilities, most of my observing is still done here at Cambridge where I am virtually the sole user of the 36-inch telescope and can observe whenever I like - weather permitting. It is therefore much easier for me to follow carefully the variations of a limited number of binary stars than it would be if I depended entirely on occasional observing runs overseas.

Last month The Observatory Magazine published my 50th paper giving binary-star orbits. Cynics point out that it may not be a coincidence that I am an Editor of The Observatory; but my own conscience is cleared by the fact that, while I have been publishing 50 orbits in my own journal, I have also published about 50 others elsewhere and no journal has yet refused one. Anyway, in looking over the Observatory orbits I couldn't help noticing that most of them have much longer periods and smaller amplitudes than most orbits published by other people. In an effort to quantify this impression I sorted the orbits of the late-type entries in the Seventh Catalogue of the Orbital Elements of Spectroscopic Binary Systems (Batten, Fletcher & Mann 1978) to form histograms in period and amplitude, using rather coarse steps of half a logarithmic unit. I used only the orbits of quality a, b and c , not trusting some of the others. The resulting histograms of P and K for the 204 Catalogue orbits satisfying the criteria for selection are shown in Fig. 1. These distributions are repeated on all the rest of my diagrams, to serve as the existing basis for comparison with the new results.

Fig. 2 shows the comparison with the 50 Observatory orbits (Griffin 1983). It will be seen that the period distributions of the Catalogue and Observatory orbits are heavily skewed in opposite directions: the Catalogue orbits have a maximum frequency in the 3-10 day bin, whereas mine show a maximum at 300-1000 days. The amplitude distributions, too, are skewed in opposite directions, but not as much - Kepler's laws show that K is proportional to the inverse cube root of P. Most of my orbits have K between 3 and 10 km/s, whereas the Catalogue orbits show a peak at 30-100 km/s.

The periods and amplitudes of all of the orbits which I have either published or am now in the course of writing up, 108 in all, are shown in Fig. 3. At the request of the Editors, I am providing in Table 1 a list of the 46 already-published orbits, other than the first 50 Observatory orbits, of which I am an author, or the sole author, with references. This Table is complementary to that giving the Observatory orbits 1-50 (Griffin 1983). The periods still show a maximum frequency in the 300-1000-day bin, but the 1000-3000-day bin is nearly as well populated. The 3-10-km/s amplitude bin is the highest, though 10-30 is not far behind.

The reasons for the differences between my orbits and the Catalogue ones are not far to seek. Most of the binaries for which I have published orbits are my own discoveries, made in the course of large survey programmes in which a second observation of a given star is often not made until years after the first, and what used to be regarded as a rather small velocity discrepancy (more than about 3 km/s) is carefully followed up. Observational selection is therefore far less fierce in my observations than in those of most previous authors. Of course, if one starts with some considerable number of binary stars and observes them all assiduously, it is the short-period ones whose orbits are determined first. A history of my own orbit publications clearly shows a tendency for orbital periods to increase as the total span of the observations lengthens, so a considerable degree of observational selection certainly remains.

In an effort to reduce as far as possible the bias against long periods, I next looked at the binaries which are in course of observation and whose orbits are sufficiently well known to be sorted into the correct bins even though they may not yet be well enough covered for publication. I found 103 of them, making a total of 211 of my orbits available for this discussion. Now in terms of number, I can more than hold my own with the Catalogue! The results appear in Fig. 4. Notice particularly the relatively enormous increase in the number of periods in the 1000-3000-day bin, bringing the maximum frequency into that range, and also the substantial number in the 3000-10000-day one. Notice also that the absolute majority of amplitudes is in the 3-10-km/s range. Clearly, longevity (or rather the lack of it) still discriminates against the longest periods; and measuring error militates against the discovery of binaries with amplitudes much less than 3 km/s. For periods of many thousands of days, too, the survey procedures that I have used become less likely to produce an obvious discrepancy between the first two observations, simply because the interval between them is not a substantial fraction of the period.

The question next arises as to whether there is any way of avoiding observational selection in any greater degree than has been done so far. Well, there are two groups of stars which

have been observed repeatedly, whether they have seemed to vary
in velocity or not, for upwards of ten years. In these groups,
discovery of velocity variations significantly greater than the
measuring error should be substantially complete up to periods
well into the 3000-10000-day range.

The first of these groups is the "Redman K stars", a set
of 80-odd seventh-magnitude stars first measured for radial
velocity by Redman at Victoria more than 50 years ago and now
reobserved photoelectrically. Observations in two seasons,
1966 and 1969, indicated that several of the objects were
spectroscopic binaries: three were pretty certain binaries,
two more were probable ones, two were "maybe" and two were
"possible" (Griffin 1970). These assessments were actually
rather conservative, as follow-up observations showed that all
nine of the objects do indeed vary in velocity. The whole set
of stars has been measured again in several subsequent seasons
and ten more binaries have been identified in it. The period
and amplitude distributions are shown in Fig. 5, where you will
see that they are even more extreme than those shown in the
previous diagrams - no doubt largely owing to a further
reduction in observational selection.

The second special group is a selected set of Hyades stars
which J.E. Gunn and I have observed in nearly every season
since 1972 with the Palomar spectrometer. Eleven of the stars
have proved to be binaries: that is hardly enough for
meaningful histograms, but I may say that there is one star
with a period in the range of 1-10 days, two of 10-100 days,
and four each of 100-1000 and 1000-10000 days. In addition,
the existence of one or two stars whose velocities so far seem
stable but deviate by one km/s or so from the expected
velocities for Hyades members might well represent the first
indication of binaries with still longer periods. However, it
is very likely that in a group of main-sequence binaries there
will be some with nearly equal components; unless their
velocity amplitudes are large enough for them to appear
double-lined at some places, they may easily be mistaken for
single-lined systems and their velocity amplitudes will be
greatly under-estimated. Indeed, their binary nature may be
discoverable only from periodic slight line-broadening, without
any detectable velocity variation at all. Improved
data-reduction programmes, written by R.E.M. Griffin, now
permit us to resolve closely blended pairs and to detect
"line-width binaries".

Because Gunn and I have made a pretty systematic study of
the Hyades, although of course we have not been able to observe
all the candidate stars as regularly as the 35 selected ones,
it seems useful to look at all the Hyades orbits so far

available. They are also of interest as referring mainly to main-sequence stars whereas most of my other orbits are of giants. There are 10 Hyades orbits in the literature, not counting our own. We have published 9 more and are writing up another 11 (see also Griffin et al. 1984). Among the many other Hyades binaries under observation, a further 16 can be assigned approximate periods and amplitudes. The distributions of these quantities are shown for all the stars in Fig. 6. Notice that the characters of the distributions are remarkably similar to the others I have shown, notwithstanding that the stars plotted here are quite differently distributed in luminosity. The one feature which is different here is the existence of several binaries with short periods (1-3 days) which are of course impossible for giant systems. It is also noteworthy that there are as many objects with periods of 3000-10000 days as with 1000-3000.

A quick summary of the conclusion to be drawn from all this is that when observational selection, which has caused overwhelming bias in the characteristics of binary orbits in the past, is reduced to a more modest level, the frequency of binaries as a function of log P is a monotonically rising function at least between 3 and 3000 days. This conclusion is in amazing contrast with the distribution of Catalogue orbits, which show a monotonically falling function for periods above 10 days. It seems quite likely that further reduction in observational bias - to be achieved by making more accurate velocity measurements extending over a longer time - would show my conclusion to be valid up to still longer periods.

Table 1

Elements of spectroscopic binary orbits

| Name | HR | HD | P days | γ km/s | K km/s | e | ω₀ | T or T₀ MJD | a sin i Gm | f(m) M☉ | Ref. |
|---|---|---|---|---|---|---|---|---|---|---|---|
| | | 7308 | 659.9 | -29.15 | 19.48 | 0.497 | 154.1 | 43375.4 | 153 | 0.331 | 1 |
| | | 9828 | 183.01 | -13.28 | 15.87 | 0.189 | 17 | 42858.4 | 39.2 | 0.072 | 2 |
| ξ Psc | 549 | 11559 | 1672.4 | +29.51 | 4.64 | 0.18 | 71 | 37651 | 105 | 0.0165 | 3 |
| 60 And | 643 | 13520 | 748.2 | -48.00 | 4.88 | 0.34 | 358 | 37886 | 47.2 | 0.0075 | 3 |
| ξ¹ Cet | 649 | 13611 | 1642.1 | -3.93 | 5.91 | 0 | - | 34985 | 134 | 0.035 | 3 |
| | | 13725 | 693.5 | -9.39 | 3.38 | 0.23 | 86 | 43928 | 31.4 | 0.0026 | 4 |
| | | 14969 | 1935 | -32.68 | 4.43 | 0.39 | 293 | 41839 | 109 | 0.014 | 5 |
| 23° 635 | | 284163 | 2.394357 | +36.76 | 66.25 | 0.057 | 279 | 43891.86 | 2.178 | 0.0719 | 6 |
| vB 162 | | 26874 | 55.130 | +38.42 | 27.76 / 30.70 | 0.393 | 71.4 / 251.4 | 43098.87 | 19.35 / 21.40 | 0.0952 / 0.129 | 6 |
| 22° 669 | | 284303 | 1.887259 | +39.31 | 95.3 / 103.4 | 0 | - | 43775.2599 | 2.472 / 2.68 | 0.1694 / 0.217 | 7 |
| δ Tau | 1373 | 27697 | 529.8 | +39.5 | 3.0 | 0.42 | 335 | 34356 | 19.7 | 0.0011 | 8 |
| vB 62 | | 28033 | 8.55089 | +38.77 | 16.46 | 0.233 | 38 | 42588.22 | 1.88 | 0.0036 | 9 |
| vB 182 | | 28545 | 358.4 | +40.75 | 13.80 | 0.364 | 139 | 43173.5 | 63.4 | 0.079 | 6 |
| vA 771 | | 286898 | 1.868017 | +39.57 | 78.8 / 111.8 | 0 | - | 43757.3508 | 2.025 / 2.87 | 0.0950 / 0.271 | 7 |
| J 318 | | 285970 | 56.449 | +14.42 | 27.14 | 0.352 | 194.4 | 43305.1 | 19.7 | 0.096 | 6 |
| vB 117 | | | 11.9293 | +41.14 | 60.9 / 64.1 | 0.516 | 90 / 270 | 42881.99 | 8.58 / 8.97 | 0.177 / 0.202 | 9 |
| vB 121 | | 30738 | 5.75096 | +42.74 | 19.70 | 0.354 | 55 | 42192.06 | 1.46 | 0.0037 | 9 |
| vB 166 | | 31181 | 7.8638 | +18.34 | 37.9 / 51.2 | 0 | - | 43749.950 | 4.10 / 5.54 | 0.0446 / 0.110 | 7 |
| | 2081 | 40084 | 219.13 | -4.09 | 32.24 / 33.24 | 0 | - | 43476.43 | 97.1 / 100.1 | 0.762 / 0.835 | 10 |
| | | 57339 | 2648 | +0.80 | 8.90 | 0.287 | 332 | 42409 | 310 | 0.170 | 11 |
| | | 60799 | 591.1 | +19.09 | 7.49 | 0 | - | 42862 | 60.9 | 0.026 | 2 |
| 81 Gem | 3003 | 62721 | 1519.7 | +83.13 | 7.21 | 0.325 | 73 | 41584 | 142 | 0.050 | 12 |
| χ Gem | 3149 | 66216 | 2438 | -8.01 | 5.19 | 0.060 | 264 | 42894 | 174 | 0.035 | 12 |
| 81 Cnc | 3650 | 79096 | 984 | +50.2 | 11.7 / 12.7 | 0.43 | 172 / 352 | 44235 | 142 / 155 | 0.119 / 0.154 | 13 |
| | 4006 | 88639 | 179.92 | +7.64 | 14.05 | 0.169 | 283 | 44199 | 32.9 | 0.048 | 14 |
| | 4474 | 101013 | 1707 | -13.71 | 6.08 | 0.196 | 301 | 42240 | 140 | 0.037 | 15 |
| | | 105341 | 194.15 | -28.54 | 13.39 | 0.027 | 263 | 43750 | 35.7 | 0.048 | 16 |
| 33° 2206 | | 1063658 | 100.260 | -9.94 | 20.26 | 0.454 | 257.7 | 44489.7 | 24.9 | 0.061 | 17 |
| | 4665 | 106677 | 64.44 | -45.29 | 36.1 / 36.8 | 0 | - | 43445.45 | 3.20 / 3.27 | 0.315 / 0.334 | 18 |
| | | 107742 | 875 | -12.69 | 6.24 | 0.099 | 293 | 44777 | 74.8 | 0.0218 | 19 |
| 26 Com | 4815 | 110024 | 972.4 | -20.26 | 10.46 | 0.590 | 102.5 | 43303.5 | 112.9 | 0.061 | 20 |
| | | 115968 | 16.1952 | -3.30 | 19.58 | 0.274 | 75.4 | 43760.70 | 4.19 | 0.0112 | 21 |
| | | 116378 | 17.7641 | -42.7 | 14.7 | 0.122 | 243 | 44978.6 | 3.57 | 0.0057 | 22 |
| | | 117064 | 2222 | -7.81 | 5.70 | 0.50 | 231 | 42495 | 151 | 0.028 | 23 |
| | 5161 | 119458 | 149.72 | -7.7 | 20.8 | 0.17 | 160 | 43785.5 | 42.2 | 0.133 | 24 |
| | | 120803 | 699.3 | -51.21 | 14.93 | 0.402 | 270.9 | 43381 | 131.4 | 0.185 | 25 |
| | | 121844 | 302.67 | -61.35 | 8.61 | 0.488 | 4.4 | 43765.0 | 31.3 | 0.0133 | 26 |
| | | 148405 | 52.453 | -33.72 | 29.48 | 0.021 | 210 | 42948.56 | 21.26 | 0.139 | 2 |
| | | 149240 | 1039.5 | -46.05 | 15.95 | 0.349 | 19.7 | 43548 | 214 | 0.360 | 2 |
| | 6626 | 161832 | 99.557 | -26.72 | 16.45 | 0 | - | 43298.76 | 22.52 | 0.0460 | 27 |
| | | 172865 | 33528 | +16.5 | 9.53 / 9.51 | 0.911 | 169.8 / 349.8 | 44546 | 1812 / 1808 | 0.211 / 0.210 | 28 |
| ν Dra | 7180 | 176524 | 258.48 | -11.11 | 5.99 | 0.21 | 298 | 41977 | 20.8 | 0.0054 | 29 |
| 29° 3805 | | | 295.36 | +12.59 | 21.68 | 0.116 | 149 | 42770 | 87.5 | 0.306 | 30 |
| | | 188753 | 155.1 | -20.56 | (2.8) | 0.26 | 58 | 42087 | - | - | 31 |
| 30 Vul | 7939 | 197752 | 2506 | +29.98 | 4.69 | 0.38 | 272 | 42511 | 149 | 0.0212 | 32 |
| | | 225292 | 954.0 | +11.54 | 7.27 | 0.37 | 233 | 43532 | 89 | 0.031 | 33 |

References to Table 1

Note: where no name is given, the author is R.F.Griffin

1.  J.R. Astr.Soc. Canada, **74**, 348, 1980.
2.  Mon.Not.R.Astr.Soc., **201**, 487, 1982.
3.  R.F. Griffin & G.H. Herbig, Mon.Not.R.Astr.Soc.,
    **196**, 33, 1981.
4.  Observatory, **103**, 284, 1983.
5.  Mon.Not.R.Astr.Soc., **190**, 711, 1980.
6.  R.F. Griffin, & J.E. Gunn, Astr.J., **86**, 588, 1981.
7.  R.F. Griffin, M.Mayor & J.E. Gunn, Astr. Astrophys.,
    **106**, 221, 1982.
8.  R.F. Griffin & J.E. Gunn, Astr.J., **82**, 176, 1977.
9.  R.F. Griffin & J.E. Gunn, Astr.J., **83**, 1114, 1978.
10. W.I. Beavers & R.F. Griffin, Publ.Astr.Soc.Pacific,
    **91**, 824, 1979.
11. Observatory, **103**, 252, 1983.
12. Mon.Not.R.Astr.Soc., **200**, 1161, 1982.
13. R. & R. Griffin, Observatory, **102**, 217, 1982.
14. R.F. Griffin & W.I. Beavers, Publ.Astr.Soc.Pacific,
    **94**, 557, 1982.
15. R. & R. Griffin, Mon.Not.R.Astr.Soc., **193**, 957, 1981.
16. J.Astrophys.Astr., **3**, 1, 1982.
17. J.Astrophys.Astr., **3**, 383, 1982.
18. B.W. Bopp, F. Fekel, Jr., R.F. Griffin, W.I. Beavers,
    J.E. Gunn & D. Edwards, Astr.J., **84**, 1763, 1979.
19. J.Astrophys.Astr., **4**, 19, 1983.
20. J.Astrophys.Astr., **2**, 115, 1981.
21. J.Astrophys.Astr., **2**, 309, 1981.
22. J.Astrophys.Astr., **4**, in press, 1983.
23. J.Astrophys.Astr., **3**, 107, 1982.
24. W.I. Beavers & R.F. Griffin, Publ.Astr.Soc.Pacific,
    **91**, 521, 1979.
25. J.Astrophys.Astr., **3**, 101, 1982.
26. J.Astrophys.Astr., **4**, 23, 1983.
27. M. Mayor & R.F. Griffin, Astr.Astrophys., **91**, 112, 1980.
28. A.H. Batten, J.M. Fletcher, W.A. Fisher, R.D. McClure,
    C.L. Morbey, R.F. Griffin & C.D. Scarfe, Publ.Astr.Soc.
    Pacific, **94**, 860, 1982.
29. R.F. Griffin, H.C. Harris & R.D. McClure, J.R.Astr.Soc.
    Canada, **77**, 73, 1983.
30. J.R.Astr.Soc.Canada, **73**, 266, 1979.
31. Observatory, **97**, 15, 1977; E.E. Bassett, Observatory,
    **98**, 122, 1978.
32. Observatory, **103**, 199, 1983.
33. J.R.Astr.Soc.Canada, **75**, 222, 1981.

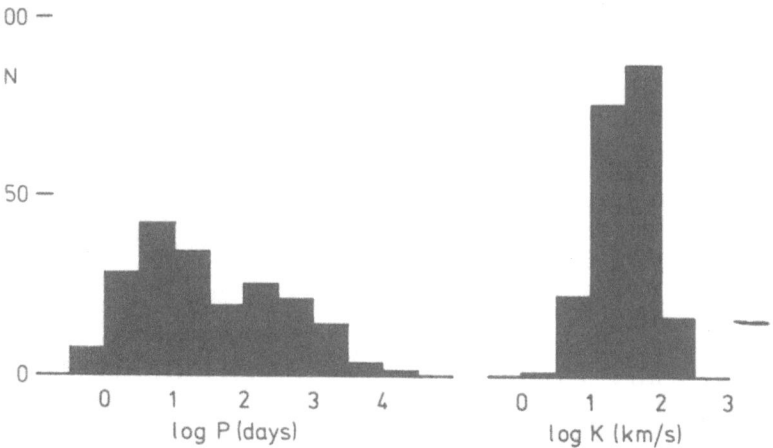

Figure 1. Distributions of periods P and radial-velocity semi-
amplitudes K for the 204 catalogued spectroscopic orbits (Batten,
Fletcher & Mann 1978) satisfying the following selection criteria:
quality a, b, or c ; spectral type of at least one component F5 or
later; and not the work of the present author. In cases where K
is given for both components of the binary the larger value is
chosen, or if only one component is of late type the value for
that component is adopted.

The histograms in this Figure are taken to represent the state
of observational knowledge up to the present time, and are repeated
in outline in Figs. 2 - 6 for comparison with the distributions
plotted there from the author's new results.

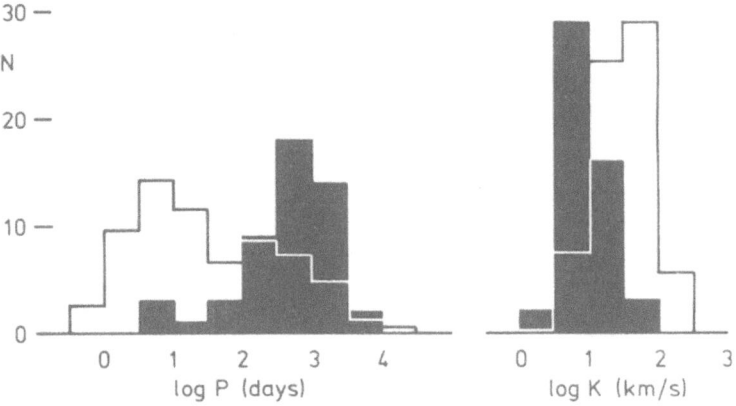

Figure 2.  Distributions of periods and amplitudes of the spectroscopic binaries treated in the first 50 papers in the author's
series in The Observatory Magazine.  The distributions of the
Catalogue orbits are repeated, in outline, from Fig. 1 (the scale
of ordinates does not apply to them, only to the Observatory
orbits).  Notice particularly that the Observatory and Catalogue
distributions are skewed in opposite senses, and that the maximum
frequencies occur at periods differing by a factor as large as 100
and at amplitudes differing by a factor of 10.

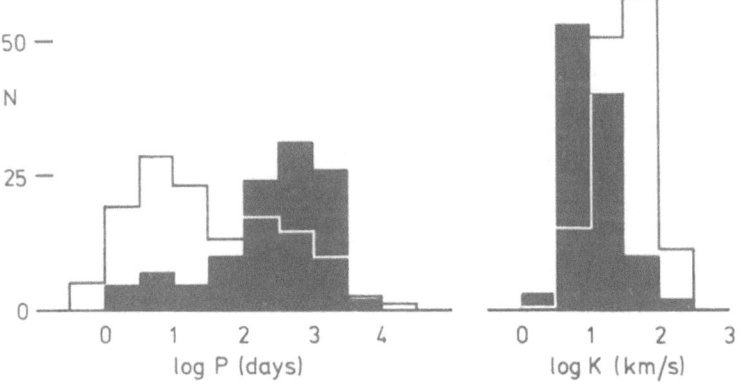

Figure 3.  Distributions of periods and amplitudes of all the 108
spectroscopic binaries with whose publication the author has been,
or is being, associated.  The Catalogue distributions are again
shown in outline.

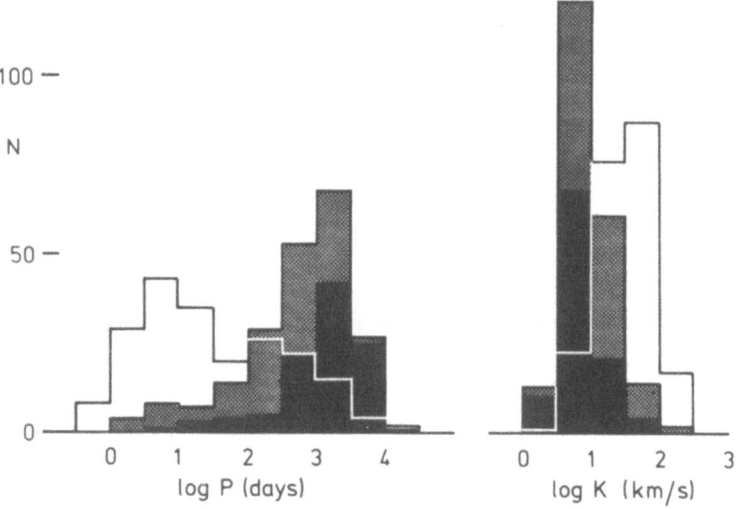

Figure 4. Distributions of periods and amplitudes for 103 unpublished orbits (solid black) as well as the 108 published ones
shown in Fig. 3 (shaded). Because the author's interest in spectroscopic binaries began seriously only several thousand days ago,
orbits with P > $10^{3.5}$ days are almost exclusively in the 'unpublished' group, being as yet inadequately covered by observations.
The steady increase of frequency with period, which is already
shown in Figs. 2 and 3 to extend as far as the 300-1000-day 'bin',
is here seen to continue to the 1000-3000 day bin; it might very
well continue further still if observational selection, in the
forms of the Observer's inappropriate discovery procedures and
lack of interest long ago, and the smallness of amplitudes at long
periods, did not discrimate against the documentation of systems
with P > 3000 days. The outlined distributions are again those of
the Catalogue orbits; the ordinates apply equally to them and to
the author's orbits in this Figure.

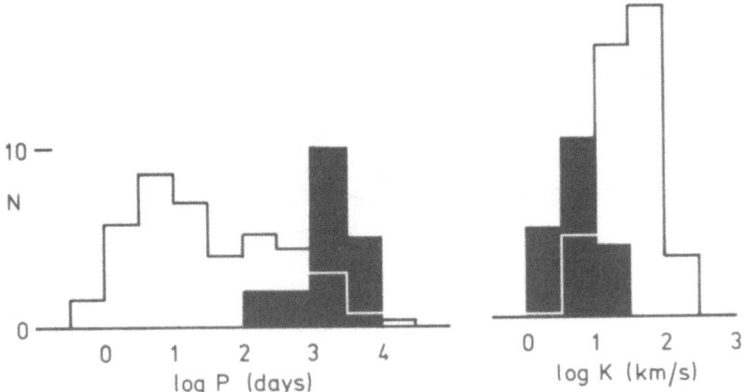

Figure 5.  Distributions of periods and amplitudes for 19 binaries among the 'Redman K stars' (see text).  This sample, which has been subject to smaller bias through observational selection insofar as the discovery of binaries is concerned, shows distributions differing even more strongly from the <u>Catalogue</u> ones than those in Figs. 2 - 4.

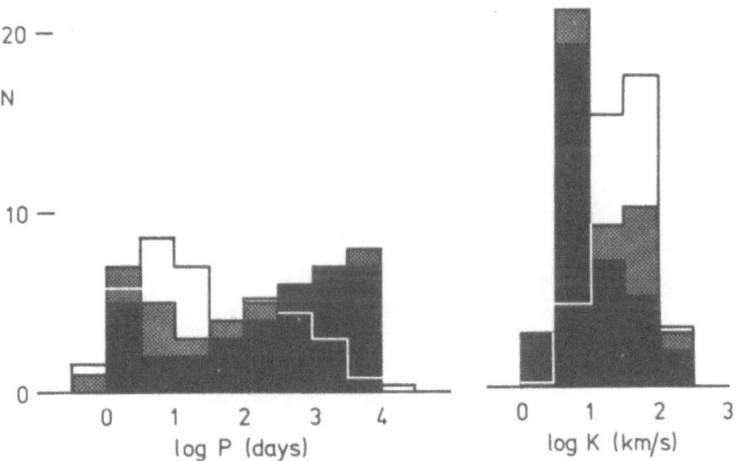

Figure 6.  Distributions of periods and amplitudes of Hyades binaries whose orbits have been obtained by the author and his collaborators (solid black) and by others (shaded), compared again with the <u>Catalogue</u> distributions (outlined).  The Hyades distributions are similar in form to those in Figs. 4 and 5 except for the existence of several orbits with short periods - periods which are necessarily restricted to dwarf stars.

REFERENCES

Batten,   A.H.,   Fletcher,   J.M.   &   Mann,   P.J.,   1978.
    Publ.Dom.Astrophys.Obs., Victoria, **15**, 121.
Griffin, R.F., 1970.  Mon.Not.Roy.Astr.Soc., **148**, 211.
Griffin, R.F., 1983.  Observatory, **103**, 273.
Griffin,  R.F.,  Gunn, J.E., Zimmerman, B.A. & Griffin, R.E.M.,
    1984, in preparation for submission to Astr.J.

# ACTIVITY OF CONTACT BINARY SYSTEMS

S.M. Rucinski

Institute of Astronomy
Cambridge
England

## 1.   INTRODUCTION

This review will consider various indicators of activity in contact bianries. We shall limit ourselves to late-type contact systems which are traditionally called the "W UMa-type" systems. This limitation results mainly from a better knowledge of properties of cooler systems (i.e. with periods below about 0.5 - 0.6 days) but is also dictated by an expectation that only late-type stars can generate any magnetic activity.

The activity of W UMa-type binaries should be studied for at least the following reasons:

a) With orbital periods as short as quarter of a day, these stars are among the most rapidly rotating late-type dwarfs.

b) Their studies might shed some light on activity of very young stars which also rotate very rapidly (e.g. K dwarfs in Pleiades : van Leeuwen and Alphenaar, 1982), but which might still accrete matter or be still contracting.

c) Their activity might be rather closely related to their evolution via magnetic wind braking (van't Veer 1979; Mochnacki, 1981; Vilhu 1981, 1982).

We shall not distinguish among these reasons, and so we describe various indications of activity using an easier, more conventional division which is related mainly to the observational techniques used. We will therefore discuss:

13

*P. P. Eggleton and J. E. Pringle (eds.), Interacting Binaries, 13–49.*
© *1985 by D. Reidel Publishing Company.*

in Sec. 2,     the W-type phenomenon and the possibly related
               photospheric activity;

in Sec. 3,     the chromospheric activity and rather fragmentary
               data on it;

in Sec. 4,     the much more systematically studied transition
               region activity;

in Sec. 5      the coronal activity observed primarily in X-rays,
               but also in lines of helium;

in Sec. 6,     the flare activity;

in Sec. 7,     the period changes;

and   in Sec 8,     a few other phenomena not described   in   other
               sections.

Very brief conclusions will be given in Sec. 9.

Of the phenomena discussed,  the  W-type syndrome (Sec. 2)
and the period changes (Sec. 6) may very well have non-magnetic
explanations.  The  rest  are  too  similar  in  character  to
activity observed in normal late type  stars  to have any other
source than magnetic fields.

It  should  be  noted  that  there  have recently appeared
reviews of activity in contact binaries by Dupree (1981, 1983).
However, they have  been  been  limited  primarily to phenomena
observable with the IUE and Einstein satellites.  The  present
review  has  a somewhat broader aim in mind and will attempt to
include all aspects  of contact binary physics which might have
something  to do with  the  stellar  activity.  The  field  is
unfortunately  not  yet  ripe  for  integration  into  a  fully
coherent  picture  so  that  only essential conclusions will be
given in Sec. 9.

An attempt  will  be made to stress weaknesses rather than
strong points in our  present  understanding,  in  the  hope of
presenting possibilities for further research of these exciting
objects.


2.   THE W-TYPE PHENOMENON AND PHOTOSPHERIC ACTIVITY

Theoretical light curves  generated using general precepts
of Lucy (1968) depend essentially  only on four parameters : q,
F, $\beta$ , i.  We comment on them briefly.

The mass ratio q is determinable by light curve fitting techniques only for totally eclipsing systems but one can obviously use a spectroscopic value, when available (this is probably the best determinable spectroscopic element). In other cases, the results for q cannot be trusted. The observed mass ratios seem to peak at about q = 0.3 - 0.4 but the real frequency distribution is difficult to estimate (van't Veer 1978). Contact binaries avoid the q = 1 situation which is frequently encountered in detached systems.

The degree of contact F and the gravity darkening exponent (locally, $T_{eff} \propto g^{\beta}$) are interdependent. Using the value suggested by Lucy (1967), $\beta_{conv.}$ = 0.08, one obtains rather shallow contacts, peaking at about F $\simeq$ 0.15 (F is defined for equipotentials V as F = $(V_1 - V)/(V_1 - V_2)$). A decrease of the exponent $\beta$ to zero, as suggested by Anderson & Shu (1977), results in a moderate increase of mean contact to about F $\simeq$ 0.25.

Inclinations i different from 90 deg. are just an inconvenience.

Theoretical light curves which result from the above models are invariably what observers call "type A" (Binnendijk 1965) : the transit eclipses are deeper by about 0.01 - 0.10 mag (Fig.1). Such systems are observed for spectral types earlier than about F8 - G0 (Rucinski 1974). For later spectral types, typical light curves are of "W-type" (with deeper occultation minimum) so that one can suspect that this phenomenon has something to do with the stellar activity which is also strongly spectral-type dependent. We shall enumerate now suggestions for explaining the W-type phenomenon, some already noted by Rucinski (1978). Each suggestion is illustrated in the corresponding part of Fig. 2. It should be noted that each suggestion is related to some peculiarity in the photospheric brightness distribution and some of these peculiarities might not be neccessarily magnetic in origin. We start with exactly such a case.

a) Large $\beta$.
It is possible to obtain theoretical light curves which are of the W-type by assuming that the gravity darkening exponent is very large, typically larger than the von Zeipel value of $\beta_{rad.}$ = 0.25. This is because the darkened area on the more massive component must be substantial and this can be obtained only when $\beta > 0.25$ (Wilson & Devinney 1973, Wilson & Biermann 1976), possibly even $\beta \geq 0.4$ (Robertson & Eggleton 1977). Such high values of $\beta$ are difficult to support theoretically, so this possibility has been abandoned.

b)    Hotter   secondary component.

It is rather easy to obtain a theoretical W-type light curve if
one assumes that the  less  massive  component  is slightly the
hotter of the two.  The value of such a temperature contrast is
small,   typically  $\Delta T/T \simeq$ 5% (Rucinski 1974).  This possibility
has some theoretical support (Whelan, 1972; Mochnacki & Whelan,
1973; Webbink, 1977), but  it  is  most  probably wrong for the
following reasons:

i) It   works well for monochromatic  light curves or curves in
   narrow    ranges  of  wavelengths  but   predicts   incorrect
   ( $\beta$ Lyr-type)   curves    for    ultraviolet    wavelengths
   (Rucinski,     1976).    Actually,   the   ultraviolet   colour
   curves  of  W UMa, m (2200)-m(3300), were used by Eaton et
   al.  (1980) to set a good  limit  on  $\Delta T/T = 0 \pm 0.009$.

ii) Some  variability  in  the  relative  depths of minima  was
   thought  to  be   explainable  within  the model  by  small
   variations in   $\Delta T/T$.  However, Stepien (1980) showed that
   such variations  are  not  compatable  with  the observed
   constancy  of  colours  in  SW Lac which  is  the  prime
   candidate for testing any   model of the  W-type phenomenon
   because  its  components  are  almost  of  the  same  size
   (q = 0.88).

c)   Dark spots on the primary component.

To have the same photometric  effect,  instead  of  making  the
secondary  component  hotter one can postulate that the primary
component is darker  for  some  reason.  Very dark starspots on
this component would contribute no light  in  the visual and UV
region and could nicely explain frequently observed asymmetries
in  heights  of  light  maxima  (the  O'Connell  effect).   In
addition,  changes  in  the  spot  coverage would result in the
shifts  of  observed moments of eclipse  minima  so  frequently
observed in W UMa  systems.  The  most important consequence,
however, is the ease with which almost  unbelievable changes in
light curves of some systems could be explained.   TZ  Boo is a
prototype of such systems which change their type from A  to  W
and  vice  versa (Hoffmann 1978a, 1978b).  In this case the
changes are particularly large, as shown in Fig. 3, but Hoffman
sees similar changes in other  stars  of late spectral type and
short orbital period.

     Spots were invoked repeatedly by Binnendijk (refer  to his
review  articles  in  1970  and  1977).  Eaton et al.  (1980)
returned to this explanation when it  became  obvious  that the
temperature  contrast  idea  failed to account for the UV data.
There is an obvious  question,  however:  why do the spots form
predominantly  on  the  more  massive component?  Mullan (1975)
explained that preference by a high sensitivity of the magnetic

field generation to gravity. However, the difference in gravity between components is so small that it would be difficult to see how the necessary preference could be achieved. Anderson & Shu (1977) speculated that magnetic fields might find it difficult to penetrate through their contact discontinuity, which in low mass systems would be expected to reside inside the less massive component.

The problem of the preference for formation of spots on the primary is now the main obstacle against full acceptance of spots as a valid explanation of the W-type phenomenon.

d) Dark belts in equatorial regions.
The dark spots do not have to be magnetic in origin. They might also result from influences of the rapid rotation on the energy transport in convective envelopes. This might be either due to switching in the mode of convection from cells to rolls (Knobloch et al., 1981) or to the existence of baroclinic winds similar to weather patterns on Earth or Jupiter. A simple minded picture would then be of dark belts extending to the same latitudes on both components, i.e. having quite different geometrical extensions (Fig. 2d). Light curves computed with this assumption can look very nicely similar to the observed W-type curves but it is difficult to apply this idea to real observations because the critical parameter - the angular entension of belts - is an entirely free one, with no constraints known at present.

e) Faculae.
Lucy (1982, private comm.) pointed out that some sort of faculae like the solar ones, which are better seen close to the solar limb, might rather easily explain the W-type phenomenon. Such faculae are expected to be rather closely related to the magnetic activity and therefore should start appearing in stars below a certain effective temperature. In addition, the faculae could cover **both** components and produce the W-type phenomenon through a difference in amounts of inclined surface visible at each eclipse (Fig. 2e). This possibility has been tested (Rucinski 1983, unpublished) assuming that the photospheric faculae are due to small magnetic tubes causing depressions, whose hot walls are seen at large angles (Spruit 1980, 1982 priv. comm.). At an extreme of full faculae coverage and brightest possible faculae, one obtains the situation of practically no limb darkening at all (but no limb brightening, either). This is exactly as in the modelling of Anderson & Shu (1977, 1978), who assumed no limb darkening for entirely different reasons. But the results are not encouraging : the light curves do tend to show a small W-type tendency at the expense of **both** minima becoming much too shallow. But one must have some degree of limb darkening to

explain deep light curves observed in many systems. Parametrization of the facular visibility by using scaled up solar data (Schatten and Sofia, 1983) gave even poorer results. Therefore, this path seems to lead to nowhere.

It turns out, however, that Lucy (1983, priv. comm.) had actually something different in mind. He thought of obtaining the necessary aspect effect by a thin slab of hot gas which would be visible only at large angles and in this way counteract the limb darkening. Calculations for such a "chromospheric" model have not been attempted so far, however.

We can summarize the brief descriptions in points a) to e) above by stating that photospheric spots have the highest chance of being the right explanation for the W-type phenomenon. The mysterious tendency for spot preference on the more massive component might find an explanation by: i) careful studies of systems which switch types, like TZ Boo, ii) mapping of binaries using spectrum deconvolution techniques. Anderson et al. (1980) showed that the method of Fourier spectrum deconvolution is capable of recovering some information on the shape of components. A subsequent study of AW UMa (Anderson et al. 1983) brought evidence of some unexplained peculiarities in line profiles which can be interpreted as manifestations of facular line emissions. This study is particularly important because it brings indications that even such a hot, extreme A-type system as AW UMa might possess areas of increased activity on its surface. Actually, small variations in brighness of this system (Hrivnak, 1982) also prove that A-type systems are not as inactive as once thought.

Possibly, peculiarities observed in the violet and ultraviolet parts of spectral distributions have something in common with the phenomena above described. Eggen (1967) discovered ultraviolet excesses $\delta$(U-B) in shortest-period systems (at a given spectral type) and interpreted these excesses as indicating decreased metal content in these systems. Rucinski & Kaluzny (1981) and Rucinski (1983) confirmed the existence of these excesses using **uvby** photometry indices (Figs. 4, 5, 6). The excesses do indeed look as if caused by a metallicity effect because they are not limited to the short wavelength side of the Balmer jump (u-filter), as would be expected for an optically thin gas emission, but affect also the violet part of the spectrum (v-filter). However, the confinement of the phenomenon strictly to the shortest period systems looks very suspicious and one cannot help thinking of some filling in of absorption lines which would operate only in the most rapidly rotating systems. Such a filling-in was described from inspection of spectra by Struve (1950). In addition, Giampapa et al. (1979) noticed that the $\delta m_1$ index of the **uvby** photometry indicates false metal under-abundances of solar active regions (Unfortunately, similar measurements of the $\delta c_1$ index do not exist). The simplest explanation of this effect is by shallowing of absorption lines due to mechanical energy being deposited not only in chromospheric layers but already in the photospheric layers of the atmosphere. The line source functions would then become less steep and, as a result, the lines should be less deep. This might be particularly important for contact systems which suffer especially strong line blending problems, but might be also relevant for explaining metal deficiency in RS CVn secondaries (Naftilan & Drake, 1980).

Unfortunately, the above explanation has one major deficiency: it does not offer a simple observational check. The point is that metallic lines become less deep irrespective of whether this is caused by an under-metallicity or by a shallowing tendency of the source function.

Finally, we should note that Linnell (1983, preprint) has recently revived the idea of Berthier (1973, 1975) that the hotter inner hemispheres of both components are responsible for the W-type systems.


3.   CHROMOSPHERIC ACTIVITY

It is surprising to realize that we know so little about the chromospheric activity of contact binaries. The best description still remains the short chapter in O. Struve's book (Struve 1950a). However, some of his conclusions can now be commented on:

a)   The lines of approaching components might not be always stronger, as described by Struve (Binnendijk 1967). Papers by spectroscopic observers subsequent to Struve do not mention that effect.

b)   "The appreciable Ca II emission in several W UMa systems" is still very poorly understood and still awaits a dedicated study. They were described only in RW Com (Struve 1950b), W UMa (Struve & Horak 1950), XY Leo (Struve & Zebergs 1959), VW Cep (Kwee 1966) and CC Com (Rucinski et al. 1977). In all cases the emission lines seem to originate in rather localized regions on the contact surfaces

c)      The statement of Struve that "The emission lines move
        with respect to the underlying absorption line and can
        be interpreted by assuming that they belong to the more
        massive component...." urgently needs a confirmation
        with the use of modern observational techniques. The
        data for four systems mentioned above are consistent
        with this statement. New observations would be
        especially important for the proper interpretation of
        the A/W type dichotomy.

d)      All observations are qualitative; we have no single
        measure of the Ca II emission intensity on any contact
        system. Any study of the Ca II emission might turn out
        rather difficult: it would require a large telescope to
        achieve both high temporal and high spectral
        resolutions, and would present serious interpretational
        difficulties in view of the extreme rotational
        broadening of lines. The existing data on Hα are meagre
        for exactly the same reasons. A low resolution
        (classification dispersion) survey of about a dozen of
        the brighter W UMa systems performed with the 72-inch
        telescope of DAO (unpublished) did not reveal any
        obvious Hα emission in any of the observed systems.
        Mochnacki (1983, private communication) did not see any
        emission in a few systems observed with 1 Å resolution;
        only TZ Boo might possibly have this line partially
        filled in.

e)      The only existing attempt to observe the Lα emission
        with the IUE satellite is that of Rucinski & Vilhu
        (1984, unpublished) for W UMa itself. They found that
        the stellar emission follows variations of the
        bolometric brightness so that $(f(L\alpha)/f(bol) = (\mathbf{3.3 \pm 0.3})$
        $\times 10^{-5}$.

In view of the above, it is rather paradoxical that we
know more about the Mg II line emission of W UMa systems. Some
brighter systems have been observed in the high resolution mode
of the IUE. As luck would have it, most of these systems are
of relatively early spectral type and, probably exactly because
of that, their Mg II emissions are either very weak or entirely
absent. The quite good quality of these spectra assures at
least good upper limits on the Mg II line emission fluxes.
Redder systems are invariably fainter and can be studied only
in the low resolution mode of the IUE. In this case the data
are of much lower quality because we cannot be sure that what
we see inside the broad absorption feature of Mg II is indeed
the emission core, unless the emission is well above the
surrounding continuum. This is rather clearly visible in the
case of SW Lac and AW UMa (Fig. 7). In AW UMa the high

resolution spectra  reveal no emission at all but the simulated
low resolution profile  gives an impression of a faint emission
core. Unfortunately, the shape  of  the low resolution profile
of the Mg II line in  W  UMa  (Fig.  8),  which  is  of a later
spectral  type  than  AW  UMa  and where we could expect a real
emission core, is very similar to  that  in  AW UMa; therefore,
the  Mg  II  emission  in W UMa (Rucinski et al.  1982) might
very well be thought to be spurious.  In  fact, it is certainly
real,   as   re-extracted  spectra  from  the  old  IUE  images
definitely show.  This illustrates particular difficulties with
measuring the Mg II emission fluxes in G-type systems which are
too faint to be observed in the high resolution mode of the IUE
satellite.

The most comprehensive collection of  data on the strength
of the Mg II emission (and  also  other activity indicators) is
contained in the paper by Vilhu (1984).  Partial data have been
given by Rucinski &  Vilhu  (1983) and Vilhu & Rucinski (1983).
Table 2 lists the existing results  which  include some not yet
published.

The present  data suggest a very strong sensitivity of the
Mg II line emission to the effective temperature.  The emission
starts appearing rather suddenly for periods shorter than about
0.3 days, i.e. for spectral types later than about G4-G5.  This
sudden turn-on will  have  to  be checked by more observations.
The fragmentary data on the Ca  II  line  seem  to  suggest the
similar  behaviour  of  this  line.  Unfortunately,  these
statements are presently based on qualitative estimates of line
emission  strengths  and  should  be  checked by measuring line
fluxes.  The  surrounding  ultraviolet  continuum falls rather
rapidly with the advancing spectral  type  and,  obviously, the
emission  lines  become  progressively better visible for later
spectral types.

Some of the chromospheric  emission  lines  originating at
similar temperatures to Ca II and Mg  II, such as lines of C I,
O I, Si II, are visible in the  short  wavelength  range of the
IUE  spectra.  They  are  somewhat  easier to measure than the
calcium or magnesium lines because they  appear in the spectral
region devoid of any continuum emission.  The relevant data for
contact binaries have been obtained by Dupree  (1981), Dupree &
Preston, (1981), Eaton (1983), Rucinski & Vilhu (1983), Vilhu &
Rucinski (1983).

An  explicit  discussion  of how some of the chromospheric
emission lines could affect  the  broad band ANS satellite data
has been published by Eaton &  Wu  (1981).  At present, we have
no  data on other chromospheric indicators such as  Lα (except
W UMa), the Ca II infrared triplet, far infrared continua, etc.

## 4.   THE TRANSITION REGION ACTIVITY

The transition region (hereinafter called TR) is a geometrically thin layer between the chromosphere and corona in which the temperature gradient is extremely steep; in the solar case the temperature rises from $20 \times 10^3$ K to about $200 \times 10^3$ K in a few hundred km. The TR is especially important from the observational point of view because with the SWP camera of the IUE we see practically all lines contributing to the radiative cooling of this region; we have therefore a good estimate of the amount of energy available to heat this region. For spectral types later than middle A, the emission lines appear well isolated with no continuum background, so that even the low resolution mode suffices in determining the line fluxes.

The first TR studies of contact binaries were published by Dupree (1981), Dupree & Preston (1981) and Rucinski et al. (1982). They contained rather general descriptions of the IUE spectra, which looked very similar to those of other very active dwarfs (Fig. 9). It was noticed also that lines originating at higher temperatures (i.e. in the upper TR) showed relatively larger enhancements in intensity relative to the Sun or other inactive stars than lines originating at lower temperatures.

Further studies were presented by Dupree (1983), Eaton (1983), Rucinski & Vilhu (1983), Vilhu & Rucinski (1983). They concentrated on the dependence of TR line fluxes on the parameters of individual contact systems. Most frequently, the strongest C IV line at 1550 Å was taken as a measure of these fluxes. Sometimes, a sum of line fluxes for N V, C IV and Si IV was used; this differs typically by a factor of about 1.5 from that of the C IV alone. Only in two cases of very similar A-type systems, $\epsilon$ CrA (Dupree 1983) and AW UMa (Rucinski et al. 1984), was it possible to obtain phase coverage, but the results were rather uninteresting: the line fluxes follow the brightness variations so that $f_{TR}/f_{bol} \simeq$ const.

The over-all trends in the TR fluxes for systems with different periods have been more revealing. They show that the ratio $f_{TR}/f_{bol}$ is almost constant within the whole group of contact binaries, irrespective of the spectral type or period. There is a small spread within the group and, on the average, the contact binaries follow the same dependence of $f_{TR}/f_{bol}$ on the period which for detached binaries starts to "saturate" at rotational periods of about 3 days (Fig. 10). The spread within the contact binary group disappears if one uses the surface fluxes $F_{TR}$: for the whole group $F_{TR} \simeq 1.8 \times 10^6$ erg $cm^{-2}$ $s^{-1}$ (within about 30%, which includes the instrumental uncertainty)[†]. It is surprising that this constancy in $F_{TR}$

seems to encompass all studied systems, starting at relatively early spectral types (middle A) and continuing to the latest systems observed.

A rather special role is played by the He II line at 1640 Å. It is formed in TR but actually can carry information on the coronal emission so it will be described in the next section.

---

†It might be appropriate at this point to stress differences between various observational quantities related to radiation fluxes. Although they are to a large extent equivalent, they are not always determinable with the same accuracy. In particular, the observed fluxes, which are called here $f_s$, where the subscript designates a spectral band "s", should be kept distinct from the surface fluxes $F_s$. The ratio $f_s/f_{bol}$ is an easily observable, distance independent quantity which is obviously identical to $F_s/F_{bol}$ or $L_s/L_{bol}$. The quantities L are, respectively, the spectral band and bolometric stellar luminosities. One can derive $F_s$ using the relation: $F_s = \sigma T_e^4 (f_s/f_{bol})$ when the effective temperature of the star is available but must know the distance to convert $f_s$ into $L_s$.

---

## 5. CORONAL ACTIVITY

The existence of coronae is one of the most important aspects of stellar activity. The stellar winds start there and their expansion, together with the downward heat conduction, constitute an important sink for energy deposited in coronae by some unknown processes (such as waves, field reconnection, etc.). The coronae are known to be highly structured due to the dominance of the magnetic pressure: the high temperature plasma, trapped in magnetic loops, escapes freely only where the magnetic structures open up to the surrounding space. Therefore, presence of the coronae tells us more about magnetic fields than about anything else. Since contact binaries rotate so rapidly, they are expected to generate strong magnetic fields and this is why observations of their coronae are so very important. The first X-ray observations of the coronal thermal emission of a contact binary were reported by Carroll et al. (1980): VW Cep was observed with the HEAO-1 to vary in X-rays on a time-scale comparable to the orbital period but in a way apparently unconnected with the brightness variations, indicating a rather large extent of the corona.

The study of temporal variability in VW Cep was repeated with the Einstein Observatory by Dupree & Cruddace (see Dupree

1983). The variations are somewhat better defined but essentially identical. So far no other system has been observed in a similar way but such observations are planned with the EXOSAT by J. Heise & O. Vilhu.

An extensive survey of 17 brighter contact binaries has been performed by Cruddace & Dupree (1984) using the Einstein Observatory. The main results for 14 detected systems were described earlier by Dupree (1981, 1983). The data are shown in Fig. 11 in the form of the distance-independent flux ratio $f_x/f_{bol}$, in a similar way as for the transition region fluxes (Fig. 10). Observations of non-contact stars with known rotational periods, obtained by other authors, are also plotted for comparison. For fuller discussion, cf. Rucinski & Vilhu (1983), Vilhu & Rucinski (1983), and Vilhu (1983). Two points can be noticed immediately:

a)  a rather clear division of stars into parallel bands corresponding to broadly defined spectral groups,

b)  an under-activity of contact binaries which is particularly striking for F-type systems (three systems, AW UMa, V535 Ara, GK Cep, which have not been detected are F-type systems).

It looks as if the W-type systems (spectral groups G and K in Figs. 10 and 11) have coronal X-ray activity slightly below expectations based on an extrapolation of results for non-contact stars into the contact binary domain. At the same time, A-type systems (spectral groups F and partly G) are very strongly under-active. This division is also present in the X-ray luminosity (Cruddace & Dupree 1984). They find that $L_x$ (0.1 - 4 keV) for W-type systems and some cooler A-type systems is (0.6 - 1.2) x $10^{30}$ erg/s, whereas a few hotter A-type systems have $L_x$ at or below $10^{29}$ erg/s. This division in $L_x$ seems to be well defined but is less striking than the difference of two orders of magnitude visible in $f_x/f_{bol}$.

Thus the picture of the X-ray activity of contact binaries looks quite different from that of the transition region, where systems seem simply to continue trends existing for non-contact binaries. However, it is possible that the under-activity of hotter contact systems might be due to a more general phenomenon of "saturation" in X-ray fluxes which has been recently noticed by Walter (1983). He found that the surface fluxes $F_x$ for an ensemble of active F-dwarfs with different (and generally unknown) periods of rotation reveal upper limits which are strongly spectral-type dependent. These upper limits were expressed as

$$\log F_x \text{ (max)} = 4.2 + 5.2 \text{ (B-V)} \quad \text{for B-V} < 0.6$$
$$\log F_x \text{ (max)} \simeq 7 \pm 0.4 \qquad\qquad \text{for B-V} > 0.6.$$

In view of this saturation tendency observed for normal stars, one can ask if contact stars of earlier spectral types are really less active than other stars. A relevant parameter might be then $F_x/F_{bol}$(max) or, as used here, $\phi_x = (f_x/f_{bol})/(f_x/f_{bol})_{max}$. The denominator in the hotter expression can be easily written as a function of (B-V) using the formulae cited above.

The plot of $\phi_x$ versus the orbital period, in a way an analogue of Figs. 10 and 11, is presented in Fig. 12. Stars belonging to the K spectral type group are not affected by the transformation from $f_x/f_{bol}$ to $\phi_x$ so that they have not been plotted. Similarly, some redder G-type stars ((B-V) > 0.6) remain in their old places (they have been marked by horizontal bars). One immediately sees that the large deviations of F-type contact binaries have disappeared in the new presentation: these systems are no longer so very under-active relative to normal stars. On the whole, the contact binaries do show some under-activity, but it is rather uniform at about $\phi_x \simeq 0.1$ to 1.0, irrespective of the spectral type. Obviously, work remains to be done in sorting out some minor problems (e.g. some normal stars in Fig. 12 have $\phi_x > 1$, due to of some systematic differences in reductions, because the original data are the same) but it is clear now that the contact binaries, especially those of earlier spectral types, are not so peculiar in their under-activity as originally suspected.

The only spectral X-ray data so far published are the Einstein Solid State Spectrometer spectra of VW Cep and 44i Boo B (Cruddance & Dupree 1984). They indicate that the coronae consist of at least two components, with temperatures about $6 \times 10^6$ K and $3.5 \times 10^7$ K, similar to other active stars so far observed (Swank et al. 1981, Golub 1983). One might hope that the new Exosat observations will bring much better spectral data than available at present.

The spectral lines of helium, He II at 1640 Å and He I at 10830 Å, occupy a special place in coronal research. Strengths of both lines are, at least partly, related to the coronal X-ray emission and can serve as a crude measure of this emission. Hartmann et al. (1979, 1982) showed that enhancements of the He II 1640 Å emission can be interpreted as the result of excess recombination following photoionization by the coronal X-ray radiation. Zirin (1975, 1976; also Zirin & Liggett 1982) advocated a close relation between the coronal radiation field in the 170-350 Å range and the population of the He I levels, resulting in the $\lambda$10830 absorption effectively measuring the coronal emission. Reservations concerning this view have been recently summarized by Smith (1983). The observational data discussed by him suggest that a good correlation between the X-ray luminosity and He I $\lambda$10830 absorption strength exists for giants but not for dwarfs, so that this matter is still not settled.

Only a few contact binaries have measured line fluxes of the He II $\lambda$1640 emission, whereas no attempt to attempt to observe contact binaries at $\lambda$10830 have been reported so far. Such observations would be rather difficult because of the severe blending problems and confusion with telluric lines. As to the $\lambda$1640 emission line: it is definitely present in contact binaries known to be X-ray emitters with strength approximately agreeing with correlations given by Hartmann et al.; it is very weak, if not absent altogether, in systems not detected in X-rays, such as AW UMa. A small feature seen in the low resolution SWP spectrum at those wavelengths (Rucinski et al. 1984) might actually be an Fe II emission (Fig. 9).

## 6.   FLARE ACTIVITY

Flare activity of contact binaries is not a new subject. Already over 30 years ago reports appeared of flares in 44i Boo B (Eggen, 1948) and U Peg (Huruhata 1952) but they were suspected of resulting from imperfections of photometers used at that time and have not been generally accepted. Kuhi (1964) observed a rather well defined flare in W UMa. It was seen only in the ultraviolet band of a scanner (center 3300 Å; width 50 Å) and lasted at least 7 minutes. More recently, Egge & Pettersen (1983) observed a flare in VW Cep of considerable duration: its rise was shorter than 5 minutes, but the half-peak decay lasted 13 minutes giving a total duration of about 35 minutes (Fig. 13). The authors estimate that the total ultraviolet energy of this flare was E = 4 x $10^{33}$ erg.

One might ask if the above properties would be consistent with those of flares in genuine flare stars. The ultraviolet colours certainly agree in this respect; Moffett (1974) gives colours of typical flares. Also, if we take a half-peak duration of flares observed in contact binaries as 10 minutes then, using the correlation between $[T_{0.5}] \equiv \log T_{0.5}$ (min) and $M_V$ (Kunkel 1973), we obtain $M_V$ = +4.1. This is quite a reasonable value for a contact binary.

We can also try to account for typical ultraviolet energies of flares $E_U$ in, say, W UMa. Using correlations between the quiescent ultraviolet luminosity $q_U$ and $E_U$ given by

Lacy et al. (1976), we obtain $E_\nu \simeq 3 \times 10^{34}$ erg/flare, in almost perfect agreement with the energy of the flare observed by Kuhi, and several orders above the strongest solar flares (q (W UMa) = $3 \times 10^{32}$ erg/s). Such flares should appear rather rarely. Using graphs in Lacy et al., one can predict a repetition time of the order of 200-400 hours, i.e. larger by a factor of about 10 than the typical time spent to obtain a decent light curve. It is therefore not surprising that so few reports of flares in contact binaries have appeared to date: apparently very few flares are expected to be observed and those which do are so large that they might be dismissed as instrumental problems.

It is worth noticing that in normal flare stars, the energy per flare is exactly proportional to the quiescent luminosity, whereas the repetition frequency decreases rather dramatically with this luminosity. The flare observed by Egge & Pettersen in VW Cep which is fainter than W UMa is roughly as strong as the absolute brightness of this system would suggest but the repetition frequency should be about 50 hours. Obviously, flares should be even more frequent in systems as cool and faint as CC Com, VZ Psc or XY Leo, so that such stars should be primarily observed for flare detection. Unfortunately, photometric observations in the ultraviolet (where the detectability of flares is highest) are frequently not done, this being entirely due to the faintness of short-period systems in this spectral region. We suggest that observers perform such observations, at least to monitor gross brightness variations in the ultraviolet.

Somewhat related to the optical flares are similar phenomena observed in the X-ray and radio bands. The slow X-ray variability detected in VW Cep (Carroll et al. 1980; Dupree, 1983) might have some flaring origin, but a similar radio flare detected recently in the same system by Hughes & McLean (1984) looks even more convincing. Using the VLA at 10.5 GHz they noticed a radio brightening of VW Cep from very low levels (see below) to $4.7 \pm 1$ mJy in about 2 hours. The total duration was certainly longer than that and could reach 4 or even more hours. The observations of Hughes & McLean were done during a radio survey of 12 contact binaries. None was detected at sensitivity levels 0.03 - 0.05 mJy (1 sigma detection limit); the only exceptions were the above mentioned flare in VW Cep and a weak, poorly defined source close to V502 Oph. The quiescent radio emission of contact binaries is therefore below about $3 \times 10^{13}$ erg s$^{-1}$ Hz$^{-1}$ (nearby systems like VW Cep or 44i Boo B) or about $10^{14}$ - $10^{15}$ erg s$^{-1}$ Hz$^{-1}$ (more distant systems), but can rise, as during the VW Cep flare, to $1.7 \times 10^{15}$ erg s$^{-1}$ Hz$^{-1}$.

## 7.   PERIOD CHANGES

The classical subject of period changes has fallen into some disfavour recently, most probably for two reasons:

a)   Proliferation of moments of minima determined by different observers, using different methods and published in different publications does not lead to an easy integration of the vast observational material,

b)   Determination of moments of minima is frequently considered an "easy observational project" and is left to unskilled observers; as a result, the observational material frequently contains simple mistakes or oversights (e.g. heliocentric corrections with a wrong sign, etc.).

It is not surprising therefore, that only three attempts to analyse that material have appeared in the last few years. Two are in abstract form, promising fuller studies (Kreiner 1977; Herczeg 1979); one is a short paper (Panchatsaram & Abhynakar 1982) and does not agree with the two former papers in important details (see below).

The abstracts of Kreiner & Herczeg, although short, are quite informative, especially the one by Herczeg. The essential results of both papers are summarized in Table 1 below. As we see, the typical behaviour consists of relatively "sudden" period changes (taking place in weeks or months), separated by longer intervals of period constancy. Some of the particularly abrupt smaller period changes might actually be shifts in photocentric moments of minima, due to surface brightness changes, sometimes a curvature in the O–C diagram within $\Delta t$ (cf. Table 1) can be seen and then the time scale of "sudden" changes can be established. This is obviously the time scale of the process of period changes itself. It comes out to be of the order of $10^4 - 10^5$ years, i.e. much shorter than any imaginable evolutionary time scale. Actually, if the "sudden" period changes were always of the same sign and happened every interval $\Delta T$ (cf. Table 1), the resulting time scales would be of the order of $10^6 - 10^7$ years. This is longer, but still rather short in comparison with, say, thermal time scales of stars observed in typical contact binaries. They are also too short to signify any evolutionary trends in contact binaries because we simply see too many of those binaries which otherwise would be quickly swept out of the contact binary domain.

Apparently, some process acts intermittently on a rather short time scale with longer intervals of quiescence; positive and negative period changes produced by this process compensate to a high degree because mean period changes averaged over long intervals are rather small. If this process is the mass-transfer instability, then typical "sudden" transfers would be of the order of $\Delta M/M \simeq 10^{-4} - 10^{-5}$. Although these amounts of mass might seem small, they nevertheless correspond in a typical system to transfering all material overlying the inner critical Roche surface. If we accept the dominance of the period lengthening events, as suggested by Herczeg, then the instability would prefer mass transfer from the less to the more massive components.

Other possibilities should be also considered. Although the period changes do not look like direct evidence of any magnetic wind braking (the changes are too large and too frequent), they actually might indicate large instabilities in this braking. Any magnetic braking dominated evolution which is reconciled with observed numbers of systems would predict relevant time scales of the order of $5 \times 10^{8} - 5 \times 10^{9}$ years (Mochnacki 1981; Rahunen & Vilhu 1982; Vilhu 1982; Mochnacki's paper in this Institute). Such long time scales can easily hide in the observed (O-C) diagrams and can be difficult to see in the prevailing "noise".

In the paper by Panchatsaram & Abhyankar (1982), the period changes in 7 systems were explained by various "classical" phenomena: light-time effects (harmonic O-C), secular changes in periods (parabolic O-C) or by period changes with noisy data. This shows how absolutely different conclusions are now possible on the basis of essentially the same observational material. This field is ripe for an in-depth study.


## 8.   MISCELLANEOUS

Direct measurements of magnetic fields of contact binaries would be of extreme value for understanding the activity of these stars. But they might be rather difficult. Measurements of the circular polarization of spectral lines in search of a poloidal component most probably will not bring much information in view of the expected high degree of cancellation in bipolar active regions. Indeed, Brown & Landstreet (1981) did not see any such field in the only W UMa system so far observed with this technique, 44 i Boo B: $B = +10 \pm 23$ G. However, this measurement is interesting in itself as it rules out strong poloidal fields of the order of 100-200 G suspected from a direct scaling up of the solar poloidal field into the contact system domain. Generally more promising methods aimed at detecting chaotic fields and based on line broadening effects (Robinson et al. 1980; Robinson 1980; Marcy 1981,

1982) can also be difficult to apply in the case of contact binaries. Possibly one might attempt a simultaneous reconstruction of the rotational and Zeeman line broadening functions, but this would require a rather advanced observational programme. Besides, current methods of spectrum deconvolution still encounter unexplained problems when applied to contact binaries (Anderson et al. 1983).

One possible manifestation of magnetic active regions in contact binaries might be rather simple to observe. If their presence really implies the appearence of a small optical polarization in normal late-type stars (Piirola 1977a; Tinbergen & Zwaan 1981), then we should also see an increase of mean polarization for late-type contact binaries. However, somewhat surprisingly, the contact binaries which so far have been observed do not show any polarization. Observations of four systems by Piirola (1977b) reveal no polarization above a very low detection limit of about 0.01-0.02%. It would be useful to enlarge the sample to see better if there is any difference in polarization between early and late-type contact systems. Such an observational project might turn to be quite ambitious if one were to aim at a measurement accuracy comparable to that of Piirola.

Another important problem seems to be the question of applicability of the Rossby number for the parametrization of activity. Noyes (1983) showed this number to be a convenient parameter combining the dependences of the Ca II activity on the period of rotation and on the spectral type for normal main sequence stars Vilhu (1984) extended this application to other indicators of activity. But, could we use this number in the same way for contact binaries? In the case of the main sequence stars, the success of the Rossby number can be plausibly explained by a chain of arguments related to homology of stars, to similar (or at least monotonic) properties of the differential rotation, to a progression in convective zone depths, etc. Contact binaries, with their period-colour correlation, and with (probably) forced depths of convective zones, might very well offer us an "orthogonal" sequence of stars. Prospects of using contact binaries for the purpose of better understanding the activity of normal stars seem to be really exciting.

## 9.   GENERAL CONCLUSIONS

a)   The W-type phenomenon remains unexplained. Dark spots have won by default but are by no means the finally accepted explanation.

b)   The chromospheric activity has been little studied. Chromospheric emissions possibly affect the broad-band colours so that line profile deconvolution attempts are of the highest importance. Surveys of Ca II, H$\alpha$, etc. emissions still wait to be done.

c)   Transition region emission seems to be very similar to that of other very active stars: it behaves as a continuation of trends "saturated" at about 3 days with very little dependence on the spectral type.

d)   Coronal activity of contact binaries, as measured by the X-ray emission, is smaller by a factor of a few than in normal, most active, non-contact stars. The very strong under-activity of F-type contact systems previously noticed is less striking when it is realized that non-contact systems of earlier spectral types also show a sharp drop in activity.

e)   A few flares observed so far in contact binaries have properties exactly as if they followed relations established for active flare dwarfs.

f)   Studies of period changes are still not conclusive. Observed changes seem to be too large in magnitude and too frequent to be related to any evolutionary processes in contact binaries.

## ACKNOWLEDGEMENTS

I would like to thank Peter Eggleton, Leon Lucy, Stefan Mochnacki, Jim Pringle, Alistair Robertson, Henk Spruit, and Osmi Vilhu for many discussions which frequently helped me to understand better my own ignorance.

Table 1

Period changes in W UMa systems

| Property | Kreiner (1977) | Herczeg (1979) |
|---|---|---|
| Number of systems studied | 15 | 35 |
| "Sudden" period changes happen during $\Delta t$ which is: | "short time" (not given) | 5-10 months (sometimes weeks - days) |
| Periods are constant for intervals $\Delta T$ which are: | 7-35 years | 3-35 years |
| Distribution of period change is: | random, both signs | random, both signs |
| Typical $\Delta P$ per "sudden" change | $\pm 0.3$ sec | $\pm(0.1-0.5)$ sec |
| Relative frequency of positive and negative $\Delta P$: $N\ (\Delta P > 0) : N\ (\Delta P < 0)$ | 1 : 1 (approx) | 2 : 1 (uncertain) |

Table 2

W UMa stars with known activity

| (1) Star | (2) Period (days) | (3) (b-y) | (4) Mg II | (5) Ly$\alpha$ | (6) f/f($bol$) TR | (7) ($10^{-5}$) COR | (8) Other |
|---|---|---|---|---|---|---|---|
| CC Com | 0.221 | 0.76 | 15 | | | | Ca II |
| 44i Boo | 0.268 | (0.6-0.7) | | | 2.7 | 23 | |
| VW Cep | 0.278 | 0.52 | | 5 | 3.3 | 25 | Ca II, f |
| XY Leo | 0.284 | 0.55 | 21 | | 3.5: | 49 | Ca II |
| TZ Boo | 0.297 | 0.40 | 10: | | | 16 | |
| TW Cet | 0.317 | 0.46 | | | | 9 | |
| SW Lac | 0.321 | 0.47 | 25 | | 4.0 | 33 | |
| YY Eri | 0.322 | 0.42 | | | 1: | 10 | |
| FG Hya | 0.328 | 0.38 | 10: | | | | |
| W UMa | 0.334 | 0.41 | 7: | 4.2 | 2.5 | 17 | Ca II, f |
| AE Phe | 0.362 | 0.41 | | 1.2 | 1.8 | | |
| AM Leo | 0.366 | 0.36 | | | | 8 | |
| AH Vir | 0.408 | 0.48 | 21 | <6: | <3.5: | | |
| V839 Oph | 0.409 | 0.34 | | | | 7 | |
| V566 Oph | 0.410 | 0.27 | 3.5 | | 3.0 | 2.6 | |
| AK Her | 0.422 | 0.34 | 14: | | 2: | 8 | |
| AW UMa | 0.439 | 0.24 | <0.5 | | 1.5 | <0.3 | |
| V502 Oph | 0.453 | (0.42) | 23 | | 3.4 | 13 | |
| OO Aql | 0.507 | 0.46 | 13: | | | | |
| $\epsilon$ CrA | 0.591 | 0.26 | | | 1.9 | 0.3 | |
| RR Cen | 0.606 | (0.23) | <0.5 | | 1.8 | | |
| V535 Ara | 0.629 | (0.25) | | | 3: | <0.3 | |
| S Ant | 0.648 | 0.21 | | | 1.3 | 0.5 | |

Comments:

Col. (3)    the observed (b-y) colour orders from Rucinski & Kaluzny (1981) or Rucinski (1983). These references give upper limits to the interstellar extinction. Data in parentheses are based on the transformed (B-V) indices.

Cols. (4)-(6) the relative emission fluxes are based on Rucinski & Vilhu (1983) and Vilhu & Rucinski (1983) with the addition of some unpublished data. Only Ly$\alpha$ in W UMa has been corrected for interstellar absorption.

Col. (7)    the relative coronal fluxes from Cruddace & Dupree (1984); or TZ Boo, from Geyer & Hoffmann (1983).

Col. (8)    presence of the Ca II emission or occurrence of flares (f) noted by different observers.

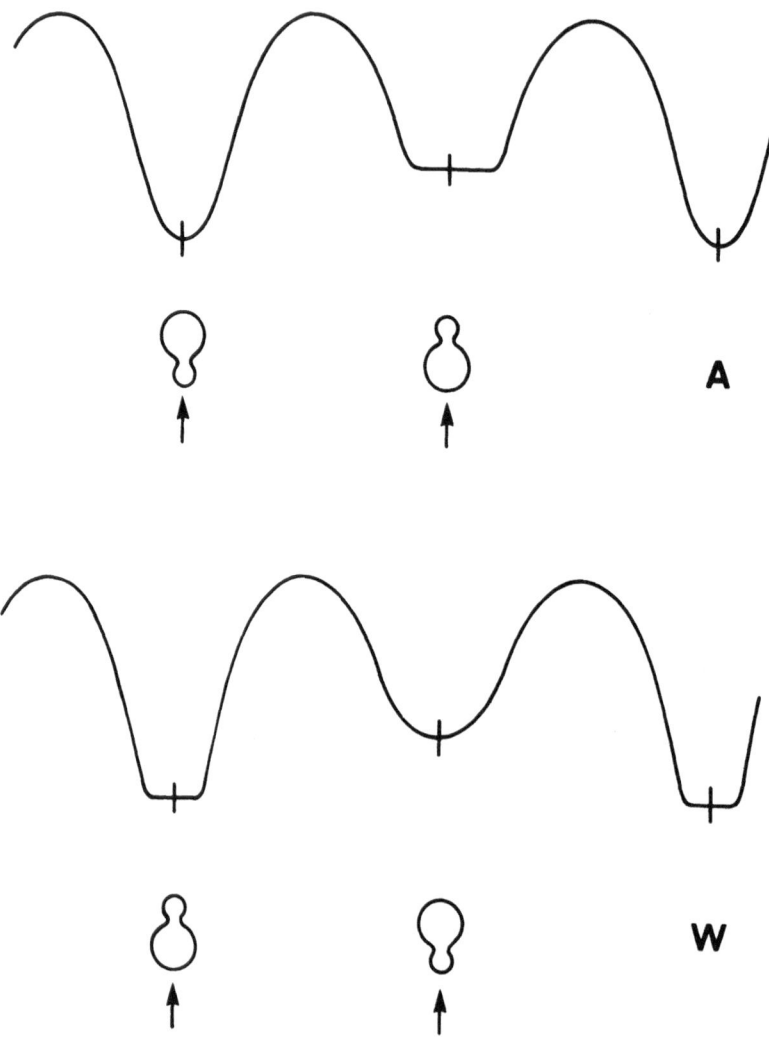

Figure 1. The basic difference between the A-type and W-type systems is shown schematically here. In the case of totally eclipsing systems, the type can be assigned from the shape of the light curve but in general one has to combine the photometric and spectroscopic data.

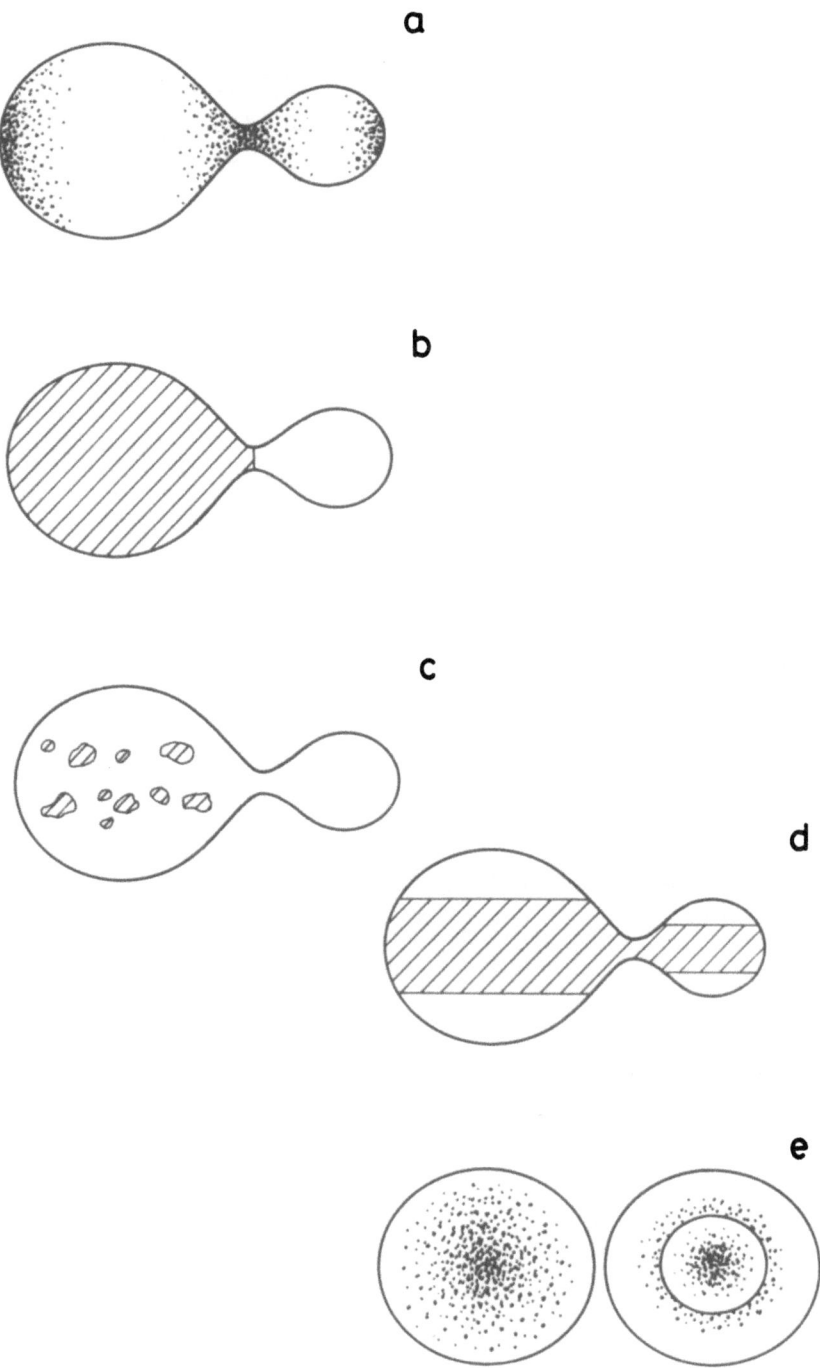

Figure 2. Possibilities of explaining the W-type phenomena, as described in the text.

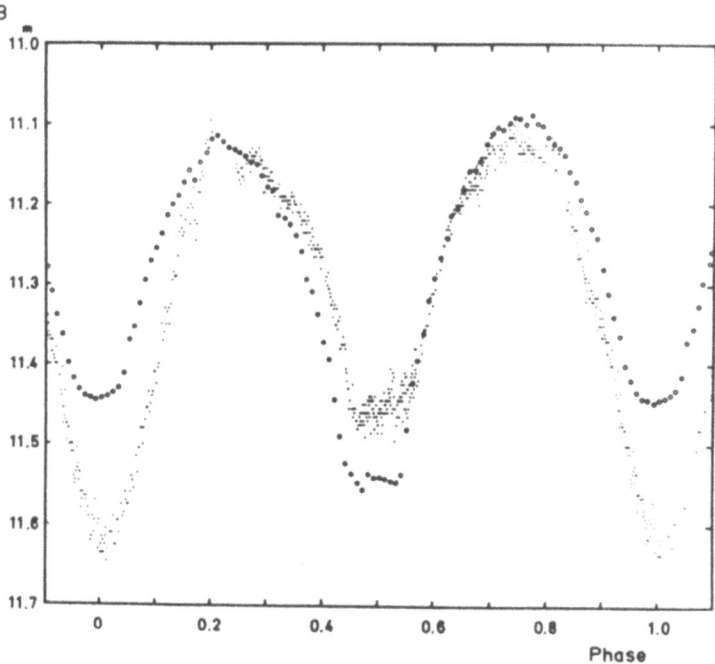

Figure 3. Striking changes of the light curve of TZ Boo have been discovered by Hoffmann (1978a).

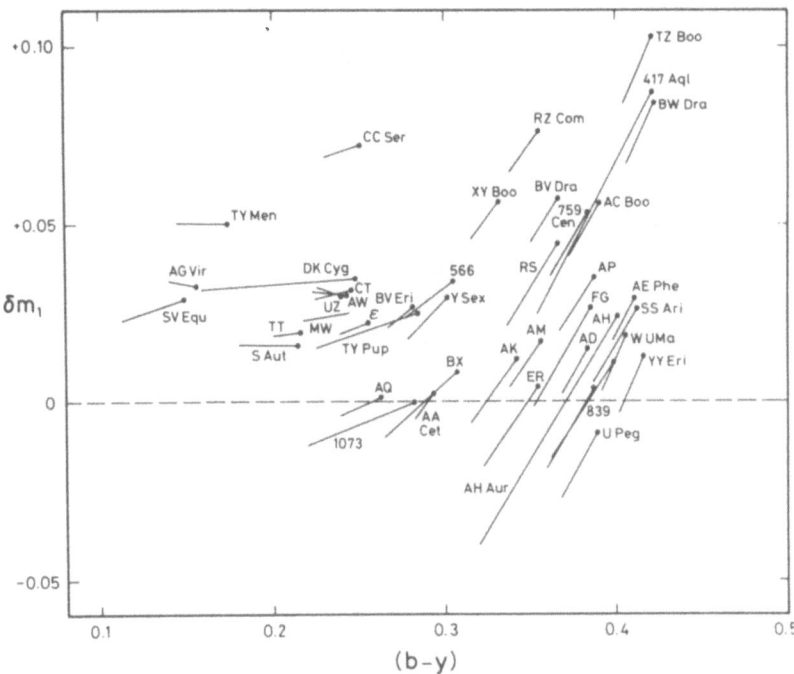

Figure 4. The $\delta m_1$ indices of W UMa systems in uvby photometry (Rucinski 1983): $\delta m_1 \equiv m_1$ (Hyades) $- m_1$(obs). Dots give the observed indices, vectors give the interstellar reddening corrections (in many cases probably exaggerated).

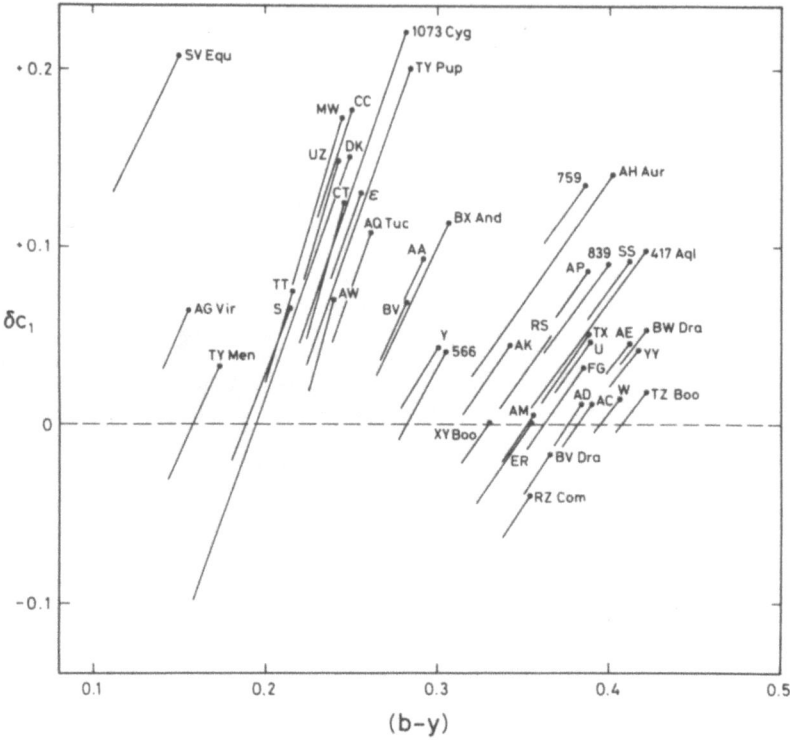

Figure 5. The $\delta c_1$ indices of W UMa systems: $\delta c_1 \lesssim c_1(\text{obs}) - c_1$ (Hyades); cf. caption to Fig. 4.

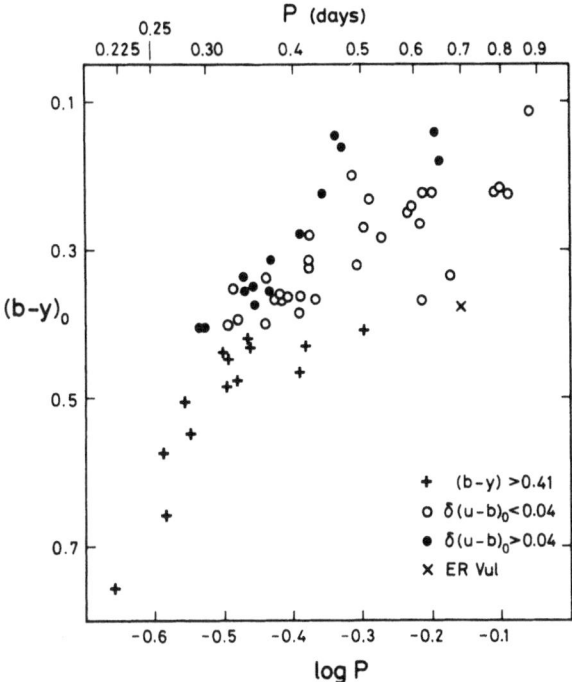

Figure 6. The period-colour diagram for W UMa systems (Rucinski 1983) observed in uvby photometry (which is well calibrated for b-y < 0.41). Systems with ultraviolet excesses $\delta(u-b) = 2\delta m$, − $\delta c_1$ are marked by filled circles. ER Vul (and possibly a few other systems in the figure) is a detached binary which is frequently misclassified as a contact system.

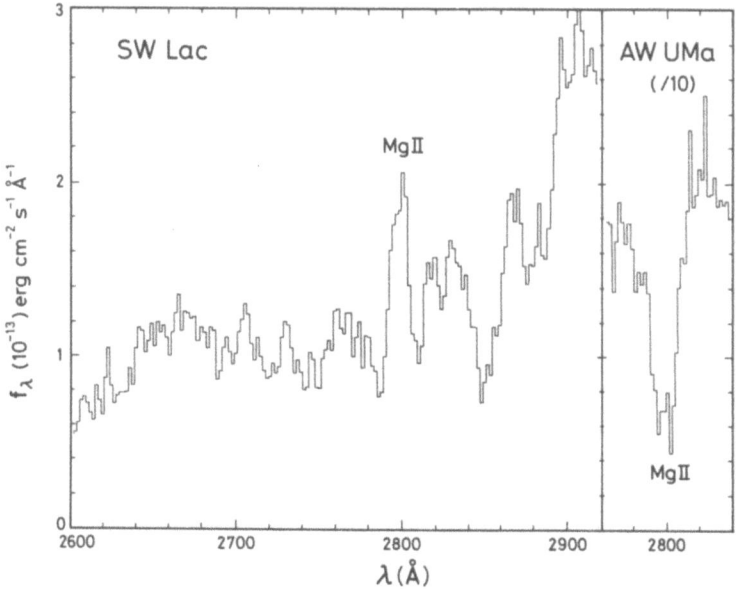

Figure 7. The low resolution ultraviolet spectra of SW Lac and of
AW UMa in the vicinity of the Mg II line at 2800 Å (Rucinski et al.
1984). The spectrum of AW UMa has been obtained by degrading a
high resolution spectrum which is entirely devoid of any Mg II
emission. The small quasi-emission in the very core actually
results from a space between two not-fully-blended absorption components, each with a deep interstellar dip. The Mg II emission in
SW Lac is one of the strongest known.

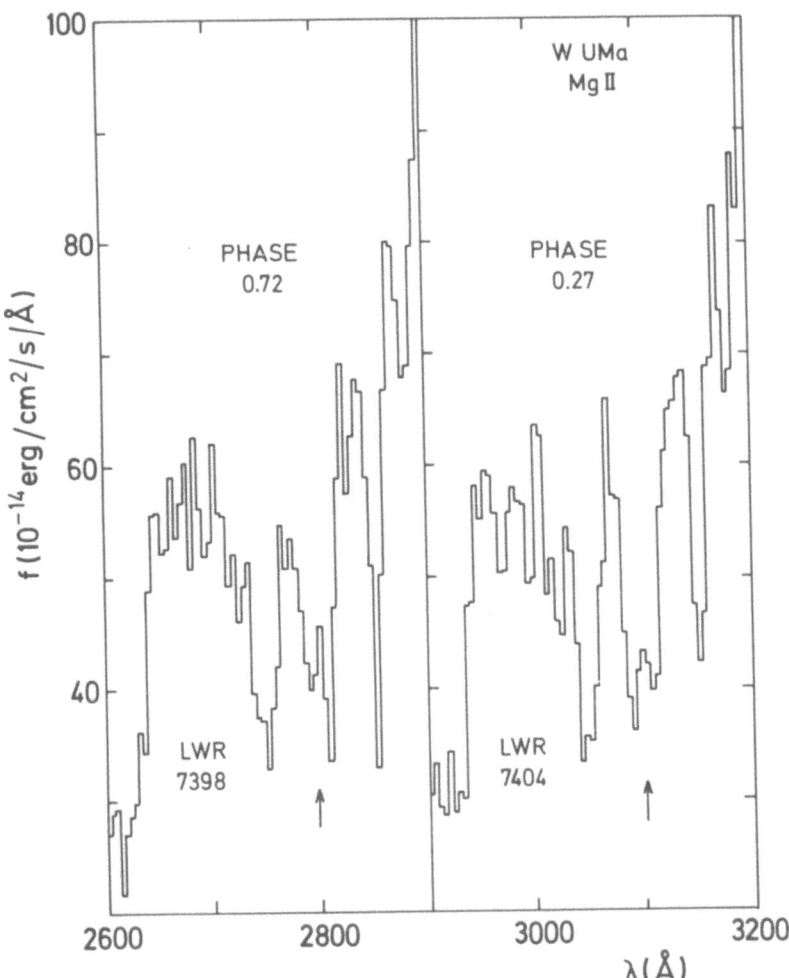

Figure 8.  The Mg II line in W UMa at opposite orbital quadratures
(Rucinski et al. 1982).  The small emission core might be genuine
but might also very well result from the presence of interstellar
extinction, which would produce two very narrow components of the
doublet.  Orbital phase-related shifts in the core speak in favour
of a real emission.

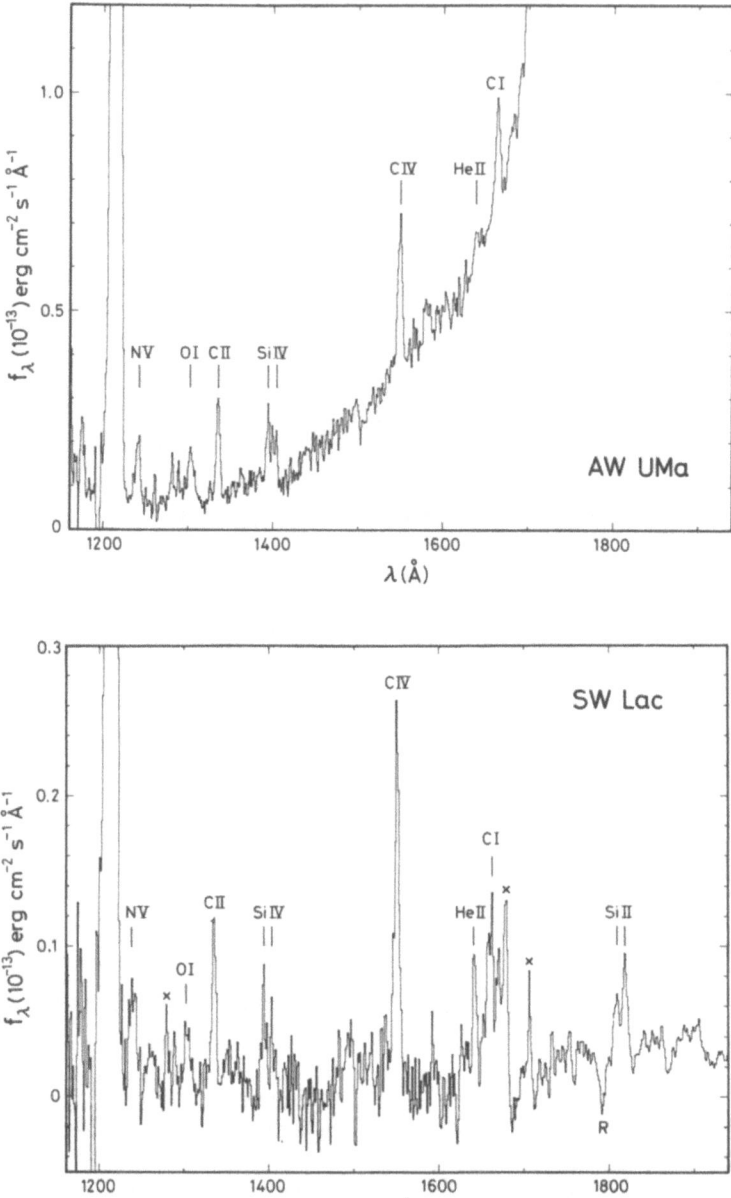

Figure 9. Low resolution IUE spectra of AW UMa (upper panel) and SW Lac (lower panel) revealing strong emission lines which originate in the transition region. Notice that in spite of very different continuum spectra (spectral type of AW UMa is F0-2; spectral type of SW Lac is G6-8), the line strengths are very similar.

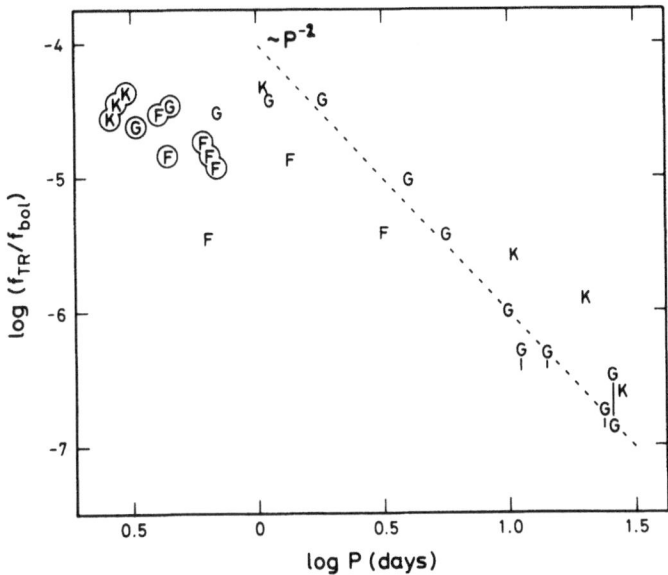

Figure 10.  The transition region emission line fluxes (summed
for the NV, Si IV and C IV lines) have been expressed relative to
bolometric fluxes and are here plotted versus the rotational period
of the star.  The contact binaries have spectral type groups
written inside circles.  The spectral type groups are rather
loosely defined: F is F0-F7, G is F8-G5, and K is G6-K5 following
Vilhu & Rucinski (1983).  Notice the hidden period-spectrum
relation for contact binaries in this figure.  The $P^{-2}$ - dependence for detached binaries and single stars is given to guide the
eye only.

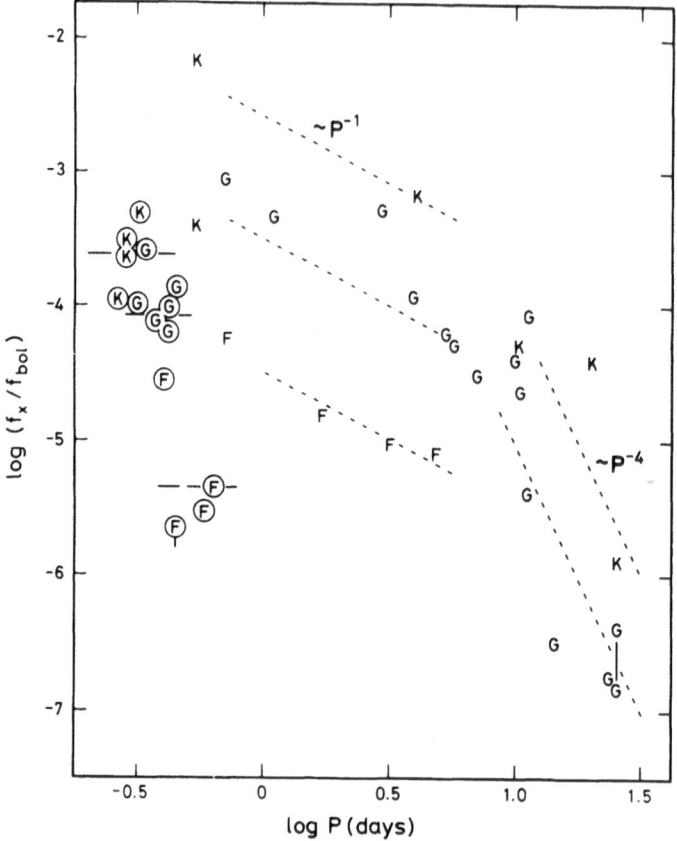

Figure 11. The coronal X-ray emission for contact binaries (circles) and non-contact stars expressed in $f_x/f_{bol}$ and plotted versus the rotational period. The mean values for the three spectral groups of contact binaries are given by horizontal dashed lines.

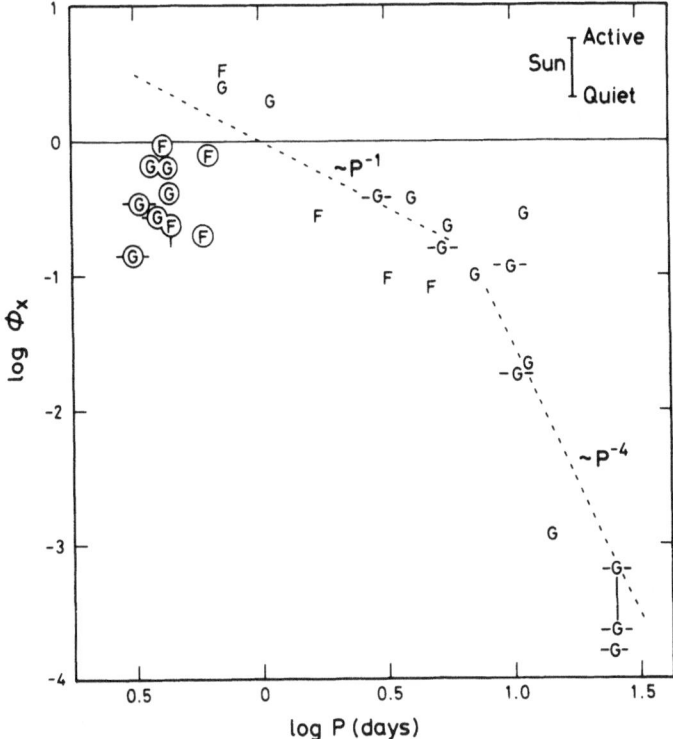

Figure 12. The relative measure of the X-ray activity $\phi'_x$ (as defined in the text) versus the rotational period. Only the stars of spectral groups F and G of Fig. 11 are shown here. To calculate $\phi_x$, the formula $\log(f_x/f_{bol}) = 6.25 (B-V)-7.25$, has been used for $B-V < 0.6$. It has been read off from the figure in Walter (1983) and does not exactly agree with that for $F_x$ given in the text. Stars marked by horizontal bars have $B-V > 0.6$ and have positions unchanged relative to Fig. 11.

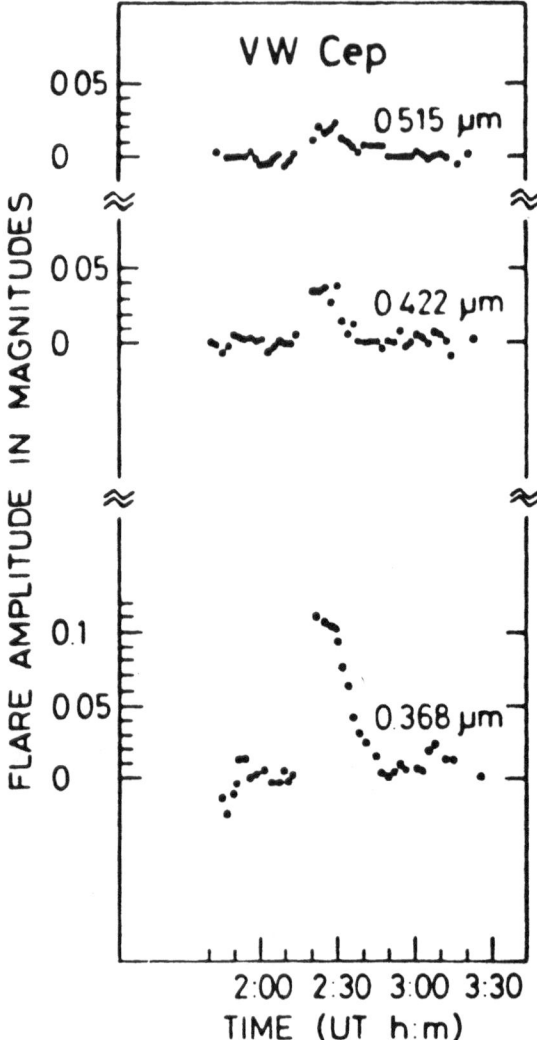

Figure 13.   The flare of VW Cep observed by Egge and Pettersen (1983).

## REFERENCES

Anderson, L., & Shu, F.H., 1977.  Astrophys.J. **214**, 798.
Anderson, L., & Shu, F.H., 1978.  Astrophys.J. **221**, 926.
Anderson, L., Raff, M., & Shu, F.H., 1980.  In "Close Binary
    Systems:  Observations and Interpretation", eds. M.J.
    Plavec et al. (Reidel), p. 485.
Anderson, L., Stanford, D., & Leiminger, D., 1983.
    Astrophys.J. **270**, 200.
Berthier, E., 1973,  Ph.D. Thesis, Paris.
Berthier, E., 1975,  Astron.Astrophys. **40**, 237.
Binnendijk, L., 1965.  Klein. Veroff. Bamberg, 40, No.40,
    p. 36.
Binnendijk, L., 1967,  Publ.Dom.Astroph.Obs., **13**, 27.
Binnendijk, L., 1970,  Vistas in Astr., **12**, 217.
Binnendijk, L., 1977,  Vistas in Astr., **21**, 359.
Brown, D.N., & Landstreet, J.D., 1981.  Astrophys.J.,
    **246**, 899.
Carroll, R.W., Cruddace, R.G., Friedman, H., Byram, E.T., Wood,
    K., Meekins, J.,Yentis,D., Share,G.H., & Chubb, T.A., 1980.
    Astrophys.J. (Lett.), **235**, L77.
Cruddace, R.G., & Dupree, A.K., 1984, Astrophys.J., **277**, 263.
Dupree, A.K., 1981,  In "Solar Phenomena in Stars and Stellar
    Systems", eds. R.M. Bonnet, A.K. Dupree, (Reidel), p. 407.
Dupree, A.K., 1983,  In "Activity in Red Dwarf Stars", IAU Coll.
    71, Catania, eds. P.B. Byrne, M. Rodono (Reidel), p. 447.
Dupree, A.K. & Preston, S., 1981,  In "The Universe at
    Ultraviolet Wavelengths - The First Two Years of IUE",
    NASA Conf. Publ. 2171, p. 333.
Eaton, J.A., 1983, Astrophys.J. **268**, 800.
Eaton, J.A. & Wu, C.-C., 1981, Astron.J. **86**, 1387.
Eaton, J.A., Wu, C.-C. & Rucinski, S.M., 1980, Astrophys.J.
    **239**, 919.
Egge, K.E. & Pettersen, B.R., 1983.  In "Activity in Red Dwarf
    Stars", IAU Coll. 71, Catania, Eds. P.B. Byrne, M. Rodono,
    (Reidel), p. 481.
Eggen, O.J., 1984, Astrophys.J. **108**, 15.
Eggen, O.J., 1967, Mem.Roy.Astr.Soc. **70**, 111.
Geyer, E.H. & Hoffmann, M., 1983.  Astrophys.Space Sci., **97**, 225.
Giampapa, M.S., Worden, S.P. & Gilliam, L.B., 1979, Astrophys.J.
    **229**, 1143.
Golub, L., 1983, In "Activity in Red Dwarf Stars", IAU Coll. 71,
    Catania, eds. P.B. Byrne, M. Rodono, (Reidel), p. 83.
Hartmann, L., Davis, R., Dupree, A.K., Raymond, J., Schmidtke,
    P.C. & Wing, R.F., 1979, Astrophys.J. (Lett.) **233**, L69.
Hartmann, L., Dupree, A.K. & Raymond, J., 1982, Astrophys.J.
    **252**, 214.
Herczeg, T., 1979, Bull.Amer.Astr.Soc. **11**, 438.
Hoffmann, M., 1978a, Inf.Bull.Var.Stars, No. 1487.

Hoffmann, M., 1978b, Astron.Astrophys.Suppl. **33**, 63.

Hrivnak, B.J., 1982, Astrophys.J. **260**, 744.

Hughes, V.A. & McLean, B.J., 1983. Preprint.

Huruhata, M., 1952, Publ.Astr.Soc.Pacific, **64**, 200.

Knobloch, E., Rosner, R. & Weiss, N.O., 1981, Mon.Not.Roy.
    Astr.Soc., **197**, 45P.

Kreiner, J.M., 1977, In "Interaction of Variable Stars with
    their Environment", eds. R. Kippenhahn et. al.
    (Veroff. Bamberg, **11**, No. 121), p. 393.

Kuhi, L.V., 1964, Publ.Astr.Soc.Pacific, **76**, 430.

Kunkel, W.E., 1973, Astrophys.J.Suppl. **25**, 1.

Kwee, K.K., 1966. Bull.Astr.Inst.Neth.Suppl., **1**, 265.

Lacy, C.H., Moffett, T.J. & Evans, D.S., 1976.
    Astrophys.J.Suppl. **30**, 85.

Lucy, L.B., 1967, Zeit. f. Astrophys. **65**, 89.

Lucy, L.B., 1968, Astrophys.J. **153**, 877.

Marcy, G.W., 1981, Astrophys.J. **245**, 624.

Marcy, G.W., 1982, Publ.Astr.Soc.Pacific, **94**, 989.

Mochnacki, S.W., 1981, Astrophys.J. **245**, 650.

Mochnacki, S.W. & Whelan, J.A.J., 1973, Astron.Astrophys.
    **25**, 249.

Moffett, T.J., 1974, Astrophys.J.Suppl. **29**, 1.

Mullan, D.L., 1975, Astrophys.J. **198**, 563.

Nafilan, S.A. & Drake, S.A., 1980, Publ.Astr.Soc.Pacific,
    **92**, 675.

Noyes, R.W., 1983, In "Solar and Stellar Magnetic Fields :
    Origins and Coronal Effects", ed. J.O. Stenflo, (Reidel),
    p. 133.

Panchatsaram, T. & Abhyankar, K.D., 1982, In "Binary and
    Multiple Stars as Tracers of Stellar Evolution", eds. Z.
    Kopal, J. Rahe, (Reidel) p. 47.

Piirola, V., 1977a, Astron.Astrophys.Suppl. **30**, 213.

Piirola, V., 1977b, Astron.Astrophys. **56**, 105.

Rahunen, T. & Vilhu, O., 1982, In "Binary and Multiple Stars as
    Tracers of Stellar Evolution", eds. Z. Kopal, J. Rahe,
    (Reidel), p. 289.

Robertson, J.A. & Eggleton, P.P., 1977, Mon.Not.Roy.Astr.Soc.,
    **179**, 359.

Robinson, R.D., 1980, Astrophys.J. **239**, 961.

Robinson, R.D., Worden, S.P. & Harvey, J.W., 1980, Astrophys.J.
    (Lett.) **236**, L155.

Rucinski, S.M., 1974, Acta.Astr. **24**, 119.

Rucinski, S.M., 1976, Acta.Astr. **26**, 227.

Rucinski, S.M., 1978, In "Nonstationary Evolution of Close
    Binary Systems", ed. A.N. Zytkow (Warsaw: Polish Sci.
    Publ.), p. 117.

Rucinski, S.M., 1983, Astron.Astrophys., **127**, 84.

Rucinski, S.M., Brunt, C.C., Pringle, J.E. & Vilhu, O., 1984,
    Mon.Not.Roy.Astr.Soc. in press.

Rucinski, S.M. & Kaluzny, J., 1981, Acta.Astr. **31**, 409.

Rucinski, S.M., Pringle, J.E. & Whelan, J.A.J., 1982, In "Binary and Multiple Stars as Tracers of Stellar Evolution", eds. Z. Kopal, J. Rahe (Reidel) p. 309.

Rucinski, S.M., Vilhu, O., 1983, Mon.Not.Astr.Soc., **202**, 1221.

Rucinski, S.M., Whelan, J.A.J. & Worden, S.P., 1977, Publ.Astr. Soc.,Pacific, **89**, 684.

Schatten, K.H. & Sofia, S., 1983, Nature, **301**, 133.

Smith, M.A., 1983, Astron.J., **88**, 1031.

Spruit, H., 1980, Solar Phys., **50**, 269.

Stepien, K., 1980, Acta.Astr. **30**, 315.

Struve, O., 1950a, "Stellar Evolution" (Princeton: Princeton Univ. Press), p. 175.

Struve, O., 1950b, Astrophys.J., **111**, 658.

Struve, O. & Horak, H.G., 1950, Astrophys.J., **112**, 178.

Struve, O., Zebergs, V., 1959, Astrophys.J., **130**, 137.

Swank, J.H., White, N.E., Holt, S.S. & Becker, R.H., 1981, Astrophys.J. **246**, 208.

Tinbergen, J. & Zwaan, C., 1981, Astrophys. **101**, 223.

Van Leeuwen, F. & Alphenaar, P., 1982, ESO Messanger, **28**, 15.

Van't Veer, F., 1978, Astron.Astrophys. **70**, 91.

Van't Veer, F., 1979, Astron.Astrophys. **80**, 287.

Vilhu, O., 1981, Astrophys.Space Sci. **78**, 401.

Vilhu, O., 1982, Astron.Astrophys., **109**, 17.

Vilhu, O., 1984, Astron.Astrophys., **133**, 117.

Vilhu, O. & Rucinski, S.M., 1983, Astron.Astropys., **127**, 5.

Walter, F.M., 1983, Astrophys.J., **274**, 794.

Webbink, R.F., 1977, Astrophys.J., **215**, 851.

Whelan, J.A.J., 1972, Mon.Not.Roy.Astr.Soc. **156**, 115.

Wilson, R.E. & Biermann, P., 1976. Astron.Astrophys., **48**, 349.

Wilson, R.E. & Devinney, E.J., 1973, Astrophys.J., **182**, 539.

Zirin, H., 1975, Astrophys.J. (Lett.), **199**, L63.

Zirin, H., 1976, Astrophys.J., **208**, 414.

Zirin, H. & Liggett, M.A., 1982, Astrophys.J., **259**, 719.

# OBSERVATIONAL EVIDENCE FOR THE EVOLUTION OF CONTACT BINARY STARS

Stefan W. Mochnacki

David Dunlap Observatory
University of Toronto
Ontario, Canada

ABSTRACT

This paper reviews data relevant to the evolution of contact binary stars. Statistics, photometric surveys, period changes and the results of light curve fitting are reviewed. These data are then compared with evolutionary models.

A new collation of data for 55 W Ursae Majoris systems is presented. Simple evolutionary models including magnetic braking have been calculated, and show good agreement with observations if contact lifetimes are long.

*P. P. Eggleton and J. E. Pringle (eds.), Interacting Binaries, 51–82.*
© *1985 by D. Reidel Publishing Company.*

## 1.    INTRODUCTION

Contact  binary  stars  by definition have both components
filling or over-filling their  inner  Lagrangian  zero-velocity
surface  (usually  called  "Roche lobes").  The light curves of
such close systems have  a  sinusoidal  appearance  due  to the
tidal  distortion  of the components.  If eclipses occur, as in
all eclipsing binaries the relative depths of the minima of the
light curve indicate  relative  surface  brightnesses and hence
effective  temperatures.  Sinusoidal-type  light  curves  with
roughly  equal minima are traditionally classed "EW", after the
prototype, W  Ursae  Majoris.  Stars with such light curves are
the main subject of the review.

The W UMa stars  are  a very distinct group, characterized
by periods of six hours to  a  day, spectral types dwarf late A
to mid  K, mass ratios not equal to unity and components having
similar  effective  temperatures.  A less distinct population of
hot contact  systems  also  exists, with spectral types O, B, A
and sharing some of the attributes of the W UMa stars.

As a consequence of their  Roche lobe geometry and similar
effective temperatures, the components of W  UMa  systems  have
luminosity  ratios  roughly  equal  to the first power of their
mass ratio rather than the  fourth  power  or  so  observed for
single main  sequence  stars.  This  paradox  of overluminous
secondaries  was solved in a classic pair  of  papers  by Lucy
(1968a,b).  A  common  convective  envelope  surrounds  both
components, leading  to  large-scale  energy  transfer from the
larger more massive component to the  less massive one, roughly
equalizing  surface  temperatures  over the entire system.  The
exact  nature of the  energy  transfer  process  is  still  not
understood (e.g.  Robertson  1980), and a contact system almost
certainly  cannot  exist  in  static  equilibrium (Lucy  1976,
Flannery  1976).  Detailed  evolutionary  models  including
instability  are difficult (Robertson &  Eggleton  1977),  and
simplified  theories  have  been  used  to  explore  various
possibilities (e.g.  Rahunen  1981,  1982,  1983;  Whyte 1982).
There  is fundamental disagreement about the interior structure
of contact  binaries,  as reviewed by Shu (1980) and briefly by
Shu & Lubow (1981).  It  is  not  my  intention  to present an
indigestible  mess of contentious theory; in this impasse it is
perhaps best to turn to observations for fresh insights.

In a  previous  paper (Mochnacki 1981) I have attempted to
show how the results  of  light-curve fitting by modern methods
can  illuminate  the  formation,  structure  and  evolution  of
contact binaries.  That work owed much  to  the earlier results
of Rucinski (1974) and Webbink (1976, 1979); the  first  author
showed  that  most  W  UMa  common envelopes lie near the <u>inner</u>

Lagrangian surface, and the second showed how this fact could be used as a "contact condition" for simplified evolutionary calculations. Webbink furthermore considered the effect of angular momentum loss on the evolution of contact binaries. This suggested that one should look for evidence of angular momentum loss as well as of nuclear evolution. In a thoughtful review, Vilhu (1981) suggested that one should also look for evidence of originally detached binaries coming into contact as a result of angular momentum loss.

This paper review statistics, photometry, period changes and the results of light curve fitting. These observational data provide useful points of comparison with theoretical predictions.

## 2. STATISTICS OF CONTACT BINARIES

Frequency: Contact binaries have long been known to be a large fraction of all eclipsing binary stars (e.g. Shapley 1948). A detailed study by Kraft (1965) show that about one in every 2000 F or G type dwarfs is a W UMa system, or about $1 \times 10^{-6}$ systems per cubic parsec in the solar neighbourhood. However, Budding (1982) suggests that this figure is too high. Furthermore, the galactic space distribution is typical of F and G dwarfs, and also resembles the distribution of cataclysmic variables (CV's), leading Kraft (1967) to conclude that the CV's may be descended from the W UMa's. More recent work (e.g. Webbink 1976) proves that W UMa systems are unlikely to be the progenitors of cataclysmic variables.

Frequency versus spectral type: Eggen (1967) claimed that the ratio of W UMa stars to all stars of the same absolute magnitude decreases with increasing luminosity. Since contact binaries are generally dwarfs, this corresponds to a decrease of contact binary frequency at earlier spectral types. This is clearly seen in Table 1, which lists the ratio of the number N(EW) of systems with EW light curves to the total number N(E) of EA, EB and EW systems in particular spectral ranges. The sample was chosen by Giuricin et al. (1983). Since restriction to a narrow spectral range defines a volume-limited sample, the ratio N(EW)/N(E) is affected only by the geometrical and period biases which favour the discovery of short-period contact systems over longer-period detached ones. Systems with EW light curves are ten times more common with respect to all eclipsing systems at spectral types later than A6 than at earlier spectral types. This means either that hot contact systems do not display EW light curves and therefore usually have different component temperatures, or

that hot contact systems are intrinsically rare. If the first
explanation is true, then hot contact systems do not transfer
energy between components as efficiently as do the cooler W UMa
stars, rather contrary to the suggestions of Lucy & Wilson
(1979) and Popper (1982). If hot contact systems are truly
rare, then their formation mechanism may not be the same as
that of the fission model of Roxburgh (1966). While the number
of O-type systems selected by Giuricin et al. is small, the
total number of eclipsing systems in the range B0-A5 is
impressively large, making the scarcity of EW light curves in
that spectral range an interesting statistic.

Frequency versus period: Lucy (1976, fig.4) plotted the
frequency of eclipsing binaries as a function of period and
light-curve type. His plot showed the previously well known 6
hour lower limit for EW systems, but surprisingly it showed a
10.5 hour lower limit for the periods of EA and EB systems.
Furthermore, the EW systems become rarer at periods longer than
10.5 hours, with a pronounced longer-period "tail" which
Lucy suggested could be due to evolution, possibly from EB to
EW.

Frequency versus period and spectral type: The study by
Giuricin et al. (1983) resolves Lucy's diagram into
spectral type ranges. A smaller sample was studied by Budding
(1981). Giuricin et al. attempt to correct the period
distributions for geometric selection effects. They classify
light curves as EA' (equal minimum depths), EA" (deep primary
minimum), EB (curved maxima, unequal minima) and EW (curved
maxima, equal minima). The EW systems of spectral type F have
longer periods on the average than EW systems of spectral type
GKM. The EW systems of spectral types GKM are much more
abundant relative to other GKM eclipsing binaries.

For the spectral range F, a substantial population of EB
systems comparable with the EW systems is distributed just
longward of the EW's, suggesting an evolutionary connection.
For the GKM stars, the EW population greatly outnumbers the EB
population with log P<0.4. The distinction between EW systems
and the EA and EB systems is therefore much more striking for
the spectral range GKM than for earlier spectral types. It is
masked in Lucy's diagram by the large number of EA and EB
systems of earlier spectral type in his magnitude - limited
sample . The distinction may be even greater because Giuricin
et al. (1983) consider only geometrical selection effects
and not photometric ones. In particular, EA" and EB systems
typically have deeper primary minima and therefore are easier
to discover than EW systems of similar period, especially in
the F spectral range.

Is the six hour cut-off real?: The lower limit of EW periods may be due to observational selection. The statistical surveys mentioned above are generally cut off at $m_B \simeq$ 12. The total number of known M dwarfs brighter than $m_B \simeq$ 12 is not much more than $10^3$ (Vyssotsky 1956). Extrapolating Kraft's frequency estimate to later spectral types, we do not expect to see many contact binaries of very late spectral type. However, there is some evidence that the frequency of spectroscopic binaries may be less at the latest spectral types (Wilson 1967), and Roxburgh (1966) argued against the formation of very low-mass contact binaries. Nevertheless, the recent discovery of young rapidly-rotating K dwarfs in the Pleiades (Alphenaar & van Leeuwen 1981) hints that more surprises may be waiting to be uncovered at the bottom of the main sequence.

Non-contact non-degenerate systems with $P < 10.5$ hours?: Lucy (1976), Flannery (1976) and Robertson & Eggleton (1977) developed the theory of thermal relaxation oscillations, according to which a contact binary should spend some fraction of its lifetime in a semi-detached configuration. It should then be observed as an EB system. Any EB with $P \lesssim 10.5$ hours is therefore extremely interesting, not just because is rare but because it may be a W UMa system in the semi-detached phase of its postulated thermal cycle.

W Crv: This system has P = 0.39 days, B-V = 0.71 and a primary component at least 600 K hotter than the secondary, unlike any known W UMa star with (B-V) $>$ 0.60. A model has been published by Lucy & Wilson (1979) and new spectroscopic and photometric observations are being analysed (D. Smith, private communication). Its light curve is EB rather than EW. Photometry and spectroscopy indicate a mass-ratio $q \sim 0.8$. Mochnacki (1981) has shown that this system has more angular momentum than typical for W UMa stars, and is therefore an unlikely candidate for a W UMa star in the semi-detached phase of its TRO cycle.

RW Dor: Mauder (1982) has noted the interesting nature of RW Dor. It has a period of 0.285 days, with a late G spectral type (Bidelman & Sanduleak 1982) and eclipse minima differing by 0.22 magnitudes (Martin & Grieco 1981).

V 1276 Sgr: A noisy light-curve for this object has been published by Mauder (1982). It has a period of 0.348 days (?) and spectral type about K3. The minima differ by 0.4 magnitudes. Mauder is re-observing this system.

FT Lupi, a system in the rapid phase of mass transfer?: The star FT Lup (P = 0.470 days) has recently been analysed by Hilditch et al. (1983). It has an F2V primary and a K2-5 secondary, with mass-ratio q = 0.45. Hilditch <u>et</u> <u>al</u>. argue that FT Lup is evolving rapidly into contact by transferring mass from the primary to the secondary. If this is so, then FT Lup is a most important system.

It is curious that the mass-losing primary is so much more massive than the secondary. Unevolved close binaries are thought to have q $\sim$1.0 (Lucy & Ricco 1979), so it is unusual to find a system undergoing its first phase of mass transfer with q = 0.45. The secondary in a system such as this is normally the more evolved component, still slowly losing mass after its initial rapid mass transfer phase. It may in fact be a good candidate for a W UMa system in the semi-detached phase of a thermal relaxation cycle.

<u>Other EB systems with 0.4 $<$ P $<$ 0.5</u>: Most of the short-period EB systems are fainter than 11th magnitude photographic, and are not well observed, especially spectroscopically. They include: CN And, V694 Aql, MT Her, WX Lib, UU Lyn, AS Ser, GR Tau, GY Tel, BS Vul. The Bamberg, Warsaw and St. Andrews groups, among others, are observing systems of this type.


3.   PHOTOMETRIC SURVEYS OF CONTACT BINARIES

The first large-scale photometric survey of contact binaries was done by Eggen (1961, 1967), in the Johnson UBV system. Eggen & Sandage (1969) published some observations of the four W UMa systems in NGC 188. Eggen (1978) surveyed hot contact systems in the uvby system; Rucinski & Kaluzny (1981) and Rucinski (1983) have observed W UMa stars in uvby.

Eggen's pioneering survey established the colour-period diagram for W UMa systems This diagram has been the principal testing-ground for all subsequent theories of contact binary structure and evolution. Lucy (1968a, fig.4) noted that two parallel sequences can be recognized in the colour-period diagram: systems with larger ultraviolet excesses $\delta$(U-B) are also bluer in (B-V) than systems of the same period with small ultraviolet excesses. A very similar effect is seen in the uvby data of Rucinski (1983). For single stars, higher ultraviolet excesses are usually attributed to lower metal abundances, implying membership in an older stellar population. Rucinski & Kaluzny (1981) have suggested that intense chromospheric activity, associated with the rapid rotation of cool stars, may also produce ultraviolet excesses. It is not obvious why this should also lead to segregation in the colour-period diagram.

## 4.   PERIOD CHANGES IN CONTACT BINARIES

Studies of period changes in W UMa stars have been carried out in recent years by Yamasaki (1975), Kreiner (1977), Herczeg (1979) and Panchatsaram & Abhyankar (1982). The most complete studies are those by Kreiner and Herczeg, but unfortunately neither study has been fully written up. Kreiner has published his diagrams, however. Both Kreiner and Herczeg have analysed the O-C diagrams of times of minima in terms of discrete jumps in period, whereas Panchatsaram and Abhyankar postulate additional stellar components. The (O-C) curves in Figure 1 of Kreiner (1977) vary quite smoothly, considering that timings of minima can be affected by photometric distortions due to "star-spot" activity and the like. If one takes $10^5$ days as a typical period for a triple system containing a W UMa binary, and assumes total masses of 1.5 $M_\odot$ in the binary and 0.5 $M_\odot$ in the third star, the W UMa star will have an orbital radius of roughly two light hours. This is similar to the semi-amplitude of the O-C curve of SW Lac, for example. Many contact binaries are known to have companions, and in the case of ADS 9537 two contact systems, BV and BW Dra, are found in one visual binary (Rucinski & Kaluzny 1982).

I am not suggesting that the only cause of period changes is the presence of third or fourth bodies. W Ursae Majoris itself has a well known discontinuous O-C curve (e.g. see Panchatsaram & Abhyankar 1982, Fig.7), but its amplitude in Kreiner's compilation is much less than the O-C amplitudes of systems such as 44 Boo B. We are probably seeing the combined effects of a third body and mass transfer in many contact binaries.

Two interesting cases: The following two stars are not W UMa systems, but are closely related. Their period changes deserve particular attention:

(a)  FT Lupi
This system has already been discussed in Section 2 as being possibly in a rapid phase of mass transfer. Its period is unmistakeably decreasing at a rate which implies a mass transfer rate of 3 x $10^{-7}$ $M_\odot$/yr (Mauder 1982), although its O-C curve could yet become sinusoidal due to a third body. Several more years of eclipse timings are needed to establish that large-scale mass transfer is indeed taking place. Although the light curve shows considerable variability, why are there no peculiarities observed in its spectra?

(b)  SV Centauri
This is a hot "contact" system, with component spectral types B1V and B6.5III. It has the most dramatically changing

period of any binary star known. The average rate $\dot{P}/P$ of period change is $-2.15 \times 10^{-5} \text{yr}^{-1}$, but the amount of variation is far from constant (Drechsel et al. 1982b). Fitting of models to light curves suggests that SV Cen has a common envelope rather close to its outer Lagrangian surface (Rucinski 1976a, Wilson and Starr 1976, Drechsel et al. 1982a), with a mass ratio of about 1.25 spectroscopically confirmed.

This bizarre system is even more puzzling when the IUE observations of Drechsel et al. (1982a,b) are considered. Firstly, resonance lines are seen blue-shifted in absorption, indicating a substantial stellar wind. Thus mass is being lost as well as transferred. Secondly, a strong UV continuum flux is seen shortward of 2000 Å, corresponding to a black-body temperature of $2 \times 10^5$ K. This flux comes from a small region corresponding to 0.6% of the total projected area of the binary. The exact location of the hot spot cannot be determined yet from the few observations published. It is possible that SV Cen is not close to outer contact but actually not in contact at all; the hot spot may distort the light-curve and mimic an over-contact envelope by adding extra light between the components. The light-curve in the visible is asymmetric and hardly fitted anyway by the near-to-outer-contact model.

The newly-discovered complications in SV Cen make untenable evolutionary conclusions based on conservative models (e.g. Nakamura & Nakamura 1982). This system deserves many more observations, particularly in the far ultraviolet.

## 5.    RESULTS OF LIGHT CURVE FITTING

Lucy (1968b) first demonstrated that the light curves of close binary systems could be directly synthesized by computer using the Roche geometry. The first fits to actual observed systems yielded surprising results, such as a mass ratio of less than 0.08 for Paczynski's Star, AW UMa (Mochnacki & Doughty 1972a). Many workers have since then produced models for contact binaries, often using the code of Wilson & Devinney (1971). Compilations of models have been published by Rucinski (1974), Wilson (1978), Mochnacki (1981), Rahunen (1981), van Hamme (1982) and Whyte (1982). Table 2 updates these lists. Only systems with $(B-V)_0 \geqslant 0.10$ are included, although this cut-off is arbitrary. The hot OBA systems are a somewhat neglected class, and need much more detailed study.

Table 2 lists observational quantities such as period and colour, and orbital elements derived from fitting models to light-curves and/or radial velocity curves. In most cases a synthetic fitting procedure has been carried out; for some, good spectroscopy is available but not a good light-curve fit (e.g. TZ Boo). The mass ratio q , the fill-out factor $F = (C_1 - C)/(C_1 - C_2)+1$ and the inclination i are listed. Quantities such as temperature differences and gravity and limb darkening coeficients are not listed since they depend too much on the model fitting procedures used by various workers. Twigg & Rafert (1980) find anomalous limb-darkening in contact and semi-detached systems they have fitted with the Wilson-Devinney code. Other workers also leave limb and gravity-darkening coefficients to be adjusted in their fitting procedures.

The sign of the apparent temperature difference between the components is contained in the A or W classification: the A-type systems have primaries with higher mean surface brightness, the W-type systems have secondaries with higher mean surface brightness (Binnendijk 1965). It now seems that the secondaries in the W-type systems are not really hotter than their primaries. Eaton, Wu & Rucinski (1980) use ultraviolet observations to argue that the apparently lower surface brightness of the primary component of W UMa itself is due to a greater density of starspots on the primary. Rucinski has discussed this effect at this Institute.

If the eclipses are partial, it is almost impossible to get the mass ratio of a contact binary without spectroscopy; furthermore, the spectroscopy has to be very good to obtain reliable masses. There are insufficient good radial velocity curves to provide a meaningful tabulation of absolute dimensions. Mochnacki (1981) has described a simple way of estimating masses and luminosities.

Derived Quantities
    Several interesting quantities can be derived from the parameters of the fitted models, as described in Mochnacki (1981):

(a) Mean density
    Kepler's law gives the mean density in abolute physical units,

$$\bar{\rho_1} = \frac{0 \cdot 079}{(1+q)V_1(F,q)P^2} \quad g \; cm^{-3},$$

(1)

where $\bar{\rho_1}$ is the mean density of the primary component, $V_1$ is the volume of the primary when the unit of length is the binary separation, q is the mass-ratio and P is the period in days. The normalized volume $V_1(F,q)$ is computed numerically for the F and q values found from fitting a model to the observations. The mean density is a probe of nuclear evolution.

(b)  Corrected temperature and colour
      According to Lucy's common convective envelope model, the primary component transfers a significant fraction of its nuclear luminosity to the secondary component, from whose surface the transferred energy is radiated. The corrected colour $(B-V)_1$ is defined as the colour the primary would have if it were not transferring energy to the secondary, without change of radius (Mochnacki 1981). The diagram of mean density versus corrected colour is a more powerful probe of nuclear evolution in contact binaries than a conventional HR diagram.

(c)  Minimum period
      The minimum possible period $P_0$ of a binary is the period the system would have if its mass were re-arranged to have components of equal mass, without change of total angular momentum. It is given by

$$P_0 = P \left\{ f(F,q)/f(1,1) \right\}^3$$

$$= \frac{P}{0 \cdot 0185} \left\{ \frac{q}{(1+q)^2} + \frac{k_1^2 + r_1^2}{1+q} + \frac{q \, k_2^2 + r_2^2}{1+q} \right\}^3,$$

(2)

where P is the observed period, q is the mass ratio, $k_1$ and $k_2$ are the radius of gyration of each component, and $r_1$ and $r_2$ are the volume radii of each component. The angular momentum J of a binary is then given by

$$J^3 = 0.0185 \, \frac{G^2 M^5}{2\pi} P_o \, ,$$

(3)

where M is the total mass of the system. Thus $P_o$ is a probe of angular momentum.

Discussion of the data

Two systems in the earlier compilation (Mochnacki 1981) have been deleted: 44 Boo B and CW Cas. The visual binary member 44 Boo B does not have a reliable colour, and its light curve shows much intrinsic variability. A model fitted to a light-curve seems to lie inside the contact surface (Maceroni et al. 1981). The system CW Cas has available only a non-Roche WINK solution, with no radial velocity curve. Four systems, FG Hya, AW UMa, DK Cyg and RR Cen are listed twice to illustrate differences between various fits; the Wilson-Devinney code seems to often give higher F values than obtained by other methods. A total of 55 systems is included in Table 2.

The stars TZ Boo and Y Sex do not have their fill-outs well determined. The value of F for TZ Boo is just a guess, being a rough average for G-type systems; the mass ratio, however, seems to be well determined spectroscopically. Y Sex was solved using a synthesis code valid only for the inner Lagrangian surface (F=1.0). The fill-outs of FG Hya and BU Vel are suspect, and better observations are required.

The system W Crv is classed as an A-type, but since it may not be in contact this classification is debatable. Its oddball nature is evident from the figures. The temperatures in Table 2 differ a little from Mochnacki (1981) due to a different interpolation procedure for the Morton-Adams colour-temperature calibration. Volume radii may be slightly different due to a more accurate numerical integration of the Roche geometry.

Figure 1 is the colour-period diagram for the systems in Table 2. Most of the $(B-V)_o$ values are from Eggen (1967), but several are from different sources or have been inferred from spectral types. The A-type and W-type systems occupy the top and bottom portions respectively of the colour-period diagram, indicating that if the W-type systems evolve into A-type systems, the evolution proceeds up the colour-period sequence.

Figure 2 shows the fill-out factor $F$ plotted against the mass-ratio $q$. This diagram extends those of Rucinski (1974) and Mochnacki (1981). The systems TZ Boo and Y Sex have not been plotted since $F$ was not determined directly for them. There is clearly some uncertainty about this parameter, especially for the A-type systems, where different authors disagree somewhat. However, the general trend is for the W-type systems to have smaller F values, averaging about 1.15, whereas the A-type systems show more scatter, with an average F of about 1.4. Figure 3 shows the fill-out F plotted against the de-reddened colour $(B-V)_o$. This diagram also shows the clear segregation of contact binaries into A and W types according to colour, even more clearly than according to mass-ratio. The only A-type systems in Table 2 with $(B-V)_o > 0.45$ are RZ Tauri, AK Her, FG HYa, TZ Boo and W Crv. Of these, AK Her, FG Hya and TZ are low-mass low-q A-type systems, RZ Tauri has uncertain reddening, while W Crv is either a non-contact system or has evolved into contact. This is more clearly seen in Figure 4, the plot of corrected colour $(B-V)_1$ against the minimum period $P_o$. The A-type and W-type systems are totally segregated except for W Crv, which sits distinctly at the long-period edge of the W-type systems. The W-type system OO Aql, seen to the right of W Crv, is clearly a system which could not have existed as a W UMa star at age zero on the main sequence. On the other hand, the semi-detached system FT Lup discussed earlier has $(B-V)_1 \simeq 0.4$ and $P_o \simeq 0.25$ days, which would put it at the boundary between the A-type and the W-type systems in Figure 4. Unlike W Crv, it is therefore a possible candidate for a W UMa system in the semi-detached phase of its TRO cycle.

An interesting possibility for future work is to plot binaries with EA, EB and EW light-curves on a diagram with minimum period versus mass ratio. Vilhu (1981, Fig.5) suggests a lack of non-contact systems with periods between 0.5 and 2.0 days in the spectral type range F5-K9. His diagram is based on the Seventh Catalogue of Spectroscopic Binaries (Batten et al. 1978), which is severely biassed by selection effects (cf. Griffin at this Institute). Nevertheless, the strong EW peak in this spectral range (Giuricin et al. 1983) does raise the possibility that short-period detached systems evolve into contact binaries, thereby depleting the number of EA and EB systems with short periods (Vilhu 1982).

Without danger of over-interpretation, one can conclude that the distinction between A and W-type systems is largely due to effective temperature, the division being at $(B-V)_o \sim 0.45$, with some dependence on mass ratio resulting in systems with $0.45 \lesssim (B-V)_o \lesssim 0.06$ and $q \lesssim 0.2$ also showing A-type light-curves. Most A-type systems also have $q \lesssim 0.3$, but some hot systems have higher mass ratios (e.g. S Ant, V1010 Oph).

Figure 5 shows the mean density of the primary component versus primary corrected colour. Again, the distinction between A and W types is clear, with many A-type and some W-type systems showing evidence of nuclear evolution.

## 6.   CONFRONTATION OF EVOLUTIONARY MODELS WITH OBSERVATIONS

The most important recent advance in understanding the evolution of contact binary stars is the realization by many workers of the importance of angular momentum loss, principally due to magnetic braking (e.g. Mochnacki 1981; Vilhu 1981, 1982; and references therein). Webbink (1976, 1979), and Robertson & Eggleton (1977) calculated evolutionary tracks including angular momentum loss due to gravitational braking, which alone is insufficient to explain how an object such as Paczynski's Star could evolve from a typical W-type system (Mochnacki 1981). All these calculations have assumed that the system is already a contact binary at age zero, and that it maintains a configuration close to its inner Lagrangian surface throughout its main sequence lifetime. This latter assumption, that $F \simeq 1.0$ is called the "contact condition". However, contact binaries may also be formed from initially non-contact systems, and the contact condition is probably thermally unstable. Rahunen (1981, 1982, 1983) and Whyte (1982) have included thermal relaxation oscillations plus a simple form of magnetic braking. Vilhu (1982) has considered the statistical consequences of short-period detached systems evolving into contact.

The theory of magnetic braking in binary stars is not well defined, mainly because we have no direct knowledge of the magnetic fields and mass loss rates in the winds from late-type dwarf stars other than the Sun. Furthermore, what indirect evidence exists is gained from slowly rotating single stars, whereas the W UMa stars are rapidly-rotating binaries. The main evidence concerning magnetic braking comes from Skumanich (1972) and Soderblom (1983), showing that the rotational velocity of single solar-type dwarfs is roughly proportional to the inverse square root of their age. Provided that the effective radius of gyration and the stellar wind expansion velocity at the Alfven radius remain constant with time, the Skumanich relation implies that the magnetic field is proportional to the rotation speed. For slowly rotating stars whose winds are accelerated primarily by the Parker mechanism, the angular momentum loss rate $\dot{J}$ depends only on the magnetic field. In rapidly rotating stars, the wind can be accelerated outwards by the magnetic field, and in this case one finds (Belcher & MacGregor 1976, Verbunt 1983)

$$\dot{J} \propto B_P^{4/3} \dot{M}^{1/3} \; ,$$

(4)

where $B_P$ is the open magnetic field averaged over the surface of the star and $\dot{M}$ is the mass loss rate (see Mochnacki 1981 for details). The full equation for the angular momentum loss rate of a binary system is

$$\frac{\dot{J}}{J} = - \frac{32 \, G^3 M^3 q}{5 c^5 A^4 (1+q)^2} - \left\{ \left( \frac{G^2 M^5}{4\pi |\dot{M}|} \right)^{1/3} \left( \frac{P}{B_P^2 R_1^4} \right)^{2/3} f(F,q) \right\}^{-1} ,$$

(5)

where the first term on the right hand side is due to gravitational radiation, and the second term is due to magnetic braking. The total mass is M , the mass-ratio is q , A is the orbital separation, P is the period, $R_1$ is the effective radius of the mass losing component and f(F,q) is the function in equation (2).

There are other ways to estimate the angular momentum loss rate, without appealing to a specific theory of stellar winds. Vilhu (1981, 1982) takes the Skumanich relation and deduces, for a binary

$$\dot{J} \simeq - 4 \times 10^{41} \left( \frac{P}{3} \right)^{-\alpha} \; g \; cm^2 \; s^{-1} \; yr^{-1} ,$$

(6)

where $\alpha$ = 3 for slow rotators. For periods $P \lesssim 3$ days, Vilhu has tried various values of the exponent $\alpha$, with $\alpha$ = 1.5 leading to a reasonable fit to the apparent period distribution of low mass binaries (Vilhu 1982).

In the following diagrams, I present the results of simple evolutionary calculations assuming long contact lifetimes, extending the models of Webbink (1976, 1979) and Mochnacki (1981) to include magnetic braking according to equation (5). I assume that only the primary has extensive surface activity (e.g Mullan 1975, Eaton, Wu & Rucinski 1980). The mass loss rate is given by Reimers' (1977) equation

$$\dot{M} \simeq -1 \times 10^{-13} \, L_1 R_1 M_1^{-1} \quad M_\odot \, yr^{-1};$$

(7)

where $L_1$, $R_1$ and $M_1$ are the luminosity, radius and mass respectively of the primary component. The magnetic field $B_P$ deduced from Skumanich's relation is $B_P = \alpha_p \Omega$, where $\alpha_p$ is a constant approximately equal to $3.5 \times 10^5$ Gauss seconds, and $\Omega$ is the angular rotation speed. However, this was found to give too strong a field, so I have arbitrarily chosen

$$B_P = \frac{B_o \, \alpha_p \, \Omega}{B_o + \alpha_p \, \Omega}$$

(8)

where $B_o$ is a maximum or saturation value of the braking field.

This approach, using equations (5), (7) and (8) is perhaps a better a priori estimate of $\dot{J}$, since estimates from single stars may be severely affected by non-uniform internal rotation. Such non-uniform rotation is now thought likely in the case of the Sun (e.g. Gough 1982), so that the effective radius of gyration may not be constant or even readily determinable. On the other hand, in a contact binary system the convective envelope is tidally coupled to the orbit, so that orbital angular momentum is extracted via the envelope and its wind. The internal spins of the component stars are important only at low mass ratios. Furthermore, the theory of angular momentum transport in a solar-type wind now rests on a firmer observational foundation (Pizzo et al. 1983), so that equation (5) should be a reasonable approximation.

The calculations were done exactly as in Mochnacki (1981), but with equation (5) for the angular momentum loss rate. It was found that $B_o = 30$ Gauss with $\alpha_p = 3.5 \times 10^5$ Gauss seconds gave the best agreement with the colour-period diagram (Fig.6), density-colour diagram (Fig.7) and colour-minimum period diagram (Fig.8). These diagrams can be compared with Mochnacki (1981), showing that magnetic braking can cover the observed distributions whereas gravitational radiation by itself is insufficient. The observational points in Figures 6, 7, and 8 are from Mochnacki (1981), but cover the same areas as the newer data in Figures 1, 4 and 5. Note, however, that the evolutionary model calculations all assume a constant fill-out $F = 1.0$.

The evolution of W-type systems into A-type systems is better understood with the inclusion of magnetic braking. The low-mass A-type systems such as TZ Boo and FG Hya appear to be the evolved analogues of low-mass W-type systems such as CC Com. Some of the hottest A-type systems are not highly evolved (compare Figure 7 with Figure 5), but the most evolved systems such as Paczyński's Star (AW UMa) can be readily explained if the long contact lifetime scenario is correct. A separate detached-to-contact sequence may exist in the colour-period diagram.

The magnetic fields $B_p$ are of the order of 20 Gauss in these models, with $B_o = 30$ Gauss. Extrapolation of Skumanich's relation $B_p = \alpha_p \Omega$ would imply $B_p \sim 100$ Gauss. These results agree with several independent lines of evidence suggesting that there is a saturation of magnetically-related activity at short orbital periods (Walter 1982, Vilhu 1982, Rucinski & Vilhu 1983, Rucinski 1983b).

It is quite likely that some contact binaries are formed from initially detached systems. However, the braking cannot be too rapid or else the contact lifetimes become too short. Since low mass-ratio A-type systems have low densities, the braking timescales cannot be much shorter than nuclear timescales. Braking timescales of $1 - 5 \times 10^8$ years proposed by van't Veer (1979) & Vilhu (1982) are probably too short to satisfy this requirement. Rahunen (1981) showed that rapid braking suppresses thermal relaxation oscillations, but if the braking is not rapid then Rahunen's explantion of the absence of detached TRO phases fails.

## 7.    CONCLUSION

I have attempted to review our present knowledge of the evolution of contact binary stars, in particular the W Ursa Majoris systems.  Since this subject is not well understood theoretically, it is vital to extract as much information as possible from observations.

Better observations are needed, especially to determine effective temperatures and mass-ratios. However, real progress is being made by many workers in fitting models to light curves, and the data collated in this review are of better quality than those I collected three years ago. It is encouraging to see some workers concentrating on certain subgroups, such as the bottom of the colour-period sequence. It would be most interesting if a distinct group can be distinguished of systems which have evolved into contact.

### ACKNOWLEDGEMENTS

This review was prepared while I was a Visitor at the Institute of Astronomy, University of Cambridge. I thank the Institute and Peter Eggleton for their unstinting hospitality. Calculations were performed on the Cambridge node of the SRC STARLINK network. I also thank Slawek Rucinski for many stimulating discussions, aided and abetted by Graham Berriman, Keith Horne, Janusz Kaluzny, Frank Verbunt and Charles Whyte.

Some of the calculations were carried out while I was a Research Fellow at the California Institute of Technology.

Table 1.

The fraction of eclipsing binaries with EB and EW light curves for various spectral ranges (adapted from Giuricin et al. 1983).

| Spectral Range | N(EB)/N(E) | N(EW)/N(E) | N(E) |
|---|---|---|---|
| O | .7 | .0 | 15 |
| B0 – B4 | .45 | .02 | 87 |
| B5 – B9.5 | .49 | .03 | 144 |
| A0 – A2 | .16 | .02 | 258 |
| A3 – A5 | .19 | .04 | 123 |
| A6 – A9 | .24 | .20 | 59 |
| F0 – F4 | .23 | .19 | 94 |
| F5 – F9 | .12 | .28 | 123 |
| G0 – G4 | .09 | .36 | 53 |
| G5 – G9 | .2 | .6 | 24 |
| KM | .15 | .4 | 13 |

Table 2

| System Name | Period (days) | (B-V)0 | q | F | I deg | r(1) | r(2) | den(1) g cm-3 | den(2) g cm-3 | T0 deg K | T1 deg K | (B-V)1 | P0 (days) | Type | Sources, notes. |
|---|---|---|---|---|---|---|---|---|---|---|---|---|---|---|---|
| CC Com | 0.221 | 1.24 | 0.521 | 1.24 | 90.0 | 0.455 | 0.342 | 2.687 | 3.294 | 4146 | 4568 | 1.08 | 0.171 | W | (1,2,3)a |
| V523 Cas | 0.237 | 1.07 | 0.610 | 1.15 | 80.8 | 0.436 | 0.350 | 2.516 | 2.973 | 4614 | 5085 | 0.87 | 0.204 | W | (4) |
| BI Vul | 0.252 | 0.94 | 0.690 | 1.29 | 78.8 | 0.437 | 0.373 | 2.099 | 2.333 | 4942 | 5404 | 0.75 | 0.236 | W | (4) |
| FS Cr A | 0.264 | 1.09 | 0.760 | 1.15 | 86.5 | 0.417 | 0.369 | 2.121 | 2.326 | 4559 | 4934 | 0.93 | 0.254 | W | (4) |
| FG Sct | 0.271 | 1.03 | 0.790 | 1.09 | 89.9 | 0.408 | 0.367 | 2.109 | 2.296 | 4721 | 5078 | 0.87 | 0.263 | W | (4) |
| VW Cep | 0.278 | 0.86 | 0.370 | 1.10 | 66.0 | 0.475 | 0.303 | 1.660 | 2.365 | 5123 | 5568 | 0.70 | 0.152 | W | (5)a,c |
| BX Peg | 0.280 | 0.66 | 0.371 | 1.16 | 87.8 | 0.479 | 0.308 | 1.597 | 2.232 | 5700 | 6197 | 0.54 | 0.154 | W | (6)b |
| XY Leo | 0.284 | 0.96 | 0.760 | 1.12 | 66.7 | 0.415 | 0.367 | 1.865 | 2.051 | 4895 | 5299 | 0.79 | 0.273 | W | (7)a |
| ER Cep | 0.286 | 0.75 | 0.590: | 1.15: | 77.8 | 0.439 | 0.347 | 1.719 | 2.056 | 5395 | 5950 | 0.59 | 0.241 | W | (8) |
| BW Dra | 0.292 | 0.63 | 0.830 | 1.10 | 69.7 | 0.405 | 0.372 | 1.823 | 1.948 | 5802 | 6176 | 0.54 | 0.287 | W | (9,10) |
| TZ Boo | 0.297 | 0.61 | 0.130 | (1.10) | 90.0 | 0.563 | 0.226 | 1.062 | 2.137 | 5885 | 6117 | 0.56 | 0.036 | A | (11)a,d |
| TW Cet | 0.317 | 0.69 | 0.581 | 1.03 | 82.8 | 0.431 | 0.336 | 1.483 | 1.821 | 5600 | 6179 | 0.54 | 0.262 | W | (12)b |
| GW Cep | 0.319 | 0.63 | 0.370 | 1.11 | 83.9 | 0.476 | 0.304 | 1.255 | 1.781 | 5800 | 6304 | 0.52 | 0.174 | W | (6)c |
| SW Lac | 0.321 | 0.75 | 0.870 | 1.37 | 81.8 | 0.427 | 0.403 | 1.254 | 1.297 | 5395 | 5651 | 0.68 | 0.327 | W | (13,10) |
| YY Eri | 0.321 | 0.66 | 0.590 | 1.18 | 79.2 | 0.441 | 0.349 | 1.343 | 1.597 | 5685 | 6269 | 0.52 | 0.272 | W | (2)a |
| FG Hya | 0.328 | 0.59 | 0.145 | 1.50 | 88.0 | 0.568 | 0.251 | 0.833 | 1.399 | 5973 | 6239 | 0.53 | 0.046 | A | (14) |
| { | 0.328 | 0.59 | 0.142 | 1.80 | 85.2 | 0.581 | 0.265 | 0.782 | 1.178 | 5973 | 6244 | 0.53 | 0.046 | A | (15) |
| AB And | 0.332 | 0.88 | 0.526 | 1.45 | 86.0 | 0.472 | 0.362 | 1.068 | 1.246 | 5078 | 5594 | 0.69 | 0.264 | W | (15,16) |
| W U Ma | 0.334 | 0.66 | 0.490 | 1.14 | 81.4 | 0.454 | 0.330 | 1.215 | 1.553 | 5685 | 6256 | 0.53 | 0.244 | W | (17) |
| RZ Com | 0.338 | 0.53 | 0.430 | 1.00 | 86.0 | 0.456 | 0.308 | 1.221 | 1.692 | 6264 | 6855 | 0.42 | 0.216 | W | (11,18) |
| V757 Cen | 0.343 | 0.62 | 0.701 | 1.13 | 69.1 | 0.423 | 0.361 | 1.248 | 1.409 | 5843 | 6368 | 0.50 | 0.319 | W | (6)c |
| BV Dra | 0.350 | 0.54 | 0.800 | 1.09 | 74.5 | 0.407 | 0.368 | 1.265 | 1.370 | 6213 | 6667 | 0.45 | 0.341 | W | (10) |
| AC Boo | 0.352 | 0.59 | 0.295 | 1.18 | 83.3 | 0.499 | 0.290 | 0.946 | 1.421 | 5973 | 6419 | 0.50 | 0.146 | W | (19)c |
| AH Cnc | 0.360 | 0.47 | 0.500 | 1.15 | 67.0 | 0.453 | 0.332 | 1.046 | 1.325 | 6589 | 7255 | 0.36 | 0.268 | W | (20) |
| RW Ps A | 0.360 | 0.75 | 0.813 | 1.08 | 77.5 | 0.405 | 0.369 | 1.210 | 1.304 | 5395 | 5771 | 0.64 | 0.352 | W | (21) |
| AE Phe | 0.362 | 0.56 | 0.400 | 1.17 | 69.0 | 0.473 | 0.315 | 0.971 | 1.319 | 6100 | 6656 | 0.46 | 0.217 | W | (22)c |
| AM Leo | 0.366 | 0.53 | 0.415 | 1.22 | 85.8 | 0.473 | 0.321 | 0.941 | 1.248 | 6264 | 6848 | 0.43 | 0.229 | W | (13) |
| HD101799 | 0.370 | 0.60 | 0.299 | 1.07 | 83.2 | 0.492 | 0.284 | 0.890 | 1.378 | 5929 | 6372 | 0.51 | 0.155 | W | (23)b |
| XY Boo | 0.371 | 0.49 | 0.185 | 1.55 | 68.4 | 0.553 | 0.273 | 0.683 | 1.046 | 6478 | 6827 | 0.43 | 0.080 | A | (11,24) |
| U Peg | 0.375 | 0.62 | 0.540: | 1.16 | 74.9 | 0.447 | 0.339 | 0.977 | 1.203 | 5843 | 6443 | 0.49 | 0.297 | W | (12)c |

Table 2 (cont.)

| System Name | Period (days) | (B-V)0 | q | F | I deg | r(1) | r(2) | den(1) g cm-3 | den(2) g cm-3 | T0 deg K | T1 deg K | (B-V)1 | P0 (days) | Type | Sources, notes |
|---|---|---|---|---|---|---|---|---|---|---|---|---|---|---|---|
| TX Cnc | 0.383 | 0.59 | 0.620 | 1.21 | 63.1 | 0.439 | 0.356 | 0.937 | 1.090 | 5973 | 6578 | 0.47 | 0.335 | W | (25,26;11)a |
| AU Ser | 0.386 | 0.82 | 0.670 | 1.13 | 87.0 | 0.426 | 0.356 | 0.982 | 1.128 | 5216 | 5725 | 0.65 | 0.351 | W | (13) |
| W Crv | 0.388 | 0.71 | 0.780 | 1.25 | 86.5 | 0.424 | 0.381 | 0.921 | 0.991 | 5514 | 5934 | 0.60 | 0.380 | A | (21)c |
| EM Lac | 0.389 | 0.68: | 0.640 | 1.23 | 74.3 | 0.438 | 0.360 | 0.904 | 1.039 | 5614 | 6173 | 0.54 | 0.348 | W | (10) |
| AH Vir | 0.408 | 0.76 | 0.342 | 1.26 | 86.5 | 0.491 | 0.307 | 0.711 | 0.993 | 5368 | 5813 | 0.63 | 0.205 | W | (7) |
| V566 Oph | 0.410 | 0.39 | 0.238 | 1.25 | 80.0 | 0.520 | 0.277 | 0.644 | 1.014 | 7051 | 7502 | 0.31 | 0.126 | A | (14;16,27,28) |
| RZ Tau | 0.416 | 0.49 | 0.372 | 1.55 | 82.9 | 0.503 | 0.336 | 0.623 | 0.780 | 6478 | 7048 | 0.39 | 0.237 | A | (18) |
| Y Sex | 0.420 | 0.43 | 0.190 | (1.00) | 76.8 | 0.528 | 0.247 | 0.611 | 1.132 | 6818 | 7175 | 0.37 | 0.090 | A | (11,29)d |
| AK Her | 0.422 | 0.53 | 0.260 | 1.16 | 81.0 | 0.508 | 0.279 | 0.639 | 1.010 | 6264 | 6689 | 0.45 | 0.147 | A | (21) |
| ER Ori | 0.423 | 0.49 | 0.757 | 1.06 | 81.1 | 0.409 | 0.360 | 0.875 | 0.970 | 6478 | 7020 | 0.40 | 0.403 | W | (12) |
| AW U Ma | 0.439 | 0.36 | 0.079 | 1.50 | 80.0 | 0.610 | 0.208 | 0.399 | 0.799 | 7228 | 7431 | 0.33 | 0.026 | A | (30,28) |
| | 0.439 | 0.36 | 0.072 | 1.74 | 79.1 | 0.623 | 0.210 | 0.378 | 0.710 | 7228 | 7422 | 0.33 | 0.023 | A | (18,31,32) |
| Y502 Oph | 0.453 | 0.61 | 0.380 | 1.24 | 71.3 | 0.482 | 0.315 | 0.596 | 0.809 | 5885 | 6406 | 0.50 | 0.257 | W | (2) |
| TY Men | 0.462 | 0.26 | 0.215 | 1.09 | 79.5 | 0.521 | 0.260 | 0.514 | 0.885 | 7775 | 8229 | 0.16 | 0.121 | A | (33) |
| DK Cyg | 0.471 | 0.28 | 0.330 | 1.35 | 84.0 | 0.500 | 0.311 | 0.512 | 0.704 | 7692 | 8316 | 0.17 | 0.228 | A | (14) |
| | 0.471 | 0.28 | 0.271 | 1.55 | 80.3 | 0.526 | 0.307 | 0.460 | 0.629 | 7692 | 8242 | 0.18 | 0.179 | A | (15) |
| OO Aql | 0.507 | 0.76: | 0.824 | 1.06 | 87.2 | 0.402 | 0.368 | 0.617 | 0.663 | 5368 | 5726 | 0.65 | 0.497 | W | (15) |
| BU Vel | 0.516 | 0.29 | 0.251 | 1.61 | 84.9 | 0.535 | 0.304 | 0.371 | 0.508 | 7635 | 8153 | 0.20 | 0.177 | A | (15) |
| Eps Cr A | 0.591 | 0.39 | 0.113 | 1.32 | 72.3 | 0.580 | 0.225 | 0.248 | 0.480 | 7051 | 7306 | 0.35 | 0.058 | A | (15,34) |
| AQ Tuc | 0.595 | 0.29 | 0.277 | 1.29 | 78.6 | 0.510 | 0.292 | 0.314 | 0.464 | 7635 | 8181 | 0.19 | 0.228 | A | (22) |
| RR Cen | 0.606 | 0.34 | 0.180 | 1.40 | 78.7 | 0.549 | 0.263 | 0.263 | 0.430 | 7345 | 7728 | 0.27 | 0.124 | A | (14) |
| | 0.606 | 0.34 | 0.179 | 1.35 | 78.7 | 0.547 | 0.260 | 0.266 | 0.442 | 7345 | 7725 | 0.27 | 0.122 | A | (15) |
| UZ Leo | 0.618 | 0.35 | 0.249 | 1.96 | 82.4 | 0.554 | 0.328 | 0.232 | 0.278 | 7257 | 7768 | 0.27 | 0.216 | A | (15) |
| V535 Ara | 0.629 | 0.10 | 0.361 | 1.13 | 82.0 | 0.479 | 0.303 | 0.319 | 0.455 | 8767 | 9517 | 0.01 | 0.334 | A | (35)b |
| S Ant | 0.648 | 0.32 | 0.590 | 1.09 | 69.5 | 0.434 | 0.342 | 0.346 | 0.418 | 7461 | 8231 | 0.18 | 0.544 | A | (12) |
| V1010 Oph | 0.661 | 0.18 | 0.489 | 1.18 | 85.1 | 0.456 | 0.332 | 0.305 | 0.387 | 6263 | 9092 | 0.06 | 0.483 | A | (36)b |
| V1073 Cyg | 0.786 | 0.10 | 0.240 | 1.13 | 69.6 | 0.514 | 0.271 | 0.182 | 0.297 | 8767 | 9326 | 0.03 | 0.243 | A | (15)b |
| MW Pav | 0.795 | 0.36 | 0.182 | 1.51 | 85.1 | 0.552 | 0.270 | 0.150 | 0.234 | 7260 | 7645 | 0.29 | 0.167 | A | (35) |
| TY Pup | 0.819 | 0.26 | 0.326 | 1.64 | 67.8 | 0.518 | 0.330 | 0.152 | 0.192 | 7800 | 8436 | 0.15 | 0.400 | A | (2)b |

## Table 2 (cont.)

### Source References

(1) Rucinski 1976b; Rucinski, Whelan and Worden 1977.
(2) Maceroni, Milano and Russo 1982.
(3) Bradstreet 1983.
(4) Bradstreet 1981.
(5) Yamasaki 1982.
(6) Kaluzny 1983.
(7) Kaluzny and Pojmanski 1983.
(8) Worden et al. 1978.
(9) Rucinski and Kaluzny 1982.
(10) Maceroni, Milano and Russo 1983.
(11) McLean and Hilditch 1983.
(12) Russo et al. 1982.
(13) Twigg 1979a.
(14) Mochnacki and Doughty 1972b.
(15) Twigg 1979b.
(16) Berthier 1975.
(17) Hilditch 1981 ("xfl").
(18) Wilson and Devinney 1973.

(19) Schleven et al. 1983.
(20) Whelan et al. 1979.
(21) Lucy and Wilson 1979.
(22) Maceroni et al. 1981.
(23) Leung 1976.
(24) Winkler 1977.
(25) Whelan, Worden and Mochnacki 1973.
(26) Wilson and Biermann 1976.
(27) Nagy 1977.
(28) McLean 1981.
(29) Hill 1979.
(30) Mochnacki and Doughty 1972a.
(31) Woodward, Koch and Eisenhardt 1980.
(32) Hrivnak 1982.
(33) Lapasset 1980.
(34) Tapia and Whelan 1975.
(35) Schoffel 1979.
(36) Leung and Wilson 1977.

### Notes

a) Spectroscopically determined mass ratio q.
b) Colour inferred from spectral type.

c) Average of parameters from one source.
d) Fill-out F assumed.

Properties of W Ursae Majoris Stars. Periods; dereddened colours $(B-V)_0$, mass ratio q; fill-outs F; inclinations i; volume radii $r_1$, $r_2$; mean temperatures $T_m$; corrected primary temperatures $T_1$, and colours $(B-V)_1$; minimum periods $P_0$; light curve type.

## Colour—Period Diagram

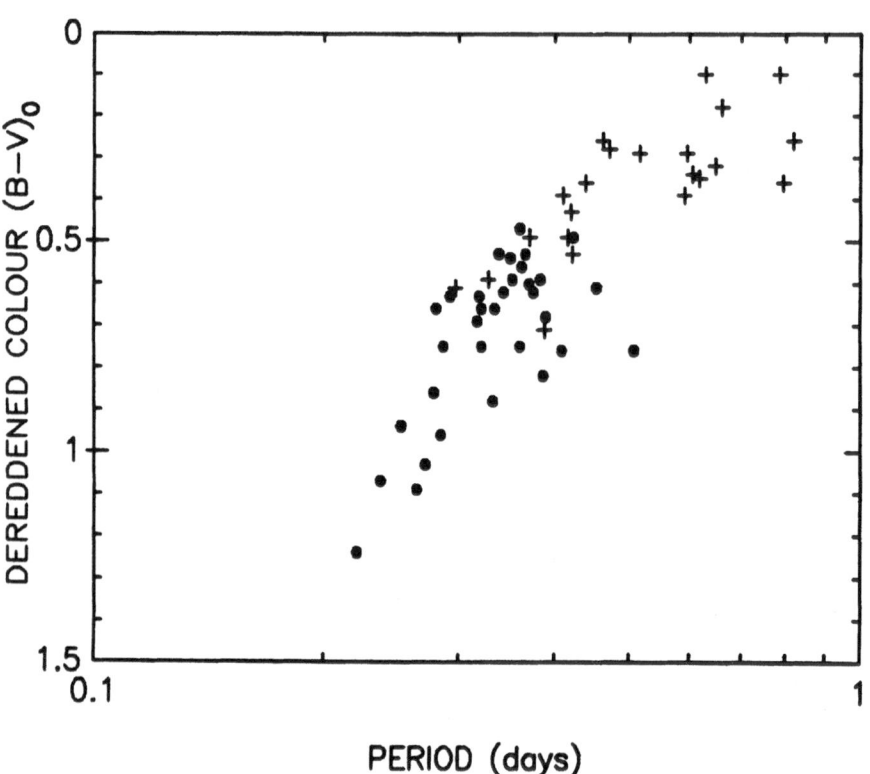

Figure 1.  Colour-period diagram for the systems in Table 2.
Data is mostly from Eggen (1967), but in a number of cases colours
have been inferred from spectral types or new observations have
been made.  Solid circles are W-type systems, crosses are A-types.

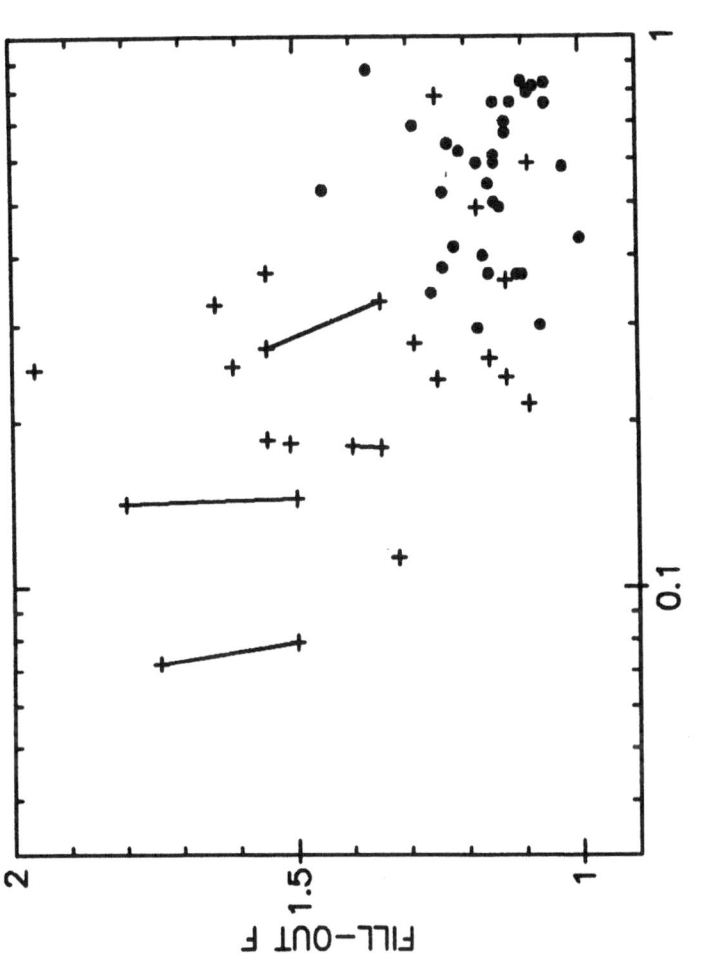

Figure 2. Fill-out plotted agains mass-ratio. Solid circles are W-types, crosses are A-types

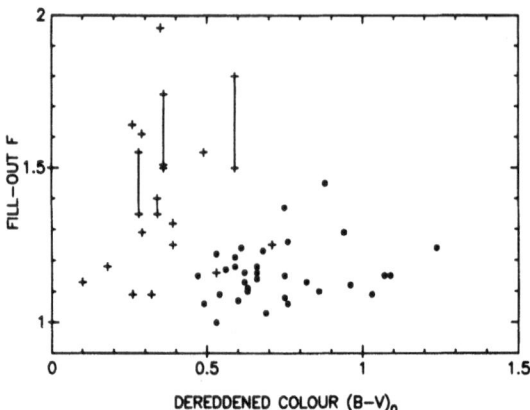

Figure 3.  Fill-out plotted against de-reddened colour index.
Solid circles are W-types, crosses are A-types.

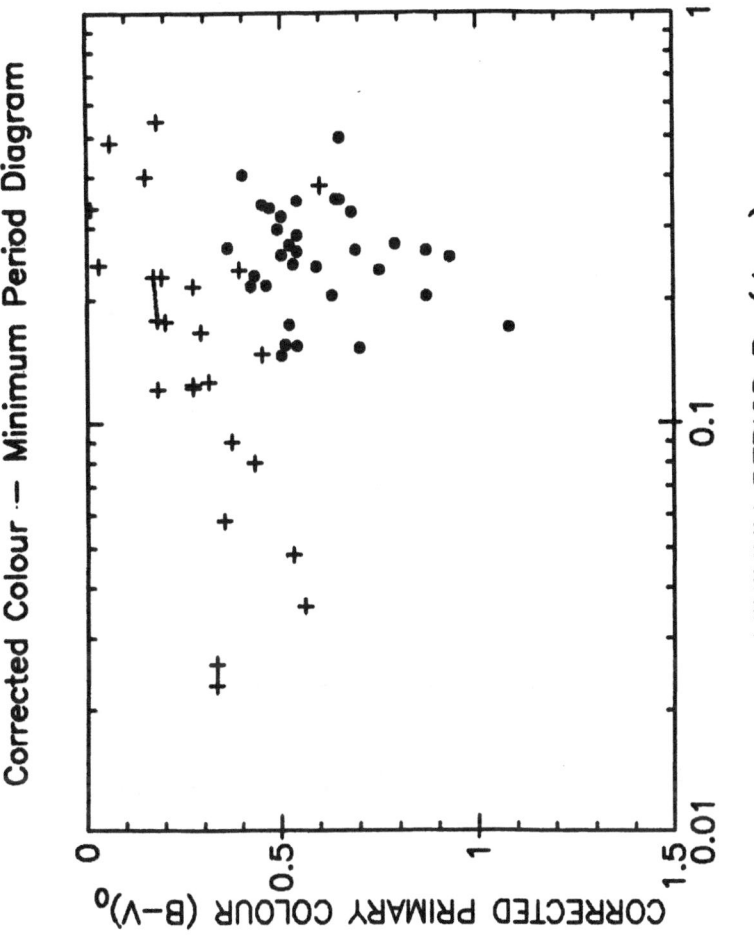

Figure 4. Corrected colour of the primary versus the minimum period. Solid circles are W-type systems, crosses are A-types.

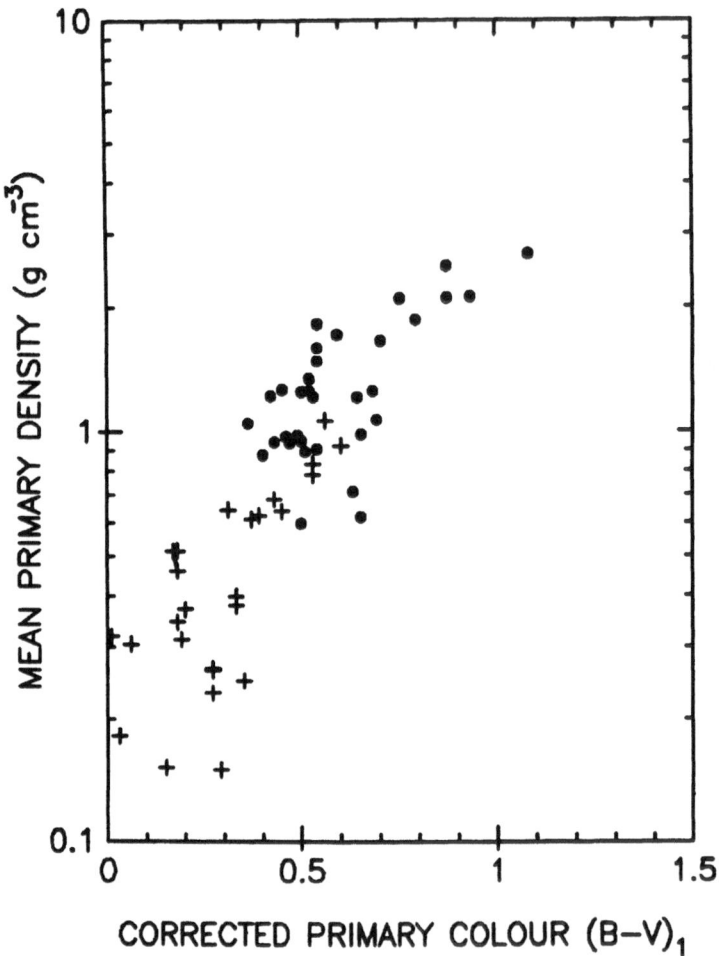

Figure 5. Mean density of the primary component versus corrected primary colour. Solid circles are W-types, crosses are A-types.

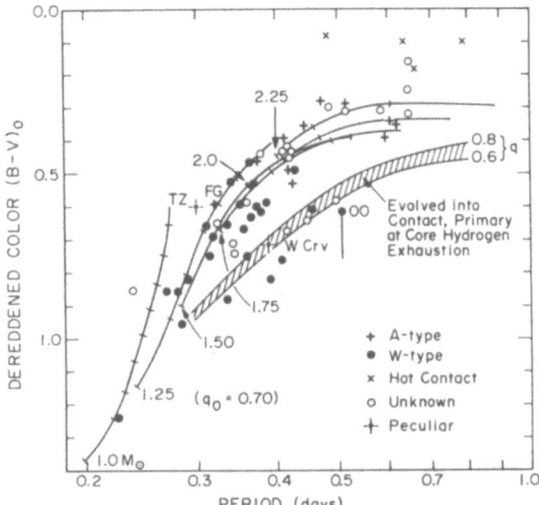

Figure 6. Colour–period diagram for contact binaries. The observed points are reliably de-reddened values from Eggen (1967), or systems listed by Mochnacki (1981). The theoretical tracks have been computed with angular momentum loss due to magnetic braking and gravitational radiation. The initial mass ratio $q_0$ is 0.7, and the metallicity Z is 0.02. Intervals of one gigayear are marked by ticks. The shaded band is where one expects to find systems with mass ratios in the range 0.6-0.8 and primaries at core hydrogen exhaustion. Such systems must have evolved into contact from initially detached configurations. The interesting systems TZ Bootes, FG Hydrae, W Corvi, OO Aquilae and AW Ursae Majoris are indicated.

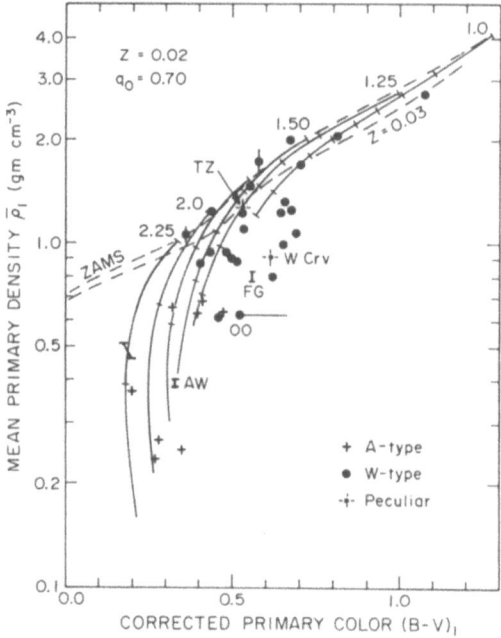

Figure 7. Mean primary density versus corrected primary colour.
The observed values are from Mochnacki (1981). Theoretical tracks
correspond to those in Figure 6. The effect of changing metallicity is shown by the zero age main sequence for Z = 0.03 (lower
dashed line).

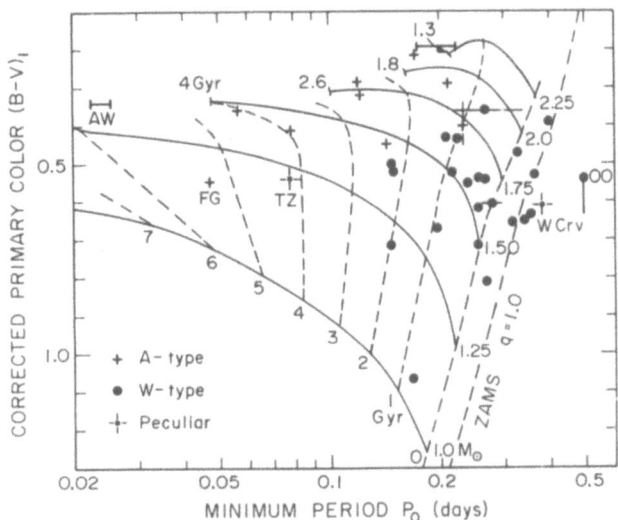

Figure 8. Corrected primary colour versus minimum period. The observed values are from Mochnacki (1981). The theoretical tracks (solid lines) are intersected by isochrones (dashed lines) labelled in gigayears, and correspond to the tracks in Figures 6 and 7. The ages at core hydrogen exhaustion of the primary are given in gigayears at the top end of each track. The dashed line marked ZAMS shows the period of systems with two identical components in contact at age zero.

Comparison with Figure 10 of Mochnacki (1981) clearly shows the better agreement with observations obtained by including magnetic braking in the evolutionary model.

REFERENCES

Alphenaar, P. & van Leeuwen, F., 1981.
    Inf.Bull.Var.Stars, No.1957.
Batten, A.H., Fletcher, J.M. & Mann, P.J., 1978.
    Publ. D.A.O., **15**, No.5.
Belcher, J.W. & MacGregor, K.B., 1976.  Ap.J., **210**, 498.
Berthier, E., 1975, Astron. Ap., **40**, 237.
Bidelman, W.P. & Sanduleak, N., 1982. Inf.Bull.
    Var.Stars, No.2122.
Binnendijk, L., 1965.  In "The Position of Variable Stars
    in the Hertzsprung-Russell Diagram", IAU  3rd
    3rd Colloquium on Variable Stars, ed. W. Strohmeier
    (Kl. Veroff. Bamberg, **40**, No.40), p.36.
Bradstreet, D.H., 1981.  A.J., **86**, 98.
Bradstreet, D.H., 1983.  Ph.D. Thesis, Univ. of
    Pennsylvania.
Budding, E., 1981.  In "Investigating the Universe",
    ed F.F. Kahn (Dordrecht: Reidel), p.271.
Budding, E., 1982.  In "Binary and Multiple Stars as
    Tracers of Stella Evolution", IAU Colloquium
    No. 69, ed. Z. Kopel and J. Rahe (Dordrecht:
    Reidel), p.351.
Drechsel, H., Rahe, J., Wargau, W. & Wolf, B., 1982a.
    In "Binary and  Multiple Stars as Traces of
    Stellar Evolution", IAU Colloquium No.69, ed.
    Z. Kopal and J. Rahe (Dordrecht: Reidel), p.205.
Drechsel, H., Rahe, J., Wargau, W. & Wolf, B., 1982b.
    Astr.Ap., **110**, 246.
Eggen, O.J., 1961.  Royal Obs. Bull., No. 31.
Eggen, O.J., 1967.  Mem. R.A.S., **70**, 111.
Eggen, O.J., 1978, A.J., **83**, 288.
Eggen, O.J., & Sandayl, A., 1969.  Ap.J. **158**, 669.
Eaton, J.A., Wu, C.-C. & Rucinski, S.M., 1980.
    Ap.J., **239**, 919.
Flannery,B.P. 1976.  Ap.J., **205**, 217.
Gough, D.O., 1982.  Nature, **298**, 334.
Giaricin, G., Mardirossian, F. & Mezzetti, M., 1983.
    Astr.Ap., **119**, 218.
Herczeg, T.J., 1979.  Bull. A.A.S. **11**, 438.
Hilditch, R.W., 1981. M.N.R.A.S., **196**, 305.
Hilditch, R.W., King, D.J., Hill, G. & Poeckert, R.,
    1983, submitted for publication.
Hill, G., 1979.  Publ. D.A.O., **15**, 297.
Hrivnak, B.J., 1982.  Ap.J., **260**, 744.
Hutchings, J.B. & Hill, G., 1973.  Ap.J., **179**, 539.
Kaluzny, J., 1983.  Preprint.
Kaluzny, J. & Pajmanski, G., 1983.  Preprint.
Kraft, R.P., 1965.  Ap.J., **142**, 1588.
Kraft, R.P., 1967.  P.A.S.P., **76**, 430.

Kreiner, J.M., 1977. In "The Interaction of Variable
    Stars with their Environment", IAU Colloquium
    No.42, ed. R. Kippenhahn and W. Strohmeier
    (Bamberg), p.393.
Lapasset, E., 1980. Astr.Ap., **82**, 225.
Leung, K.-C., 1976. P.A.S.P., **88**, 936.
Leung, K.-C., & Wilson, R.E. 1977, Ap.J., **211**, 853.
Lucy, L.B., 1968a. Ap.J. **151**, 1123.
Lucy, L.B., 1968b. Ap.J. **153**, 877.
Lucy, L.B., 1973. Ap.Space Sci., **22**, 381.
Lucy, L.B., 1976. Ap.J., **205**, 208.
Lucy, L.B. & Ricco, E., 1979, A.J., **84**, 401.
Lucy, L.B. & Wilson, R.E., 1979, Ap.J., **231**, 502.
Maceroni, C., Milano, L., Russo, G. & Sollazzo, C.,
    1981. Astr.Ap.Suppl., **45**, 187.
Maceroni, C., Milano, L., & Russo, G., 1982.
    Astr.Ap.Suppl., **49**, 123.
Maceroni, C., Milano, L., & Russo, G., 1983.
    Astr.Ap.Suppl., **51**, 435.
Marton, S.F. & Grieco, A., 1981. Inf. Bull.Var.
    Stars No. 1960.
Mauder, H., 1982. In "Binary and Multiple Stars as
    Tracers of Stellar Evolution", IAU Colloquium No.69,
    ed. Z. Kopal and J. Rahe (Dordrecht: Reidel), p.275.
McLean, B.J., 1981. M.N.R.A.S., **195**, 931.
McLean, B.J. & Hiditch, R.W., 1983. M.N.R.A.S., **203**, 1.
Mochnacki, S.W., 1981, Ap.J., **245**, 650.
Mocknacki, S.W. & Doughty, N.A., 1972a. M.N.R.A.S.,
    **156**, 51.
Mocknacki, S.W. & Doughty, N.A., 1972b. M.N.R.A.S.,
    **156**, 243.
Mullan, D.J., 1975. Ap.J., **198**, 563.
Nagy, T.A., 1977. P.A.S.P., **89**, 366.
Nakamura, M. & Nakamura, Y., 1982, Ap.Space Sci.,
    **83**, 163.
Panchatsaram, J. & Abhyankar, K.D., 1982. On "Binary
    and Multiple Stars as Tracers of Stellar
    Evolution", ed. Z. Kopal and J. Rahe (Dordrecht:
    Reidel), p.47.
Pizzo, V., Schwenn, R., Marsch, E., Rosenbauer, H.,
    Muhlhauser, K.-H. & Neubauer, F.M., 1983.
    Ap.J., **271**, 335.
Popper, D.M., 1982. Ap.J., **262**, 641.
Rahunen, T., 1981. Astr.Ap., **102**, 81
Rahunen, T., 1982. Astr.Ap., **109**, 66.
Rahunen, T., 1983. Astr.Ap., **117**, 135.
Reimers, D., 1977. In "The Interaction of Variable
    Stars with their Environment", IAU Colloquium
    No.42, ed. R. Kippenhahn, J. Rahe &
    W. Strohmeier (Bamberg), p.559.

Robertson, J.A., 1980. M.N.R.A.S., **192**, 263.
Robertson, J.A. & Eggleton, P.P., 1977. M.N.R.A.S., **179**, 359.
Roxburgh, I.W., 1966. Ap.J., **143**, 111.
Rucinski, S.M., 1974, Acta Astr., **24**, 119.
Rucinski, S.M., 1976a. P.A.S.P., **88**, 244.
Rucinski, S.M., 1976b. P.A.S.P., **88**, 777.
Rucinski, S.M., 1983a. Submitted for publication.
Rucinski, S.M., 1983b. The Observatory, submitted for publication.
Rucinski, S.M. & Kaluzny, J., 1981. Acta Astr., **31**, 409.
Rucinski, S.M. & Kaluzny, J., 1982. Ap.Space Sci., **88**, 433.
Rucinski, S.M. & Vilhu, O., 1983. M.N.R.A.S., **202**, 1221.
Rucinski, S.M., Whelan, J.A.J. & Worden, S.P., 1977. P.A.S.P., **89**, 684.
Russo, G., Sollazzo, C., Maceroni, C. & Maceroni, C. & Milano, L., 1982. Astr.Ap.Suppl., **47**, 211.
Schieven, G., Morton, J.C., McLean, B.J. & Hughes, V.A., 1983. Astr.Ap.Suppl., **52**, 463.
Schoffel, E., 1979. Astr.Ap.Suppl., **36**, 287.
Shapley, H., 1948. In "Centennial Symposia", Harvard Obs. Monograph No.9, p.249.
Shu, F.M., 1980. In "Close Binary Stars: Observations and Interpretation", IAU Symposium No.88, ed. M.J. Plavec, D.M. Popper and K.K. Ulrich (Dordrecht: Reidel), p.477.
Shu, F.M. & Lubow, S.M., 1981. Ann.Revc.Astr.Ap., **19**, 277.
Skumanich, A., 1972. Ap.J., **171**, 565.
Soderblom, D.R., 1983. Ap.J. Suppl., **53**, 1.
Tapia, S. & Whelan, J.A.J., 1975. Ap.J., **200**, 98.
Twigg, L.W., 1979a. Private communication.
Twigg, L.W., 1979b. M.N.R.A.S., **189**, and microfiche MN 189/2.
Twigg, L.W. & Rafert, J.B., 1980. M.N.R.A.S., **193**, 775.
van Hamme, W., 1982. Astr.Ap., **116**, 27.
van't Veer, F., 1979. Astr.Ap., **80**, 287.
Verbunt, F., 1983. In preparation.
Vilhu, O., 1981. Ap. Space Sci., **78**, 401.
Vilhu, O., 1982. Astr.Ap., **109**, 17.
Vyssotsky, A.N., 1956. A.J., **61**, 201.
Walter, F., 1982. Ap.J., **253**, 745.
Webbink, R.F., 1976. Ap.J., **209**, 829.
Webbink, R.R., 1979. Ap.J., **227**, 178.
Whelan, J.A.J., Worden, S.P. and Mochnacki, S.W., 1973, Ap.J., **183**, 133.

# TIDAL INTERACTIONS OF CLOSE BINARY SYSTEMS

G.J. Savonije

Astronomical Institute
"Anton Pannekoek"
University of Amsterdam      and
The Netherlands

J.C.B. Papaloizou

Dept. of Applied Mathematics
Queen Mary College
Mile End Road
London E1 4NS, U.K.

## 1.     INTRODUCTION

Observations show that close binary systems (in the sense that the orbital separation D is not much larger than the stellar dimensions) tend to have small orbital eccentricities (e.g. Young & Koniges 1977, Koch & Hrivnak 1981, Middelkoop & Zwaan 1982). There appears to be no such correlation for wide binary systems, which show a wide scatter of orbital eccentricities. This phenomenon is usually explained in terms of tidal interaction between the two stars, which is only effective in close binary systems. The mutual gravitational attraction between the two stars in a binary system induces a "tidal" deformation. In the case of an eccentric orbit, or when the rotation of the binary stars is not synchronized with the orbital revolution, each star in the binary feels a varying potential as its companion revolves around in its orbit. The varying potential excites tidal oscillations in the stars. Inevitably, part of the orbital energy fed into the tidal oscillations is dissipated, causing a secular variation of the orbital parameters.

Eqn. (1) relates the orbital eccentricity e to the orbital angular momentum H and orbital energy E:

$$e = \left[ 1 + \frac{2(M_p + M_s)}{G^2 M_p^3 M_s^3} E H^2 \right]^{\frac{1}{2}} \tag{1}$$

*P. P. Eggleton and J. E. Pringle (eds.), Interacting Binaries, 83–102.*
© *1985 by D. Reidel Publishing Company.*

where G is the constant of gravity and $M_P$ and $M_s$ are the masses of the primary and secondary component. In general both E and H change as a result of tidal effects. This is illustrated (Fig. 1) for the case of a circular orbit and stars aligned perpendicular to the orbital plane, but with a secondary that rotates slower than synchronously. In this situation a tidal wave is excited that runs along the surface of the secondary with the relative orbital angular velocity of its companion. As a result of the inevitable dissipation in the tidal flow, the two tidal bulges lag behind the orbiting companion by a small angle $\delta$. This phase shift gives rise to a tidal torque because the distance to the perturbing companion differs slightly for the two bulges. The tidal torque converts orbital angular momentum to spin angular momentum of the slowly rotating secondary, so that this star is spun up towards corotation. Thus dissipation of orbital energy is accompanied by exchange of orbital and spin angular momentum in the binary. In the particular case of e = 0 discussed above Eqn. (1) requires $EH^2 < 0$ to remain constant, while both E and H decrease as a result of tidal effects.

Tidal evolution tends in general to a state in which the tidal deformation becomes stationary and the dissipation vanishes. The tidal deformation of the stars is stationary if none of the following conditions is violated:

i.          the orbit is circular
ii.         the rotation axes of the two stars are aligned
            perpendicular to the orbital plane
iii.        the rotation of the two stars is synchronized with
            the orbital revolution.

It can be shown by a linear perturbation analysis (e.g. Counselman 1973, Hut 1980) that such a tidal equilibrium state can only be attained if the orbital angular momentum of the binary remains larger than three times the sum of the stellar spin angular momenta. When this condition is violated tidal effects will cause the two stars either to spiral in until they collide or, conversely, to spiral out indefinitely, depending on the initial conditions.

It is of interest to estimate the timescale on which the tidal evolution towards tidal equilibrium, described above, takes place. Papaloizou & Pringle (1982) studied the damping of precessional motion in slightly non-aligned binary stars and concluded that this damping occurs on a timescale that is (usually much) shorter than the circularization timescale. One expects that (for binary systems that fulfill the above mentioned stability condition) both alignment and corotation will in general be attained before the orbit is circularized.

Exceptions to this rule may occur for systems that are close to the tidal instability mentioned above, i.e. when the orbital angular momentum of the binary is not much larger than the total spin angular momentum that corresponds to synchronized binary stars. This is, for example, the case with massive X-ray binary systems which have extreme mass ratios (typically q $\simeq$ 20). The spin angular momentum of the massive star in these systems (when synchronized) is comparable to the orbital angular momentum and circularization can be achieved while the massive star is still far from corotation (Section 6).

Observations indicate that there are binary systems that manage to evolve towards e $\simeq$ 0 within the nuclear lifetime of their stellar components. But it is difficult to infer whether a given binary star is synchronized. This is because studies of rotation rates (for example Slettebak 1970) inform us only about the rotation rate of the stellar surface layers. The bulk of the stellar mass may rotate at a very different rate. This seems especially true for close binary systems since tidal effects are strongly concentrated in the outer deformed layers of the stars. It is thus dangerous to conclude that a binary star (as a whole) has been synchronized when the observed rotation rate is (roughly) equal to the orbital revolution rate.

In this article we shall discuss some recent results of tidal evolution calculations (Sections 5 and 6). But first, in Section 2, we give expressions for the perturbing potential and, in Section 3, we discuss the simplified hydrostatic approximation to the tidal response. In Section 4 we give general expressions for the rate of tidal evolution, assuming that the tidal interaction is weak enough to allow a linearized approach.

## 2.    THE PERTURBING POTENTIAL

Let us assume for simplicity that the rotation axes of the two stars are aligned and that the orbit has a moderately small eccentricity e. Introducing spherical polar coordinates r, $\theta$ , $\phi$ , based at the centre of the secondary component, we can write the lowest order, dominant spherical harmonic $l$ = m = 2 of the perturbing potential $\phi_T$ due to the primary component as

$$\phi_T = -\left(\frac{GM_P}{4D^3}\right) r^2 \, Re\left\{P_2^2(\mu)exp[2i(\omega t-\phi)]+\phi_e\right\},$$

$$(2)$$

where $\phi_e = \phi_1 + \phi_2 + \phi_3$ and

$$\phi_1 = 6 e\, P_2(\mu)\, exp\,(i\omega t)$$

$$\phi_2 = \frac{1}{2} e\, P_2^2(\mu)\, exp\left[2i\left(\frac{\omega}{2}t - \phi\right)\right]$$

$$\phi_3 = -\frac{7}{2} e\, P_2^2(\mu)\, exp\left[2i\left(\frac{3\omega}{2}t - \phi\right)\right].$$

Here $M_p$ is the mass of the perturbing star, D the orbital separation of the binary system, $\mu$ = $\cos\theta$, $\omega$ the orbital angular velocity and $P_2(\mu)$, $P_2^2(\mu)$ the usual Legendre and associated Legendre functions. In the above formulation the deformation of the primary has been neglected as we are only interested in the lowest order effects. For circular orbits $\phi_e$ = 0.

3.    THE HYDROSTATIC RESPONSE; THE WEAK FRICTION MODEL

For sufficiently slow tidal oscillations the response of the star approaches the hydrostatic or "equilibrium tide". It is simple to derive an order of magnitude for the (radial) surface amplitude $\xi_r$ of the tidal wave in this case by noting that (Eqn. 2)

$$\left(\frac{G M_s}{R_s^2}\right)\xi_r \simeq \left(\frac{G M_p}{4 D^3}\right)R_s^2 .$$

This yields for the equilibrium tide

$$\xi_r \simeq \frac{1}{4}\left(\frac{M_p}{M_s}\right)\left(\frac{R_s}{D}\right)^3 R_s .  \qquad (3)$$

The mass in the tidal bulges can be written as

$$M_t \simeq k\left(\frac{M_s}{R_s}\right)\xi_r \simeq k\, M_p\left(\frac{R_s}{D}\right)^3 , \qquad (4)$$

where the dimensionless constant k ≪ 1 depends on the mass distribution of the secondary.

Let us now crudely approximate the mass distribution of the deformed secondary by three colinear point masses: $M_s$ at the centre of the secondary and $M_t$ ≪ $M_s$ at a distance $R_s$ on both sides of the centre. If the line through the two point masses $M_t$ makes an angle $\delta$ with the line through the two stellar centres and if the spin angular velocity of the secondary is less than the orbital angular velocity (Fig. 1), the tidal torque follows from an elementary calculation as

$$\frac{dH}{dt} \simeq - \frac{3 G M_t M_p}{R_s} \left(\frac{R_s}{D}\right)^3 \sin(2\delta).$$

Substituting Eqn (4) and assuming $\delta \ll 1$, the tidal torque on the secondary can be written as

$$\frac{dH}{dt} \simeq - 6 k \left(\frac{G M_s^2}{R_s}\right)\left(\frac{M_p}{M_s}\right)^2 \left(\frac{R_s}{D}\right)^6 \delta. \tag{5}$$

It is apparent from the large exponent of $(R_s/D)$ in Eqn. (5) that, indeed, the tidal torque decreases very strongly with increasing orbital separation. Furthermore, the tidal torque and thus the rate of tidal evolution is seen to be proportional to the small phase angle $\delta$, which depends on the dissipation rate in the tidal flow.

Up to now most analyses of tidal evolution in binary stars (for example Alexander 1973 and Hut 1981) were based on the so-called weak friction model (Darwin 1879). That is, it is assumed that the tidal response is hydrostatic and that the phase lag $\delta$ is proportional to the relative orbital angular velocity of the perturbing companion. The constant of proportionality in the latter relation is usually treated as an undetermined free parameter, so that the time scale of tidal evolution remains unknown in this type of analysis.

Let us stress here that the weak friction model has no physical basis: firstly, the actual frequency of tidal oscillations in close binary systems of interest does not in general allow a hydrostatic approximation to be made for the tidal response; and secondly, there seems no physical basis for the assumption that $\delta$ is linearly proportional to the forcing frequency. The weak friction model seems to have been applied to stellar tidal evolution merely because of its mathematical simplicity. In the following we will ignore the crude equations derived in this section. Eqn. (5) is, however, illustrative for the functional dependence of the tidal torque on the binary parameters.


4.    EQUATIONS FOR TIDAL EVOLUTION

Following Savonije & Papaloizou (1983), let us define a characteristic tidal timescale $t_0(\omega)$ such that the tidal torque on the non-rotating secondary can be written as (assuming e = 0)

$$\frac{dH}{dt} = - I\omega \left(\frac{M_P}{M_s}\right)^2 \left(\frac{R_s}{D}\right)^6 t_o^{-1}(\omega).$$

(6)

Here H is again the orbital angular momentum of the binary system, I the secondary's moment of inertia and $\omega$ the orbital angular velocity. $M_s$ and $R_s$ are the mass and radius of the secondary component. Because the scaling factors $(M_P/M_s)^2$ $(R_s/D)^6$ (compare Eqn. 5) have been factored out of $t_o^{-1}(\omega)$, this latter quantity expresses an intrinsic tidal timescale independent of the particular binary configuration. $t_o(\omega)$ depends on the non-adiabatic tidal response of the secondary which varies with the forcing frequency.

It follows, by comparison with Eqn. (5), that the weak friction model requires $t_o$ to be a constant independent of $\omega$, which is not very plausible.

Note further that $t_o(\omega)$ is defined for a circular orbit and so expresses (see Eqn. 2) only the response to the tide with time dependence exp $[2i(\omega t - \phi)]$. If the secondary rotates uniformly with spin angular velocity $\omega_s$, one should replace in Eqn. (6) by the relative angular frequency $\omega - \omega_s$. We will not consider the appreciably more complicated case of a differentially rotating secondary star.

For small tidal deformations one can apply a linearized treatment in which the effect of the different tidal components ($\phi_1$, $\phi_2$, $\phi_3$ etc.) can be caluculated separately and then superposed. In this way it follows that the torque on a uniformly rotating secondary with spin angular velocity $\omega_s$ can be written as (Savonije & Papaloizou 1984)

$$\frac{dH}{dt} = - I\omega \left(\frac{M_P}{M_s}\right)^2 \left(\frac{R_s}{D}\right)^6 \left\{ \frac{\left(1 - \frac{\omega_s}{\omega}\right)}{t_o(\omega - \omega_s)} \right.$$
$$\left. + \frac{e^2\left(1 - 2\frac{\omega_s}{\omega}\right)}{8t_o\left(\frac{\omega}{2} - \omega_s\right)} + \frac{147e^2\left(1 - \frac{2\omega_s}{3\omega}\right)}{8t_o\left(\frac{3\omega}{2} - \omega_s\right)} \right\}$$

(7)

In a similar way the orbital circularization rate for a moderately small eccentric orbit can be expressed as

$$-\frac{1}{e}\frac{de}{dt} = \frac{I}{mD^2}\left(\frac{M_p}{M_s}\right)^2\left(\frac{R_s}{D}\right)^6\left\{\frac{3}{8t_o\left(\frac{w}{2}\right)} - \frac{\left(1-2\frac{w_s}{w}\right)}{16t_o\left(\frac{w}{2}-w_s\right)}\right.$$

$$\left. - \frac{\left(1-\frac{w_s}{w}\right)}{2t_o(w-w_s)} + \frac{147\left(1-\frac{2w_s}{3w}\right)}{16t_o\left(\frac{3w}{2}-w_s\right)}\right\}, \qquad (8)$$

where $m = M_p M_s/(M_p + M_s)$ is the reduced mass of the system.

Note that de/dt can be either positive or negative, depending on the ratio $w_s/w$ and on the values of $t_o$ for the different tidal components. For the weak friction model $t_o$ is independent of $w$, so that it follows at once from Eqn. (8) that for this approximation de/dt < 0 only if $w$s/$w$ < 18/11, a result obtained already by Darwin (1879). In reality $t_o$ is not a constant and the situation is more complicated. Equations (7) and (8), together with the constraint that the total angular momentum of the binary system should be conserved, enable the tidal evolution to be followed in detail, once we know the function $t_o(w)$ and the initial conditions. We have thus reduced the determination of the tidal evolution of a binary system to the calculation of the intrinsic tidal timescale $t_o$ as a function of tidal frequency $w$. Because tidal dissipation processes, and thus $t_o(w)$, seem to differ appreciably for stars with convectively stable and convectively unstable envelopes, we will consider both in turn.

## 5.   TIDAL DISSIPATION IN STARS WITH CONVECTIVE ENVELOPES

Unfortunately, there is no good theory of convection available to describe the effects of turbulent viscosity. We can therefore only guess about the nature of the presumably complicated coupling between the turbulent convective motions and the global tidal oscillations of the star. One usually adopts the extremely simplified mixing length theory to treat the transport properties of the convection (for example Tennekes & Lumley 1974, Cox & Giuli 1968). The crudest approximation for turbulent viscosity is then given by the so-called eddy viscosity, which is defined as

$$\nu_t = \frac{1}{3}\ell V$$

$$(9)$$

where $\ell$ is the mixing length or typical size of large scale eddies and V the average convective velocity. In stars the mixing length is usually adopted to be of order of the pressure scale height $H_P$ (except near the stellar centre where the scale height diverges). When most of the stellar energy flux is carried by convection the typical convective velocity is given by

$$V = \left( L / 20\pi \, r^2 \rho \right)^{\frac{1}{3}}$$

where L is the stellar luminosity and $\rho$ the local mass density. The eddy viscosity is thus of order

$$\nu_t \simeq \frac{1}{3} H_P \left( \frac{L}{20\pi r^2 \rho} \right)^{\frac{1}{3}}.$$

Averaging over the star we obtain an (extremely) crude estimate

$$\nu_t \simeq f \, R \left( \frac{LR}{M} \right)^{\frac{1}{3}}$$

$$(10)$$

where f is a dimensionless factor that accounts for the structure of the star. The corresponding crude viscous timescale for the star is then

$$T_{ev} = \frac{R^2}{\nu_t} \simeq f^{-1} \left( \frac{MR^2}{L} \right)^{\frac{1}{3}}$$

$$(11)$$

which is of the order of one year, roughly. If we now approximate the intrinsic tidal timescale $t_0$, as defined in Eqn. (6), by this crude viscous timescale $T_{ev}$, we deduce from Eqns. (7) and (8) the following expressions for the tidal spin up rate $t_{su}^{-1}$ and orbital circularization rate $t_c^{-1}$ due to dissipation effects in the secondary:

$$t_{su}^{-1} = \left( \frac{-1}{I\omega} \frac{dH}{dt} \right)_{e=0} = f \left( \frac{M_P}{M_s} \right)^2 \left( \frac{R_s}{D} \right)^6 \left( \frac{L_s}{M_s R_s^2} \right)^{\frac{1}{3}}$$

$$(12)$$

$$t_{\varepsilon}^{-1} = \left(-\frac{1}{e}\frac{de}{dt}\right)_{\omega=\omega_s} = \frac{7}{2} f\left(\frac{I}{mD^2}\right)\left(\frac{M_p}{M_s}\right)^2\left(\frac{R_s}{D}\right)^6\left(\frac{L_s}{M_s R_s^2}\right)^{\frac{1}{3}}. \quad (13)$$

The derivation of Eqns. (12) and (13) as given here is, of course, rather formal since we have not specified the structural constant f. Although all estimates made in the framework of the mixing length theory are inevitably rather crude, it may be worthwhile to estimate the factor f in Eqns. (12) and (13) from a somewhat more detailed analysis.

Neglecting stellar rotation, Zahn (1977) derived a crude estimate $f = 6 k_2$, where $k_2$ is the apsidal motion constant of the secondary. Note that some of the numerical factors in Zahn's equations for $t_{su}$ and $t_c$ for constant $t_o$ (Zahn 1977, 1978) are incorrect. The correct coefficients can easily be derived from Eqns. (7) and (8) above.

It is sometimes argued that tidal dissipation by means of turbulent viscosity is an example of a case where the weak friction model (Section 3) applies, because the tidal timescale $t_o$ appears in that case to be independent of the orbital angular velocity $\omega$. This argument is, of course, dubious because $t_o$ appears independent of $\omega$ merely as a result of our ignorance regarding the details of the turbulent dissipation, which forced us to adopt the crudest possible approximation (Eqn. 9). Moreover, we shall see below that even in the crude framework of the mixing length theory the tidal dissipation can be frequency dependent.

Campbell & Papaloizou (1983) calculated the tidal synchronization rate in somewhat more detail than Zahn. They studied the synchronization of rotating, fully convective stars with application to the cataclysmic binary systems. They approximated convective stars by n = 3/2 polytropes and assumed zero orbital eccentricity. Neglecting all dissipation, they derived the non-synchronous velocity field inthe linear approximation. They then determined the tidal dissipation from the action of eddy viscosity on the inviscid non-synchronous velocity field. From a two-dimensional integration over the deformed star they derived a synchronization rate equivalent to Eqn. (12) with $f \simeq 0.4$, in rough agreement with the result obtained by Zahn (1977). They argued on the basis of a variational principle that their linearized calculations for small tidal deformations may, in view of the general crude level of approximation, be reasonably well extrapolated to critical lobe filling conditions.

Campbell & Papaloizou noted that the factor f in Eqns. (12) and (13) needs a correction factor to account for the reduced efficiency of turbulent viscosity once the convective timescale is longer than the tidal oscillation period. Suppose that the secondary's degree of non-synchronism is $\Delta\omega/\omega$, so that the relative forcing frequency (in a frame that is on average corotating with this star) is $2\Delta\omega$ (Eqn. 2 with e = 0). Adopting the approach by Goldreich & Keeley (1977), who ignored the contribution of all eddies with turnover time longer than $(2\Delta\omega)^{-1}$, they adopted the following expression for the reduced, now frequency dependent turbulent viscosity:

$$\nu_t^* = \eta^{-2}\nu_t \tag{14}$$

where

$$\eta = 2\Delta\omega\left(\frac{M_s R_s^2}{L_s}\right)^{\frac{1}{3}}.$$

Their numerical results for the synchronization of low mass convective main sequence stars can be represented approximately by substituting into Eqn. (12)

$$f = \min\left(0.4, 30\eta^{-2}\right). \tag{15}$$

The limiting value of $\eta$ for which the turbulent viscosity thus starts to be reduced is $\eta \simeq 9$. For lobe filling low mass main sequence stars this corresponds to a deviation from synchronism not greater than $\Delta\omega/\omega \simeq 10^{-4}$. It then follows that for very small $\Delta\omega/\omega$ the synchronization time is of order $10^2$ yr, whereas for $\Delta\omega/\omega \gtrsim 10^{-1}$ this time is greater than $10^7$ yr. It is interesting to compare the above synchronization timescale with the timescale $\tau_\gamma$ on which the orbit of a close binary system decays by the emission of gravitational radiation (Landau & Lifshitz 1970)

$$\tau_\gamma = 1.6\times10^8\left(\frac{M_p M_s(M_s+M_p)}{M_\odot^3}\right)^{-1}\left(\frac{D}{R_\odot}\right)\ yr.$$

We conclude that for a moderately small initial degree of non-synchronism, gravitational radiation is not likely to drive the secondary in cataclysmic binaries (with $M_p \simeq 1\ M_\odot$, $M_s \lesssim 1\ M_\odot$ and $D/R_\odot \gtrsim 0.5$) out of synchronism. But we should, of course, remember that all results based on a mixing length theory have large intrinsic uncertainties. As Campbell & Papaloizou note, convection is likely to be inhibited by rotational effects in nearly synchronous stars in close binary systems. Rotation tends to make the convective motions, and thus the eddy viscosity, anisotropic (for example Wasiutynski 1946, Kippenhahn 1963 and Tayler 1973). It is therefore not at all obvious that a synchronous state of uniform rotation can actually be obtained in convective binary stars.

RS Canum Venticorum binary systems may be an interesting group in this respect. The slowly migrating quasi-sinusoidal wave in the lightcurves of these systems have been interpreted (Hall 1976) as evidence for differential rotation of the extended convective envelope of the cool component in the system, superposed on nearly synchronized rotation enforced by tidal coupling. Scharlemann (1981, 1982) studied the tidal interaction in these systems for isotropic turbulent viscosity.

We have so far neglected the interaction between the periodic tidal forces and the natural oscillation modes of the star. Papaloizou & Pringle (1981) studied the oscillation spectrum of slowly, uniformly rotating polytropes of index n = 3/2, which represent stars that are convectively neutral. By a linear and adiabatic treatment they demonstrated the existence of a spectrum of toroidal modes which they called r-modes with frequencies in the range $-2\Omega < \sigma < 2\Omega$, where $\Omega$ is the rotational angular velocity. When the star is a member of a close binary system, resonances may occur when the tidal oscillation frequency coincides with a natural oscillation frequency of the star. Tidal interaction can then obviously be strongly enhanced. To determine ' the resulting tidal torque requires a non-adiabatic treatment of these forced non-radial oscillations, which has not yet been carried out for rotating stars.

The resonance effects can also influence the apsidal motion in eccentric binary systems. The classical expression for the apsidal advance rate was derived for the equilibrium (hydrostatic) tide (Cowling 1938). Resonance effects may give qualitatively different results (Papaloizou & Pringle 1981), but have as yet not been calculated in detail.


## 6. TIDAL DISSIPATION IN STARS WITH RADIATIVE ENVELOPES

That tidal dissipation is also efficient in stars with convectively stable envelopes can be inferred from the observations of massive X-ray binary systems. These systems consist of a massive, early type star accompanied by a neutron

star (or perhaps a black hole, see contribution by N. White),
that can be regarded as a point mass.

In several cases the compact companion is an eclipsing
short period X-ray pulsar, which allows for a very accurate
determination of the orbital elements of the system. Two well
studied members of this group, SMC X-1 and Cen X-3, appear to
have orbital eccentricities smaller than 0.0008 (for example
Primini et al. 1977). This is surprising because the
neutron star is thought to originate from a violent supernova
explosion, which must have made the orbit eccentric (Wheeler et
al. 1975 and Flannery & van den Heuvel 1975). The orbit
appears to have been circularized within the lifetime of the
early type star, that is within some $10^7$ yr.

This conclusion led to a revived interest in tidal
interaction of close binaries (for example Sutantyo 1974,
Chevalier 1975, Lecar et al. 1976 and references below).
Atomic or radiative viscosity yields a characteristic viscous
timescale $\tau_v \simeq R^2/\nu \gtrsim 10^{16}$ yr, much too long for any tidal
relaxation to occur on interesting timescales. In order to
explain the efficient circularization it was speculated (Horedt
1975 and Press et al. 1975) that the shearing tidal flow
might itself generate turbulence with efficient dissipation of
tidal energy. But it was pointed out by Sequin (1976) that the
growth rate of small perturbtions of the tidal flow is much
smaller than the typical tidal frequencies, so that
self-generated turbulence is probably not of importance.

However, for negligibly small viscosity the tidal
oscillations are still damped by radiative diffusion. Cowling
(1938, 1941) studied the adiabatic non-radial oscillation modes
of a non-rotating free polytrope of index n = 3. He classified
the natural oscillation modes into p-modes and g-modes. The
p-modes form a class of high frequency oscillations that cannot
be resonantly excited by the tidal oscillations. The g-modes,
on the other hand, form a set of low frequency oscillations for
which the restoring force is mainly due to (negative) buoyancy.
These gravity modes have oscillation periods of the order of
days and can therefore interact resonantly with the periodic
tidal forces in close binary systems (for example Unno et
al. 1979).

Zahn (1975, 1977) was the first to attempt to calculate
the tidal interaction caused by radiative damping of the forced
non-radial oscillations of stars in close binary systems. He
used an asymptotic theory for low orbital frequencies for which
the resonant interaction with g-modes is heavily damped by
radiation. It is, however, not obvious that his results can be
applied to close binary system of interest (which do not have

very low orbital frequencies).

Savonije & Papaloizou (1983) improved on this by solving the linearized hydrodynamic equations that determine the radiatively damped tidal response of non-rotating zero age main-sequence stars by numerical calculations. They wrote the hydrodynamic equations as a set of finite difference equations which were solved by matrix inversion according to the procedure adopted by Henyey, Forbes & Gould (1964) for stellar evolution calculations. Because of the very high order g-modes involved in the response for lower forcing frequencies they used a large number ($>$ 1000) of zones which were carefully spaced. In this way they determined the tidal response time $t_o(\omega)$ for zero-age-main sequence stars of 5, 10 and 20 $M_\odot$ for all orbital frequencies greater than $\omega = 0.05\,\omega_c$. Here $\omega_c = (G\,M_s/R_s^3)^{1/2}$ is the Kepler frequency of a particle in orbit at the stellar surface.

As expected, the deformed star shows a strong response, when the forcing frequency is close to one of the Eigen-frequencies of the free g-mode oscillations. These resonances can be denoted by $g_1$, $g_2$, $g_3$... in order of decreasing frequency, where the index denotes the order and corresponds to the number of nodes in the imaginary part of the tidal response. For low frequencies, $\omega/\omega_c \lesssim 0.1$, the radiative damping becomes severe and the resonances become weak. As a result of damping the nodes in the real part of the tidal response disappear and for $\omega \to 0$ the radial tidal amplitude approaches the hydrostatic value $\xi_{eq} = 1/4\,(M_p/M_s)\,(r/D)^3\,r$, where r is the radial distance from the stellar centre. In a following paper Savonije & Papaloizou (1984) extended their numerical calculations to evolved stars with a non-homogenous chemical composition. They calculated the frequency dependent tidal timescale $t_o(\omega)$ for radiatively damped forced tidal oscillations of a 20 $M_\odot$ star at different phases of core hydrogen burning, starting at the zero age main-sequence for which the central hydrogen mass fraction $X_c = 0.7$.

The unperturbed spherical models were obtained with a stellar evolution code developed by Eggleton (1972). (Semi-) convection was treated by solving the standard stellar equations for the chemical composition simultaneously with a diffusion equation for the chemical composition on a moveable mesh. While matter was fully mixed in the convective core, it was only partially mixed in the growing semi-convective region in between the shrinking convective core and the outer radiative envelope. The evolution code mixes the matter in the semi-convective region until it is almost convectively neutral according to the Schwarzschild criterion (c.f. Kato 1966). By applying the same method of matrix inversion as before,

Savonije & Papaloizou were able to determine $t_o$ for a whole range of forcing frequencies. Strong resonances were found for $\omega/\omega_c > \alpha$ , where $\alpha$ varies from 0.1 for $X_c$ = 0.1 to 0.2 for $X_c$ = 0. As the star evolves more and more resonances, corresponding to higher and higher order g-modes, can be seen in the response. These resonances become more densely packed on the $\omega/\omega_c$ axis when the central hydrogen content $X_c$ decreases. The reason for this is that the resonant frequency $\omega_k/\omega_c$ for a mode $g_k$ increases for fixed k as the star evolves. This is a consequence of the increasing central condensation in the star. Hence, due to stellar evolution higher and higher order resonances are shifted into the weakly damped high frequency range.

Another result of these calculations is that for fixed $\omega$ the values of $t_o$ in general tend to decrease as the star evolves. Hence, tidal effects become stronger as the star evolves. This is in conflict with the results of Zahn (1975, 1977) who found that tidal effects decrease rapidly in importance when stellar evolution proceeds. Zahn performed an asymptotic calculation for low frequencies in which he fitted solutions for an adiabatic interior to solutions for a non-adiabatic exterior.

Because of the very short wavelength response at low forcing frequencies, the strongly damped low frequency regime could not be studied satisfactorily with the above mentioned numerical scheme. But, even for low forcing frequencies the stellar response is still very nearly adiabatic throughout most of the stellar interior. Therefore, Savonije & Papaloizou (1984) analysed the behaviour at low frequencies by studying the adiabatic equations. The radiative damping was approximately taken into account by a fully outgoing energy condition at the interface between the adiabatic interior and the non-adiabatic surface layers. In this way $t_o(\omega)$ can be estimated without matching adiabatic and non-adiabatic solutions. When $F_E$ is the outward energy flow at the interface connected with the tidal perturbation, the rate of change of orbital angular momentum is given by

$$\frac{dH}{dt} = -\omega^{-1} F_E$$

and $t_o(\omega)$ can be calculated from Eqn. (6). $F_E$ can be expressed as an integral over the adiabatic interior. The integrand of this integral oscillates rapidly as a function of radial distance r for low forcing frequencies, so that cancellation effects occur in large parts of the star. For a main sequence star subject to a forcing frequency

$$\frac{\omega}{\omega_c} \ll \frac{R_c}{R_s},$$

(16)

where $R_c$ is the radius of the star's convective core, most of the contribution to $F_E$ is expected to come from the region just outside the convective core where cancellation effects are less important. For forcing frequencies that satisfy inequality (16) $t_o(\omega)$ can be estimated as

$$t_o(\omega) = \frac{K_g}{\omega_c} \left(\frac{M_c}{M_s}\right)^{4/3} \left(\frac{R_c}{R_s}\right)^{-9} \left(\frac{\omega}{\omega_c}\right)^{-\frac{5}{3}},$$

(17)

where $K_g$ is the secondary's radius of gyration and $M_c$ the mass of its convective core. Eqn. (17) has the same frequency dependence as found by Zahn (1977) and makes similar numerical predictions. For the 20 $M_\odot$ zero age main sequence model (with $R_c/R_s = 0.29$, $K_g = 0.08$ and $\omega_c = 0.19 \times 10^{-3}$ s$^{-1}$) Eqn. (17) yields

$$t_o(\omega) = 0.32 \, (\omega/\omega_c)^{-5/3} \ yr.$$

For a suitable low frequency $\omega/\omega_c = 0.05$ this gives $t_o = 47$ yr, in reasonable agreement with the detailed numerical calculations by Savonije & Papaloizou. From an expression similar to Eqn. (17) Zahn (1977) predicted that tidal effects would weaken dramatically with stellar evolution (because of the shrinking convective core radius). However, at the same time the necessary condition (16) for Eqn. (17) to be valid becomes violated, so that Zahn's conclusion is incorrect.

For example, when $X_c = 0.1$, $\omega_c$ has decreased by a factor of 3, $K_g$ by a factor of 2, while $R_c/R_s$ has decreased by a factor of about 3.

From this it follows that

$$t_o(\omega) = 3.0 \times 10^3 \, (\omega/\omega_c)^{-5/3} \ yr$$

which, for $\omega/\omega_c = 0.15$ yields a $t_o$ three orders of magnitude
larger than the result of the detailed numerical calculations.
The discrepancy arises because the condition (16) is violated
and regions external to the convective core contribute
appreciably to $F_\epsilon$. It was found that an approximate empirical
estimate of $t_o(\omega)$ for low frequencies can be obtained by using
Eqn. (17), not with parameters appropriate to the convective
core, but appropriate to the convective core plus
semi-convective region.

The calculations by Savonije & Papaloizou indicate that
the tidal evolution of massive X-ray binaries can speed up
considerably as the massive star approaches the end of core
hydrogen burning. This is partly caused by the rapid shift to
higher frequencies of the g-mode spectrum, so that resonance
passages become frequent. Besides that the tidal interaction
becomes more effective, not only as a result of a general
decrease in the tidal timescale $t_o$ with stellar evolution, but
also because of the increasing value of (R/D) as the star
expands during its evolution. It was found that as a result of
this binary systems consisting of 20 $M_\odot$ + 1 $M_\odot$ and an orbital
period of up to 6 days can have their orbits circularized
during the core hydrogen burning phase. However, in all cases
studied the massive star could not achieve complete corotation
during the core hydrogen burning phase.

A point of interest is the effect of stellar wind ejection
from massive stars (for example Cassinelli 1979) as mass loss
affects the profile of chemical composition in the star and
thereby influences the tidal interaction.


7.   CONCLUDING REMARKS

From the discussions of recent results of tidal evolution
calculations in the previous two sections it must be clear that
the tidal problem is still far from being solved. For binary
stars with convective envelopes simple expressions for the
tidal spin-up and circularization rates are given in Section 5.
But these expressions are based on the greatly simplified
mixing length theory for convection and must be handled with
caution.

For massive binary stars with convectively stable
envelopes the situation is somewhat more promising. When
stellar rotation is neglected the intrinsic tidal timescale
$t_o(\omega)$ can be calculated rather straight-forwardly, at least
when tidal effects are assumed sufficiently weak to allow a
linearized treatment. When the linear approximation breaks
down, for example during resonance passage, the situation

becomes less simple. For example, the non-linear motions may induce some mixing of stellar matter which affects not only the nuclear evolution of the stars, but also their non-radial oscillation properties (for example Unno et al. 1979). This effect was not taken into account by Savonije & Papaloizou (1984). A further complication arises because of stellar rotation and its influence on the tidal oscillations. The calculations by Savonije & Papaloizou (1983, 1984) indicate that, since the tidal torque can change sign many times as a function of radial distance from the stellar centre and is much enhanced near the stellar surface, there is a strong tendency to set up a complicated form of differential rotation in the star. The tidally induced differential rotation is likely to be unstable. For a discussion of the instabilities in rotating stars, see for example Tassoul (1978) and references therein. Furthermore, the tidally induced changes of the stellar rotation will react back on the tidal coupling, because the oscillation properties of the star depend on stellar rotation. This is, of course, also true for convective envelope stars for which toroidal oscillation modes are of importance. It is clear that much work remains to be done in the field of tidal evolution.

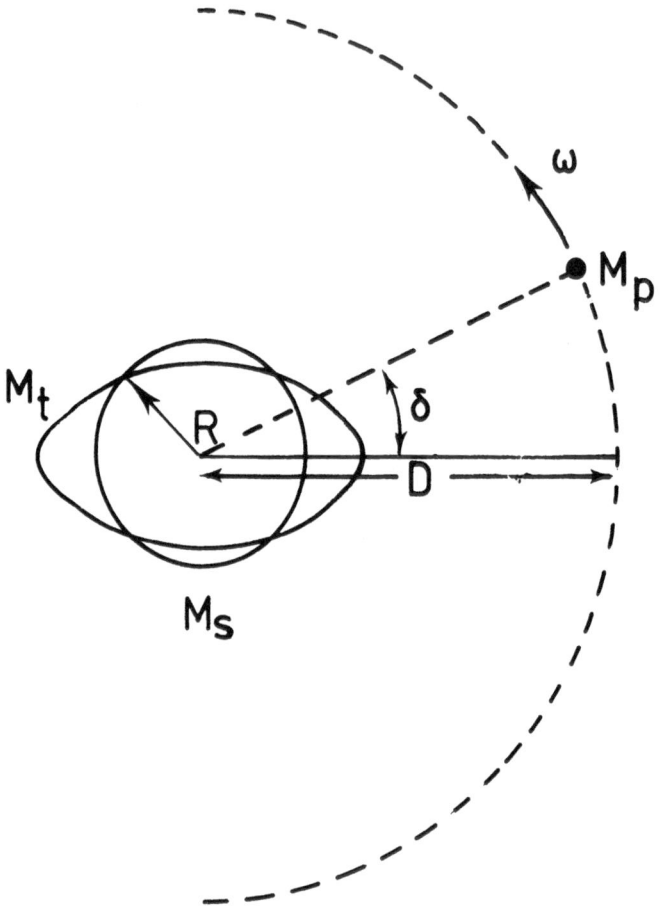

Figure 1. Schematic illustration of the tidal response in case of a circular orbit and a secondary (with mass $M_s$) that rotates slower than the orbital revolution of its companion (with mass $M_p$). As a result of dissipation the tidal bulges (with mass $M_t$) raised on the secondary lag the orbital revolution of the primary by a small angle $\delta$. This creates a gravitational torque that retards the primary in its orbit and spins up the secondary towards corotation.

REFERENCES

Alexander, M.E., 1973. Astrophys. Space Sci., **23**, 459.
Campbell, C.G. & Papaloizou, J., 1983. Mon.Not.Ro.Astron.Soc.,
    **204**, 433.
Cassinelli, J.P., 1979. Ann.Rev.Astr.Astrophys., **17**, 275.
Chevalier, R.A., 1975. Astrophys.J., **199**, 189.
Counselman, C.C., 1973. Astrophys.J., **180**, 307.
Cowling, T.C., 1938. Mon.Not.Roy.Astron.Soc., **98**, 734.
Cowling, T.C., 1941. Mon.Not.Roy.Astron.Soc., **101**, 367.
Cox, J.P. & Fiuli, R.T., 1968. Principles of Stellar
    Structure Vol. 1, Gordon and Breach.
Darwin, G.H., 1879. Phil.Trans.Roy.Soc., **170**, 1.
Eggleton, P.P., 1972. Mon.Not.Roy.Astron.Soc., **156**, 361.
Flannery, B.P. & van den Heuvel, E.P.J., 1975. Astron.
    Astrophys., **39**, 61.
Goldreich, P. & Keeley, D.A., 1977, Astrophys.J., **211**, 934.
Hall, D.S., 1976. In IAU Colloqium No. 29, Multiple Periodic
    Variable Stars, ed. W.S. Fitch (Reidel) p. 287.
Henyey, L.G., Forbes, J.E. & Gould, N.L., 1964. Astrophys.J.,
    **139**, 306.
Horedt, G., 1975. Astron.Astrophys., **44**, 461.
Kato, S., 1966. Publ.Astr.Soc. Japan, **18**, 374.
Hut, P., 1980. Astron.Astrophys., **92**, 167.
Hut, P., 1981. Astron.Astrophys., **99**, 126.
Kippenhahn, R., 1963. Astrophys.J., **137**, 664.
Koch, R.H. & Hrivnak, B.J., 1981. Astron.J., 86, vol. 3, 438.
Landau, L. & Lifshitz, E., 1970. The Classical Theory of
    Fields, Pergamon Press, p. 105.
Lecar, M., Wheeler, J.C., McKee, C.F., 1976. Astrophys.J.,
    **205**, 556.
Middelkoop, F. & Zwaan, c., 1982. Astron.Astrophys., **107**, 31.
Papaloizou, J. & Pringle, J.E., 1981. Mon.Not.Roy.Astron.Soc.,
    **195**, 743.2
Papaloizou, J. & Pringle, J.E., 1982. Mon.Not.Roy.Astron.Soc.,
    **200**, 49.
Press, W.H., Wiita, P.J. & Smarr, L., 1975. Astrophys.J.,
    **202**, L135.
Primini, F., Rappaport, S. & Smarr, L., 1975. Astrophys.J.,
    **217**, 543.
Savonije, G.J. & Papaloizou, J.C.B., 1983. Mon.Not.Roy.
    Astron.Soc., **203**, 581.
Savonije, G.J. & Papaloizou, J.C.B., 1983. Mon.Not.Roy.
    Astron.Soc., **206**, 000.
Scharlemann, E.T., 1982. Astrophys.J., **253**, 298.2
Scharlemann, E.T., 1981. Astrophys.J., **246**, 292.
Seguin, F.H., 1976. Astrophys.J., **207**, 848.
Stettebak, A. ed., 1970. "Stellar Rotation", Gordon and
    Breach, New York.
Sutantyo, W., 1974. Astron.Astrophys., **35**, 251.

Tassoul, J.L., 1978. "Theory of Rotating Stars",
    Princeton University Press, U.S.A.
Taylor, R.J., 1973. Mon.Not.Roy.Astron.Soc., **165**, 39.
Tennekes, H. & Lumley, J.L., 1974. "A first course in
    turbulence", the M.I.T. Press, Cambridge, MA, U.S.A.
Unno, W., Osaki, Y., Ando, H., Shibahashi, H., 1979.
    "Non-Radial Oscillations of Stars", University of
    Tokyo Press.
Wassiutynski, J., 1946. Astrophys.Norv., **4**, 1.
Wheeler, J.C., Lecar, M. & McKee, C.F., 1975. Astrophys.J.,
    **192**, L171.
Young, A. & Koniges, A., 1977. Astrophys.J., **211**, 836.
Zahn, J.P., 1975. Astron.Astrophys., **41**, 320.2
Zahn, J.P., 1977. Astron.Astrophys., **57**, 383.
Zahn, J.P., 1978. Astron.Astrophys., **67**, 162.

# WOLF-RAYET STARS AND BINARITY

Allan J. Willis

Department of Physics and Astronomy
University College London
Gower Street
London, U.K.

## 1. INTRODUCTION

Studies of Wolf-Rayet (WR) stars in binary systems play a significant role in improving our overall understanding of these unusual objects, yielding information on stellar masses, the motions and physics of interacting material streams connecting the stellar components and providing tests for binary stellar evolution models. The past decade has witnessed some significant changes to the picture of what constitutes a WR star and as to what role binarity plays, which is of particular relevance to this meeting. The most recent advances in the WR field can be found in the proceedings of IAU symposium No.99 - "Wolf-Rayet stars: Observations, Physics and Evolution" (de Loore & Willis 1982), whilst summaries of earlier work are contained in previous symposia devoted to studies of WR and/or O-type stars (Gebbie & Thomas 1981, Bappu & Sahade 1973, Conti & de Loore 1979 and Chiosi & Stalio 1981). In this review I will address the question of numbers and statistics of WR + OB systems, and our knowledge of WR masses that come from orbit solutions for such systems, and discuss the proposed existence of WR stars with low mass (collapsed) companions. I will not discuss investigations of interaction effects in individual systems although my own involvement with WR binarity has concentrated in this area. In addressing the overall question of WR binarity in this review I have relied heavily on the recent results from specialists in the binarity of WR stars, notably Peter Conti, Phillip Massey, Virpi Niemela and Tony Moffat who, together with their colleagues, have made an immense contribution to this field.

*P. P. Eggleton and J. E. Pringle (eds.), Interacting Binaries, 103–126.*
*© 1985 by D. Reidel Publishing Company.*

In order to put into focus the question of WR binarity, it is pertinent to provide a brief resume of current thinking concerning the physical and chemical nature of the WR stars, and this is given in Section 2. Section 3 discusses the detection methods for WR binaries and their statistics in the Galaxy and Magellanic Clouds. Section 4 summarises current knowledge of WR masses and orbital parameters with inferences for evolutionary mechanisms involved in producing a WR star. Finally the question of whether or not WR + collapsar systems exist is addressed in Section 5.

## 2.    THE PHYSICAL AND CHEMICAL NATURE OF WOLF-RAYET STARS

WR stars exhibit optical spectra dominated by the presence of strong, broad emission lines superimposed on a relatively weak stellar continuum indicative of a hot object. The prototypes were discovered by Wolf and Rayet (1867) and their spectral characteristics make these stars fairly easy to pick up on low dispersion, objective prism or filter-photometry sky surveys. Consequently such surveys of the Galaxy and Magellanic Clouds have yielded reasonably complete samples of WR stars in these galaxies down to appropriate magnitude limits. The latest catalogue of galactic WR stars lists 161 objects (van der Hucht et al. 1981); 101 are known in the LMC (Breysacher 1981) and (only) 8 are known in the SMC (Azzopardi & Breysacher 1979). In addition some 70 WR stars have been found in M33 (Wray & Corso 1972, D'Odorico & Rosa 1981, Conti & Massey 1982) and about 30 are known in M31 (Corso 1975, Shara & Moffat 1982).

Most WR stars can be divided spectroscopically into two well defined sequences: WN stars show emission lines of predominantly He and N ions, although a few carbon lines are seen in most WN spectra, whilst WC stars exhibit emissions in He, C and O ions with virtually no evidence for any N lines. A very few examples are known which show both WN and WC features. Recently a new WO sequence has been proposed in which very strong O lines are seen in an otherwise WC-type spectrum (Barlow & Hummer 1982). Each WN and WC sequence is divided into spectral subtypes, WN2-WN9 and WC4-WC10, based on relative line ratios in appropriate N or C ion transitions, and reflects a decreasing degree of spectral excitation and/or ionisation going from low to high subtype numbers, analogous to spectral typing in OB stars.

The emission lines are broad, reflecting velocity widths in the range of hundreds to thousands of km/s. Within the same stellar spectrum different ions can display different velocity widths, presumed to result from atmospheric stratification in either excitation (viz. density) and/or ionisation (viz. temperature): see Willis 1983 for a discussion on this point. P-Cygni profiles are often seen in the optical spectra of excited transitions as well as in the usual UV resonance lines illustrative of mass loss in early-type stars. The stellar winds of WR stars are much more extended and dense than in OB supergiants and are thus capable of supporting P-Cygni absorptions in highly excited transitions. However it has recently been recognised that some WR stars can also show intrinsic absorption lines implying that in these cases the wind is sufficiently transparent to allow the underlying photosphere to be observed. The basic model of a WR star (Beals 1933) thus envisages a central hot stellar core emitting a photospheric continuum + line radiation field surrounded by a very extended outflowing dense stellar wind producing the predominant emission line spectrum.

The effective temperatures and radiative luminosities of the WR stars are still somewhat uncertain. Most recently deduced values of temperature using UV-visible energy distribution analyses have yielded values in the range 30000-50000 K, implying luminosities in the range $\log L/L_{\odot}$ = 4.5 - 5.5 (Underhill 1980, 1981, Nussbaumer et al. 1981, Barlow et al. 1983). However, given that the WR winds are likely to be optically thick in both lines and continuum, the validity of such simple energy distribution analyses is in question, and it is possible that the temperatures and luminosities could be higher.

Recent detections of many WR stars at radio wavelengths coupled with UV P-Cygni profile measurements of wind terminal velocities have enabled mass loss rates to be determined using the simple approach, generally regarded to be reliable, of modelling the observed radio (or IR) free-free emission assuming a steady spherically symetric outflow (Wright & Barlow 1975). By making use of the radio data from Abbott et al. (1982) and Hogg (1982) together with IR 10 $\mu$m data available in the literature, mass loss rates have now been deduced for 38 galactic WR stars (Willis 1982). The results show a mean value of $3.8 \pm 2.0 \times 10^{-5}$ $M_{\odot}$/yr, and have a remarkably small spread in the sample given the gross spectral differences apparent between WN and WC stars. There appears to be little difference in mass loss rate with: (i) WN and/or WC subclass, (ii) single or binary WN and/or WC stars. Moreover the rates appear much larger than can be accommodated on the basis of current radiation pressure-driven wind models, implying that other mechanism(s) may play a major role in driving the WR mass outflows.

In general most WR stars show little or no evidence for any atmospheric hydrogen. Perry & Conti (1982) have derived H/He ratios for 60 WR stars in the Galaxy and LMC which are generally less than unity. A few stars, usually WN7 and WN8 classes (denoted WNL stars) have values of H/He $\geq$ 2, but most WNE stars have H/He = 0.0. An important recognition is that there is not a one-to-one correspondence of the H/He ratio with WN subtype. The hydrogen content for WC stars is less well determined, but it seems likely that it can be taken as zero (Willis 1982). A recent analysis of the UV and visible spectra of 10 galactic and 9 LMC WR stars by Smith & Willis (1982, 1983) has yielded the following estimated C/N ratios: $2-6 \times 10^{-2}$ for WNE stars; $2-14 \times 10^{-2}$ for WNL stars and $\geq$ 60 for the WC types. These results demonstrate the chemical separation of the WN and WC sequences and are consistent with the C/N values expected from the exposure of CNO-burning products near the end of the core hydrogen burning lifetime (WN stars) or He-burning products (WC stars), if the nuclear processed material can be exposed to observation by mass removal. This is in line with the "peeled onion" model for a WR star in which successive layers of outer atmospheric material are removed by extensive mass loss in single stars and/or by mass transfer via Roche lobe overflow in binary systems. The latter mechanism was regarded by Paczynski (1973) as the most plausible, with the implication that binarity could be a necessary condition for the development of a WR star, a topic discussed below.


3.   THE BINARY FREQUENCY OF WR STARS - DETECTION METHODS &
     STATISTICS

The question of whether all WR stars are binaries was addressed by Kuhi (1973) who discussed the detection methods and criteria employed to establish or infer binarity, viz: (i) the observation of periodic radial velocity variations, (ii) observation of eclipses, (iii) the presence of a second spectrum (generally OB-type absorption lines), and (iv) anomalous emission line-continuum ratios implying extra light from a companion star. Kuhi based his assessment of WR binary frequency on a deliberate (and at the time quite reasonable) restriction on statistics to a consideration of only northern hemisphere stars brighter than $m_v \leq 10$ (15 stars), for which 11 or 73% were quoted as being binaries in the Smith (1968) catalogue, generally using methods (iii) and (iv). On allowing for the difficulty of picking up low mass/luminosity companions, Kuhi concluded that it was very likely that all WR stars are binaries, with the implication that binarity is a necessary condition for the development of a WR star. However, more recent studies have radically altered this picture.

Firstly, it is now recognised that the WR stars form a highly heterogeneous group of objects - stars of a given subtype can exhibit a wide spread ($\geq$ 2 mag) in absolute visual magnitude (Conti 1982) as well as showing a wide range of up to a factor of ten in intrinsic line strengths (Leep 1982). It is therefore highly unsafe to use method (iv) to infer binarity. Vanbeveren & Conti (1980) made use of an extremely large body of uniformly exposed 26Å/mm spectra of most WR stars in the Galaxy and the Magellanic Clouds to search for absorption lines in stars much fainter than the $m_v$ cutoff used by Kuhi (1973). The main thrust of their results is given in Figure 1, showing histograms (for each galaxy) of stars with and without absorption lines as a function of apparent magnitude. Maintaining the assertion that the presence of absorption lines is a sufficient criterion for duplicity, Vanbeveren & Conti (1980) find that about 40% of the 91 stars studied in the Galaxy and the 48 stars in the LMC are binaries and all 8 stars in the SMC are such. From arguments based on theoretical stellar evolution expectations they concluded that an equal percentage of the remaining stars should be WR + low mass (collapsed) systems, with the remaining 20% being truly single WR stars. However, the use of the presence of absorption lines as a binarity criterion has also now come under serious question.

Although good quality spectra have now become available for most WR stars in the Galaxy and Magellanic Clouds, detailed and long-term spectroscopic monitoring to establish binary orbits has been carried out for only a small sample of about 20 stars. In some cases no significant radial velocity variations are seen, implying that the observed absorption spectrum is intrinsic to the WR star itself, or possibly attributable to a widely separated companion star. Examples are HD 193077-WN5 (Massey 1980), HD 9974-WN3 (Massey & Conti 1981), HD 156327-WC7 and HD 192641 - WC7 (Massey, Conti & Niemela 1981) and HD 193793-WC7 (Conti & Roussel-Dupree 1981). Thus it is now felt that the simple presence of an OB absorption spectrum cannot be taken per se as direct evidence for binarity and criterion (iii) above becomes invalid. Massey et al. (1981) have re-assessed the WR binary frequency based on these new spectroscopic monitoring programmes which have covered all the 12 northern hemisphere ($\delta > -35°$) and sufficiently bright ($m_v \leq$ 11) WR stars exhibiting absorption lines. Of this sample, 5 stars show no radial velocity variations ($K \leq 15$ km/s) implying no close ($P \leq 500$ days) companions. The statistical arguments then run as follows: 60% of the absorption-lined stars are close binaries with massive luminous companions, and since Vanbeveren & Conti (1980) found 40% of all WR stars in the Galaxy and LMC to have absorption lines, the fraction of

WR+OB binaries is at most 25%. To correct for the occurence of
the expected second WR stage in the binary evolution scenario
proposed by Van den Heuvel (1976), one simply doubles the
latter figure to obtain an overall binary frequency of $\lesssim$ 50%.
This value is similar to that found for progenitor O-type stars
(Garmany et al. 1980). Massey et al. (1981) stress
the uncertainties still present from the small number
statistics involved and from the correction for the "unseen"
low mass companions. Nevertheless it now seems certain that
single WR stars do exist, probably in comparable numbers to
those in binaries. The overall picture is consistent with the
Conti (1976) scenario that single O-type stars evolve into
single WR stars, with stellar wind mass removal providing the
required atmospheric stripping, whilst binary O-type stars can
evolve into binary WR stars by mass loss and/or Roche lobe
overflow. Binarity does **not** appear to be a necessary
condition to produce a WR star.

## 4. WR + O BINARIES - ORBITAL SOLUTIONS AND STELLAR MASSES

Massey (1981, 1982) has reviewed the status of orbital
solutions for the comparitively small number of WR + O systems
for which the required spectroscopic monitoring has been
obtained. The discussion below summarises the most salient
conclusions arrived at in this recent work. Table 1 lists the
WR binaries, both double lined (SB2) and single lined (SB1)
systems, for which the derived masses or mass functions
indicate an early-type companion. These data are taken
principally from Massey (1981) but also include the recent
orbit determinations for HDE 320102 and CD $-45^{\circ}$ 4482
Niemela (1982) HD 5980 in the SMC Breysacher (1982), HD 90657
(Niemela & Moffat 1982), HD 94305 (Niemela et al. 1983) and
HD 63099 (Niemela 1981).

For the SB2's, for which the minimum masses, $m \sin^3 i$, and
the mass ratios can be determined, Massey (1981) estimates the
mass of the WR component, $M_{WR}$, for the eclipsing systems by
deducing sin i from assumed values of stellar radii and the
projected orbital separation. For non-eclipsing systems, the
O-star masses are estimated using their spectral types (giving
$T_{eff}$ and $M_{Bol}$) and placing them on evolutionary tracks including
the effects of mass loss. It is estimated that these techniques
involve an uncertainty of $\lesssim$ 5$M_{\odot}$ in the deduced WR masses.
Massey points out that with these new results it is no longer
proper to think of a canonical WR mass of $\sim$10$M_{\odot}$, since many of
the stars clearly have masses $\gtrsim$ 20$M_{\odot}$. Moreover the use of the
mass-losing evolutionary tracks, in fact, gives a lower limit to
$M_{\odot}$ and hence also to $M_{WR}$. Massey also stresses that it is
inappropriate to assume that all WR stars of a given subtype have

the same mass, since the available results indicate a spread of
at least a factor of two. Moreover, as Massey & Niemela (1981)
point out, WC stars do not appear to have an average mass lower
than that for WN stars, implying that not all WN stars will
evolve into WC types through continual mass removal. This
latter conclusion is rather different to that asserted by Moffat
(1981), who considered a qualitative analysis of the mass ratios
$M_{WR}/M_{\odot}$ derived from orbital solutions for SB2's both in the
Galaxy and the MC's. Moffat prefers to use the mass ratios
since these are derived directly for SB2 systems, rather than
assess minimum masses for individual components since these can
suffer uncertainties in assumed inclination angles and/or
assumed values of $M_{\odot}$. Moreover, Moffat asserts that, to a
good first approximation, the values of $Q = M_{WR}/M_{\odot}$
should be expected to reflect the relative total mass loss of
the WR star during that evolutionary phase, regardless of the
original large absolute masses, since the initial Q-values are
expected to be near unity for $O + O$ progenitors (Garmany et
al. 1980, and below) and the current mass loss of the O
companion is much less than that of the WR component. Figure 2
taken from Moffat (1981) shows the mass ratios determined for
his WR + O sample plotted against WN and WC subtype. A good
correlation is apparent, such that in both sequences the higher
excitation subclasses exhibit the lower values of Q. With the
realisation that the WR mass loss rates are very large, Moffat
concludes that this result suggests an evolution along the
sequence WN8-WN3 and WC8-WC4. It is unclear at what point a WN
star might become a WC star, since this will depend on the
mass loss rate and on the galactic metallicity Z. It is
noteworthy that in the SMC, with its much lower value of Z than
either the Galaxy or the LMC, only one (low mass) WC4 star is
known, whilst the rest of the WR population are WN3 stars,
which are also postulated to be of low mass ($\lesssim 4M_{\odot}$ if a typical
O-star companion mass of, say, $40M_{\odot}$ is assumed). Massey (1981)
questions some of these conclusions, pointing out that the mass
ratios Q will only reflect WR masses if the values of $M_{\odot}$
are the same - an unlikely circumstance. Nevertheless, I feel
that the Q vs. WR-subtype relations found by Moffat are highly
suggestive - they must be telling us something about the
evolutionary schemes involved in producing different WR
subtypes!

Massey (1981) has used the new and improved orbit
solutions for the 13 SB2's in his galactic sample to clear up
an old enigma highlighted by Smith (1968) in that the mass
ratios for WR + O binaries seemed to depend on the orbital
inclination - a physical impossibility. The Smith result was
in fact based on the hypothesis that all WR masses are
effectively the same, whilst in reality, as Massey has shown,
the WR masses are not identical and moreover exhibit a

considerably larger relative mass range than do their O-type companions. Massey also finds that, assuming an initial value of Q = 1 and taking the current masses of the relatively unevolved O-companions back to their main sequence values, at least 40% of the WR progenitor's initial mass must have been removed in forming the WR star.

Massey (1981) used the WR + O SB2 solutions together with complementary data now available for a large number of O + O systems (Garmany et al. 1980) to investigate period-mass-ratio relationships and throw light on the evolutionary aspects of massive binary systems. Figure 3, taken from Massey (1981), shows the orbital period P plotted against the mass ratio Q for both WR + O and O + O systems, and spot-lights the two stars which have recently been identified as likely intermediately evolved systems: BD + 40°4220 (Bohannon & Conti 1976) and HDE 228766 (Massey & Conti 1977). For the O + O systems the range of Q is very limited and most systems have values near unity (Garmany 1979). Values of Q for WR + O binaries generally fall in the range 0.3 - 0.5, but can be larger and up to 1.19 for CQ Cep. This overall result is understood since the initially most massive component will have evolved fastest through mass loss and Q should decrease. Massey notes that while there is a fairly clear separation in Q, there is really no significant difference in period between WR + O and O + O binaries. Restricting consideration to systems with $P \lesssim 20$ days (to minimise selection effects) gives an average value of $\overline{P}_{O+O} = 5.3$ days and $\overline{P}_{WR+O} = 5.7$ days.

To investigate dynamical effects that have been involved in the binary evolution, Massey uses the following equations relating the initial (subscript i) and final (subscript f) values of the orbital period P, component mass m, total masses M, and total orbital angular momentum J, at each stage in: (1) conservative evolution in which the total orbital angular momentum and mass of the system is constant in time, and (2) non-conservative evolution in which mass loss is assumed to be spherically symmetric, following the formulation of Vanbeveren & de Greve (1979):

(1) Conservative:

$$\frac{P_f}{P_i} = \left(\frac{J_f}{J_i}\right)^3 \left(\frac{M_f}{M_i}\right)^{-5} \left(\frac{q_f}{q_i}\right)^{-3} \left(\frac{1 + q_f}{1 + q_i}\right)^6 \qquad (1)$$

# (2) mass-loss

$$\frac{P_f}{P_i} = \left(\frac{M_f}{M_i}\right)^{-2} \left(\frac{1 + q_i^2/q_f}{1 + q_f}\right)^3 \left(\frac{1 + q_f}{1 + q_i}\right)^6 \tag{2}$$

where $q = m_2/m_1 = Q^{-1}$.

From the available data, typical values of Q for O + O systems and WR + O systems are 1.2 and 0.4 respectively, and we would thus expect $P_{WR}/P_O = 1.8$ in the conservative case (Eqn. 1 with the first two terms equal to unity on the r.h.s.), whilst in the mass loss case (Eqn. 2) adopting $M_f/M_i = 0.5 - 0.8$, we would expect $P_{WR}/P_O = 10-4$. The observations suggest $P_{WR} = P_O$, implying that more angular momentum must have been removed for the same relative M and q than that accommodated by spherically symmetric mass loss, or indeed by conservative binary stellar evolution.

From an analysis of a smaller number of WR + O systems, Smith (1973) suggested that WC + O systems exhibit larger orbit separations, a sini, than WN + O binaries. Massey has noted that with the larger binary sample now available this apparent difference is no longer statistically significant. Moreover, on plotting a sini vs. period using available data (Fig 4), he finds no clear separation between WR + O and O + O systems. This suggests that their total masses are similar, since from Kepler's laws one expects the two quantities to be related by

$$\log a = \frac{2}{3} \log P + \frac{1}{3} \log M + 0.625. \tag{3}$$

Since the O + O systems in the available sample contain mainly O8 or O9 stars, and only a few high mass O + O systems are known, Massey concludes that, in the expected non-conservative evolution, only the more massive O + O systems can become WR + O binaries. This assertion supports the views of Vanbeveren and de Loore (1980) that the progenitors of the WR stars in binaries must be more massive than $50M_\odot$. Finally, Massey notes that many of the long period, high a sini, WR binaries (and also O+O systems) exhibit high orbital eccentricities (e = 0.4-0.6) suggesting that mass transfer has not played a dominant role in their evolution, since such material transfer, either by stellar wind accretion or Roche lobe overflow, would tend to circularize the orbit.

The results of these recent studies of WR + O systems can be summarised as follows:

(a)   there is no one typical mass for a WR star, since the known masses span a large range from 10 - 50M$_\odot$, and span a larger relative range than do their O-companions.

(b)   the average mass appears to be about 20M$_\odot$.

(c)   the masses of WR stars of similar subtype can be quite different and in the galaxy there does not seem to be a systematic connection between mass and subtype. Moffat (1981) suggests a correlation exists between the mass ratio $M_{WR}/M_\odot$ and WR subtype, implying an evolutionary trend WN8 → WN3 and WC8 → WC4.

(d)   the masses of WC stars are not statistically lower than WN stars.

(e)   although the mass ratios for WR + O and O + O systems are quite different, the orbital periods are similar, suggesting that orbital angular momentum losses are greater than can be accounted for by either conservative or simplistic wind models. It appears that WR members of binaries have lost at least 40% of their initial mass.

(f)   the orbital separations of WC binaries appear to be statistically similar to those of WN systems, contrary to earlier beliefs.

(g)   it appears that only fairly massive O + O systems evolve into WR + O systems, and that mass transfer has not played a dominant role.

## 5.   SINGLE-LINED WR BINARIES WITH LOW MASS FUNCTION - WR + COLLAPSARS?

During the past few years extensive searches have been undertaken for WR binary systems that might contain collapsed object secondaries, since these systems are expected to occur naturally as one phase in the overall evolutionary history of massive binary systems in which mass removal from the component stars has played an important role. The evolutionary scheme, discussed by Van den Heuvel (1976), was developed to account for the existence of massive X-ray binary systems, and is schematically shows as follows:

$$OB_1 + OB_2 \longrightarrow WR_1 + OB_2 \longrightarrow n.s. + OB_2$$

(a)                    (b)                      (c)

$$n.s. + n.s. \longleftarrow n.s. + WR_2$$

(e)                      (d)

Stages (a), (b), (c) and (e) are all observed in nature, representing respectively massive OB binaries, WR + O systems, OB X-ray binaries and binary pulsars. Only stage (d) corresponding to the predicted WR + neutron star binary phase has not been so observationally obvious.

Moffat (1982) has summarised the current status of attempts to isolate such WR + collapsar systems, highlighting the difficulties involved in the requirement to undertake extensive observational programmes aimed at detecting small amplitude periodic variations in radial velocities and/or continuum light related to the orbital motion of the WR component. The compact object in these systems is not expected to be readily observable at optical wavelengths, nor in X-ray emission since all but the hardest, lower luminosity, X-rays would be expected to be absorbed in the overlying dense WR stellar wind (but see below!). Currently optical surveys are under way to try to find these systems in samples selected because of their apparently high galactic z-distances (asserted to be the result of a recoil in a SN explosion) which might enhance the detection possibility, as well as in more complete samples of WR stars down to a specified magnitude or space-volume limit.

To date 14 candidates have been found to exhibit low amplitude radial velocity variations, and these are listed in Table 2, taken principally from Moffat (1982) but with the addition of HD 143414 recently proposed as a WR + collapsar system by Isserstedt et al. (1983). In most cases the K-values are small, $\lesssim$20 km/s, and there is by no means consensus that the apparent velocity variations are in fact real. For instance, Massey (1982) states that he finds no significant velocity variations for either HD 177230 or HD 193077, and argues that those seen by Moffat and colleagues are dominated by measurement errors - a view contested by Moffat. More credence can be given to the higher K-value systems and in particular to those objects that exhibit both velocity and light variations with the same periodicity, eg. HD 50896 and HD 143414. Indeed the former star also shows similar periodicity in optical polarisation (McLean 1980) and probably represents, at present, one of the most promising WR + collapsar candidates.

Moffat (1982) argues that the low mass functions indicate unseen companions of mass $M_2 = 0.5-2$ $M_\odot$ in most cases, compatible with the presence of a neutron star. On making this assumption and adopting $M_2 = 1.6$ $M_\odot$, he deduces the WR masses listed in Table 2, which are seen in most cases to be compatible with the WR masses deduced for SB2 systems. The exceptions are HD 97950, HD 38268 and HD 197406 for which the deduced values of $M_{WR}$ appear too low. The nature of HD 97950 and, in particular, HD 38268 (the central object of 30 Doradus) is unclear and there are some who suggest that they may be supermassive stars, which confuses the issue. For HD 197406, Moffat prefers to adopt a "typical WR mass of 50-60 $M_\odot$" thence deriving $M_2$ in the range 5-15 $M_\odot$, suggesting that the unseen companions in these cases may be black holes.

Moffat (1982) notes that the current WR + neutron star candidates show a preponderence of WN stars, particularly WN7 and WN8 stars. This however may be a straight forward selection effect, since it is easier to pick up small velocity variations in their narrower emission lines compared to other WR subclasses. Using the current sample of candidates, he reassesses the WR binarity statistics, limiting consideration to WNL stars of which there are about 15 galactic objects brighter than $12^m.3$ for which optical studies for binarity have been completed. Notwithstanding the small number statistics involved, Moffat concludes that the relative frequency of WNL + c (= collapsar) and WNL + OB systems is 5:3, which is sufficiently close to parity to be consistent with the expected two phases of WR stars in massive binary evolution in which initial mass functions are close to unity.

The WNL+c candiates have a mean z-distance of 482 pc, compared to the mean value of 75 pc for other WNL stars and WNL + OB binaries. Moreover the former sample have different radial velocity components of peculiar velocity of 50 km/s compared to a value of 9 km/s for the other samples. Moffat argues that this result supports a runaway picture for the proposed WR+c systems. However, again it must be noted that there is no consensus that the apparent high z-values are real, since in deducing such values one has to adopt an absolute magnitude for the WR component (and thus a distance) and it is now recognised that there is a large range in $M_v$ for individual WR subclasses. Massey (1982b) suggests that the adoption of more modest absolute magnitudes than used by Moffat and colleagues would give more normal z-distances. In their discussion of the system HD 143414, Isserstedt et al. 1983 point out that if the runaway status is correct one can expect a corresponding z-component of proper motion of about 0.006" x 5/d (d in Kpc), which although small, should become measurable with future astrometric satellites such as HIPPARCOS. The advent of such facilities will clearly be of great importance in confirming (or denying) the runaway proposition.

A further problem encountered when examining the case for the reality of the WR + c systems relates to the low levels of X-ray emission observed from these objects. One might expect to see a similar level of X-ray emission as for OB X-ray binary systems, say $L_x \sim 10^{36}$ -$10^{38}$ erg/s, given that a similar X-ray production mechanism of accretion on to a neutron star will be operating. However, unlike the X-ray binaries, the WR+c systems were not seen with the early satellites like UHURU, and indeed with the advent of the EINSTEIN satellite we now know that the values of $L_x$ are typically $\lesssim 10^{33}$ erg/s in the 0.2 - 4 KeV range for single WR stars, WR + O binaries and proposed WR + c objects (Sanders et al. 1982). Moffat et al. (1982) report some variability in the observed X-ray emission from HD 50896 as well as the SB2 system V444 Cyg, but these are attributed to orbital-linked variations of colliding winds, with no evidence of harder X-rays emanating from accreting material onto a compact object. Indeed the typical values of WR X-ray emission are found to be similar in level and spectral shape to those seen in O-type stars of comparable radiative luminosities. It has long been qualitatively argued that the WR winds are so dense and extensive that any neutron star companion would in effect be buried in the wind material, and the expected X-ray production would suffer so severe attenuation as to be masked from observation. Initial computations by Van den Heuvel (1976) and Moffat & Seggewiss (1979) supported this view, and the picture does seem at first glance to be reasonable since we know that the WR mass loss rates are typically a factor of 10 larger than for O stars. However the question has recently been reassessed by Vanbeveren et al. (1982) with some disturbing consequences for the WR + c proponents.

Vanbeveren et al. (1982) have calculated the expected X-ray luminosity from WR + neutron strar systems, considering in their models both the production of hard X-rays by accretion and subsequent absorption of this by the WR stellar wind. They find that the wind is not sufficiently opaque to extinguish the X-rays beyond the observational limits of instrumentation prior to EINSTEIN. Although the enhanced WR mass loss and wind extent does increase the absorption of the X-rays, it also enhances the production of the X-ray emission. For WR mass loss rates of $3 \times 10^{-5}$ (known now to be realistic) and reasonable values of the stellar wind velocity at the separation of the neutron star (deduced from measured values of wind terminal velocities and reasonable velocity laws) they are able to compute the expected X-ray emission in, say, the 2-6 KeV (UHURU) range as a function of binary period and a range of values of kT (5-20 KeV) describing the X-ray spectrum. Typically for periods of about 5 days they find values of $L_x$ (2-6 KeV) in the range $10^{35} - 10^{37}$ erg/s, considerably above that measured with EINSTEIN. This result casts considerable doubt on the reality of the proposed WR + c systems, but may not be a killer blow. As Vanbeveren et al. (1982) point out, one can conceive of other ways to reconcile the data. For instance the accretion may take place at supercritical rates such that the expected accretion disk may become optically thick, or the production of X-rays may proceed ineffectively as a consequence of an abnormally large wind velocity at the neutron star separation coupled with a large period. Clearly there is scope here for further detailed modelling of expected X-ray production and attenuation mechanisms.

As a further test on the reality of the proposed WR + c systems, one can look for spectral effects induced by the presence of the neutron star companion which are known to be common phenomena in massive X-ray binaries, and one such approach is to utilise UV observations. IUE observations of X-ray binaries have confirmed the theoretical predictions of Hatchett & McCray (1977) that UV P-Cygni profiles (say in CIV $\lambda$ 1550, NV $\lambda$1240, SiIV $\lambda$1400) should show binary phase-linked changes in shape resulting from the anisotropic stellar wind ionisation produced by the X-ray emission from the neutron star companions. In such cases one expects an anomalously high region of ionisation around the neutron star (the extent depending on the level of X-ray emission) which will affect the wind material in its immediate environment, producing changes in the P-Cygni absorption at binary phases near 0.5 (neutron star in front in the line of sight), whilst at phases near 0.0 the emission part of the P-Cygni profile is affected and the absorption component reverts to its undisturbed strength. This is precisely what is observed in the IUE spectra of Cyg X-1 (Treves et al. 1980), Vela X-1 (Dupree et al. 1980) and the sources SMC X-1 and LMC X-4 (van der Klis et al. 1982), and is clearly a common signature of the neutron star/X-ray emission in such systems. Clearly searches for comparable effects in the UV P-Cygni profiles for WR + c candidates would assist in confirming their nature. So far only one candidate, HD 50896 has been extensively observed with IUE, with numerous exposures recorded during 1978-1981.

Together with colleagues at JILA, I have extracted these
available data and we have indeed found significant UV spectral
variations, particularly in the unsaturated P-Cygni profile to
the excited NIV $\lambda$1718 line. Example spectra are shown in Fig
5a, and in Fig 5b the equivalent width of the P-Cygni
absorption component in NIV $\lambda$1718, measured from these data,
is plotted as a function of proposed binary phase. There is
considerable scatter in the data, but clearly gross changes are
evident. Most of the available spectra were acquired over many
different binary cycles, and some of the scatter may be the
result of secular changes. Those data obtained over a single
cycle do appear to show a cleaner pattern of variations, which
is broadly consistent with that seen in the X-ray binary
spectra - at phases near 0.5 the P-Cygni absorption is reduced
over that seen at phases near 0.0. Thus available UV data
appears consistent with the proposed collapsar candidature for
HD 50896, but more spectra are still required to tidy up the
variation curve. Ideally one would also like further X-ray
observations of this and other systems, preferably simultaneous
with UV observations, in order to allow a detailed modelling of
the UV/X-ray variability, and with the availability of EXOSAT
this is at least potentially possible.

The current position regarding the reality of WR + c
systems is thus unclear. There is increasing evidence for low
mass function SB1 systems, although much more optical data is
required on individual systems to clean up the RV and light
curves. Forthcoming satellites like HIPPARCOS should confirm
or deny the runaway status of such systems, and more extensive
X-ray observations are needed to search for variability as a
function of binary phase and examine the shape and level of the
hard X-rays ($\geqslant$ 4 KeV). More theoretical work on the expected
accretion processes involved when a neutron star is embedded in
a dense, fast flowing WR stellar wind are required to attempt
to reconcile the low levels of X-ray emission observed from
these systems. Clearly the topic is still in its infancy and
is worthy of a concerted attack on both observational and
theoretical fronts.

TABLE 1 : DOUBLE-LINED WR+O SYSTEMS AND SINGLE-LINED WR BINARIES WITH HIGH MASS FUNCTIONS

| STAR | Sp | P (days) | a sini (R$_\odot$) | M$_{WR}$ sin$^3$i (M$_\odot$) | M$_{WR}$/M$_O$ | M$_{WR}$ (M$_\odot$) | eclipsing | |
|---|---|---|---|---|---|---|---|---|
| CQ CEP | WN7+O? | 1.6 | 20 | 23 | 1.19 | 23 | yes | |
| GPCEP | WN6+O | 6.7 | 40 | 3.6 | 0.22 | 10-25 | yes | |
| HDE 311884 | WN&+O5V | 6.34 | 64 | 40 | 0.84 | 50 | no | |
| CX CEP | WN5+O8V | 2.13 | 18 | 5 | 0.43 | 5-11 | yes | |
| V444CYG | WN5+O6 | 4.21 | 35 | 9.3 | 0.40 | 11 | yes | |
| HD 190918 | WN4.5+O9I | 112.8 | 130 | 0.7 | 0.26 | 9 | no | |
| HD 186943 | WN4+O9V | 9.55 | 55 | 9-11 | 0.52 | 13 | no | |
| HD 94546 | WN4+O7 | 4.9 | 38 | 8 | 0.34 | ≥8 | no | Massey |
| HD 5980 | WN4+O7I | 19.6 | 100 | 8 | 0.30 | 8 | yes | Massey |
| HD 90657 | WN4-04-6 | 8.2 | 50 | 8.4 | 0.52 | ≥8 | no | Niemela & Moffat 1982 |
| HDE 320102 | WN3+O5-7 | 8.83 | 35 | 1.8 | 0.33 | 20 | no | Niemela 1982 |
| HD 68273 | WC8+O9I | 78.5 | 285 | 17 | 0.54 | 21 | no | |
| CV SER | WC8+O8V | 29.7 | 120 | 11 | 0.48 | 13 | no | |
| HD 152270 | WC7+O5 | 8.89 | 34 | 1.8 | 0.36 | 20 | no | |
| HD 97152 | WC7+O7V | 7.89 | 36 | 3.6 | 0.59 | 20 | no | |
| HD 94305 | WC6+O-8 | 18.8 | 108 | 15 | 0.47 | ≥15 | no | |
| HD 63099 | WC5+O7I | 14.7 | | 4.7 | 0.25 | ≥4.7 | no | |
| | | | | | | | | |
| HD 92740 | WN7 abs | 80.35 | f = 1.75 | | | | | |
| CD-45°4482 | WN7 | 23.9 | f = 5.5 | | | | | |
| HD 193928 | WN6 | 21.64 | f = 4.3 | | | | | |
| HD 113904 | WC6 | 18.34 | f = 9.9 | | | | | |

for references, see text.

TABLE 2 : SINGLE-LINED WR BINARIES WITH LOW MASS FUNCTION ($f \leqslant 0.3M_\odot$) - WR + COLLAPSAR CANDIDATES*
(taken from Moffat (1982))

| STAR | Sp | Z (pc) | P (days) | e | $K_{wR}$(KM/s) | $M_{wR}(M_\odot)$* |
|---|---|---|---|---|---|---|
| HD 50896 | WN5 | -356 | 3.8 | 0.34 | 36 | 12 |
| HD 143414 | WN6 | -744 | 7.7 | 0 | 21 | 23 |
| HD 192163 | WN6 | +53 | 4.5 | 0.3 | 20 | 27 |
| HD 193077 | WN6 (OB) | +29 | 2.3 | 0 | 16 | 51 |
| HD 97950 | WN6 (OB) | -68 | 3.8 | 0 | 54 | (5) |
| HD 38268 | WN6 (OB) | | 4.4 | 0 | 43 | (7) |
| HD 197406 | WN7 | +799 | 4.3 | 0.11 | 90 | (1.3) |
| HD 96548 | WN8 | -342 | 4.8 | 0 | 10 | 71 |
| HD 86161 | WN8 | -181 | 10.7 | 0 | 6 | 105 |
| 209 BAC | WN8 | +264 | 2.4 | 0 | 13 | 68 |
| HD 177230 | WN8 | -834 | 1.8 | 0 | 22 | 35 |
| HD 164270 | WC9 | -228 | 1.8 | 0 | 20 | 41 |

NB: * - $M_{wR}$ ESTIMATED ASSUMING M(ns) = 1.6 M and i = 60°.

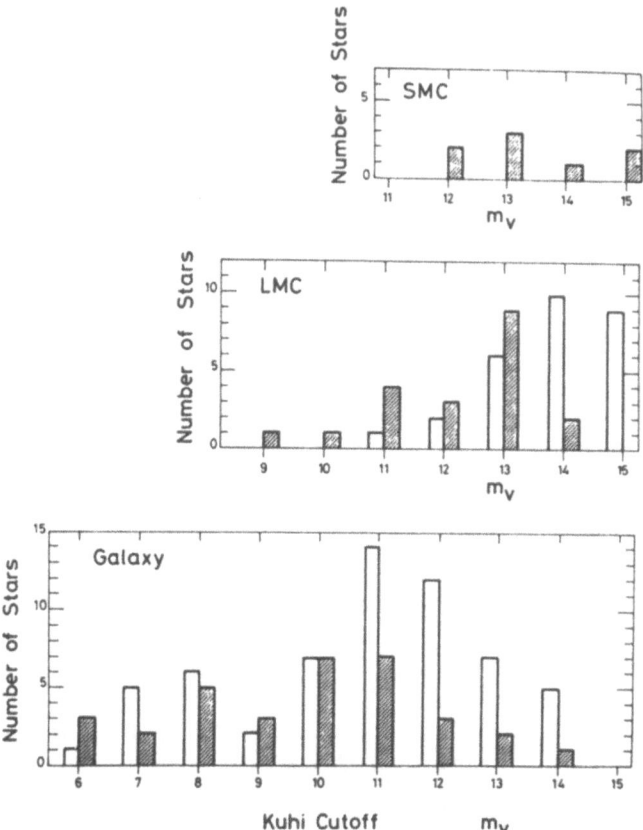

Figure 1. The frequency distribution with apparent visual magnitude of WR stars in the Galaxy and Magellanic Clouds which (a)
show absorption lines (shaded boxes) and (b) those that do not.
These data imply 40% of WR stars in the Galaxy and LMC are binaries,
if the presence of an absorption spectrum is taken as a criterion
for binarity (taken from Vanbeveren & Conti 1980).

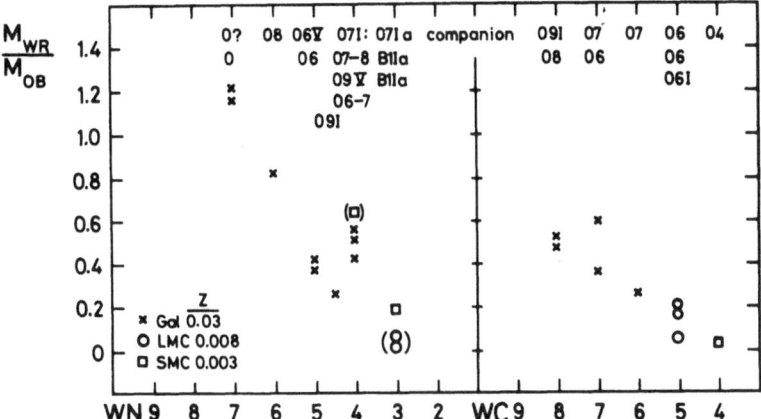

Figure 2. The mass ratio, $M_{WR}/M_O$, for SB2 components of galactic
and MC systems plotted as a function of WN and WC subtype. These
data imply an evolutionary sequence WN8-WN3 and WC8-WC4 (taken
from Moffat 1981).

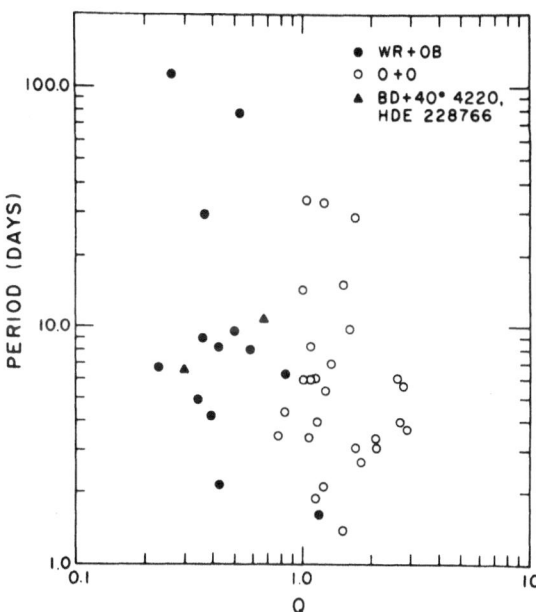

Figure 3. A comparison of the orbital periods and mass ratios
derived for O-type and WR-type SB2 systems in the Galaxy (taken
from Massey 1981).

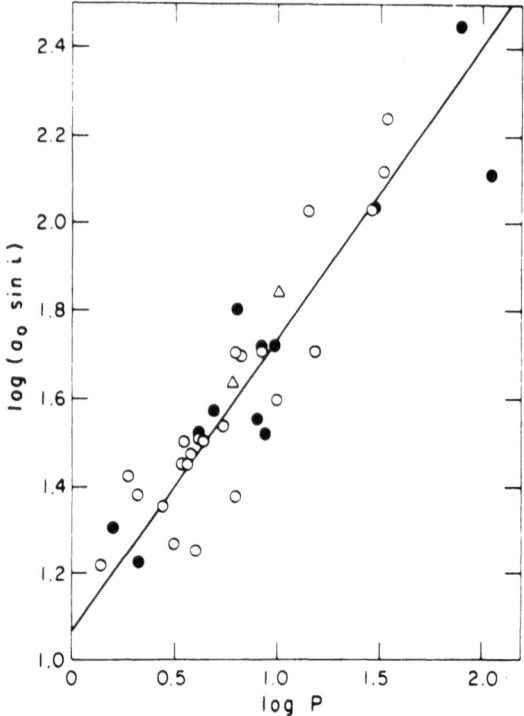

Figure 4.   The correlation of projected orbital separation and
orbital period for O+O and WR+O SB2 systems in the Galaxy (taken
from Massey 1981).

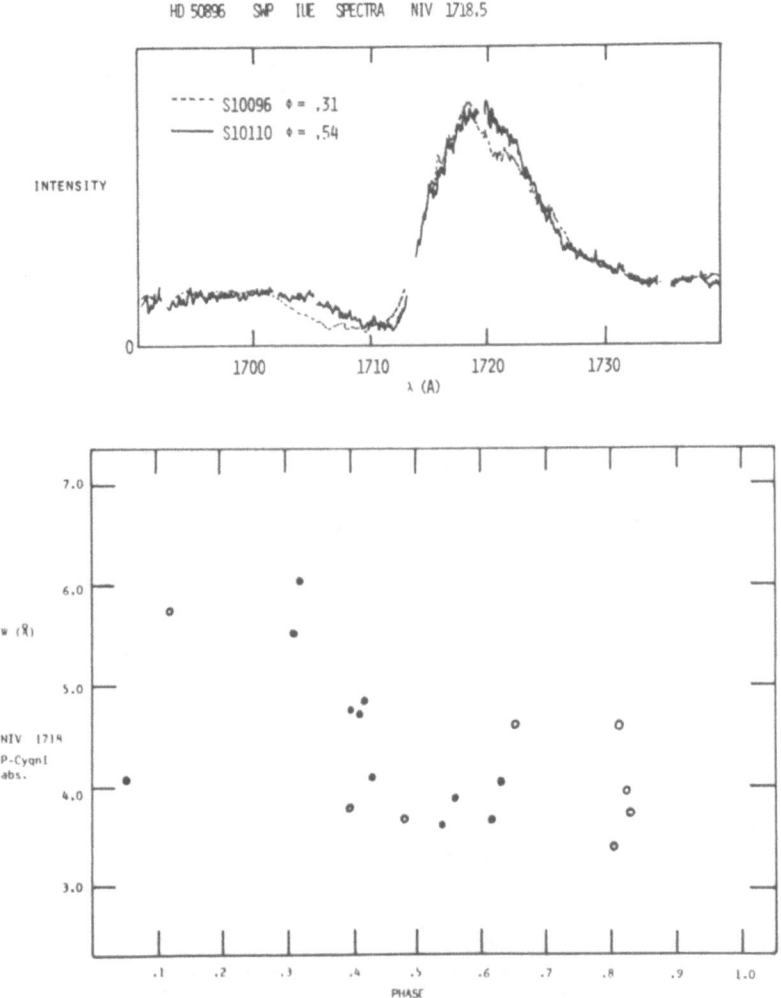

Figure 5. (a) Sample IUE spectra of the NIV λ1718 P-Cygni profile at two binary phases in HD 50896 (a prime WR+collapsar candidate).

(b) The equivalent width of the absorption component to the NIV λ 1718 P-Cygni profile as a function of phase in HD 50896. Filled circles represent data obtained during a single binary cycle, open circles are data acquired over a variety of epochs and cycles. The large variations apparent and their rough dependence of phase are similar to those seen in the UV spectra of X-ray binary systems.

REFERENCES

Abbott, D.C., Bieging, J.H. and Churchwell, E., 1982. IAU
    Symp. 99 (eds. C.W.H. de Loore & A.J. Willis), D. Reidel
    Pub. Co., Holland, p 215.
Azzopardi, M. and Breysacher, J., 1979, Astron.Astrophys.,
    **75**, 120.
Bappu, M.K.V. & Sahade, J., 1973. Eds. IAU Symp. 49, D.
    Reidel Pub. Co.    Holland.
Barlow, M.J., Smith, L.J. & Willis, A.J., 1981. Mon.Not.
    R.Ast.Soc., **196**, 101.
Barlow, M.J. & Hummer, D.G., 1982. IAU Symp. 99 (eds. C.W.H.
    de Loore &    A.J. Willis), D. Reidel Pub. Co.,
    Holland, p 387.
Beals, C.S., 1933. Pub. Dom. Ast. Obs., **6**, 95.
Bohannon, B. & Conti, P.S., 1976. Astrophys.J., **204**, 797.
Breysacher, J., 1981. Astron.Astrophys.Suppl., **43**, 203.
Breysacher, J., Moffat, A.F.J. & Niemela, V., 1982. IAU
    Symp. 99 (eds. C.W.H. de Loore & A.J. Willis), D. Reidel
    Pub. Co., Holland, p 317.
Chiosi, C. & Stalio, R., 1981. Eds. IAU Coll. 59,
    D. Reidel Pub. Co., Holland.
Conti, P.S., 1976. Mem.Soc.Roy.Sci.Liege, 6 Ser, **9**, 193.
Conti, P.S. & de Loore, C.W.H., 1979. Eds. IAU Symp. 83,
    D. Reidel Pub. Co. Holland.
Conti, P.S., 1982, IAU Symp. 99 (eds. C.W.H. de Loore &
    A.J. Willis), D. Reidel Pub. Co., Holland, p 3.
Conti, P.S. & Massey, P., 1982. Astrophys.J., **249**, 471.
Conti, P.S. & Roussel-Dupree, 1982. Preprint.
Corso, J., 1975. Ph.D. Thesis, North Western University.
de Loore, C.W.H. & Willis, A.J., 1982. Eds. IAU Symp. 99,
    D. Reidel Pub. Co. Holland
D'Odorico, S. & Rosa, M., 1981. Astrophys.J., **248**, 1015.
Dupree, A.K., Gursky, H., Black, J.H., Davis, R.J.,
    Hartmann, L., Matilsky,T., Raymond, J.C.,
    Hammerschlag-Hensberge, G., van den Heuvel, E.P.J.,
    Burger, M., Lamers, H.J.G.L.M., Vanden Bout, P.A.,
    Morton, D.C., de Loore, C.W.H., van Dessel, E.L.,
    Menzies, J.W., Whitelock, P.A., Watson, M., Sanford, P.W.
    & Pollard, G.S.G., 1980, Astrophys.J. **238**, 969.
Garmany, C.D., 1979. IAU Symp. 83 (eds. P.S. Conti & C. de
    Loore), D. Reidel Pub. Co., Holland, p 261.
Garmany, C.D., Massey, P.M. & Conti., P.S., 1980. Astrophys.
    J., **242**, 1063
Gebbie, K. & Thomas, R.N., 1968. Eds. Wolf-Rayet Stars, NBS
    SP-307.
Hatchett, S.P. & McCray, R., 1977. Astrophys.J., **211**, 552.
Hogg, D.E., 1982. IAU Symp. 99 (eds. C.W.H. de Loore &
    A.J. Willis), D. Reidel Pub. Co., Holland, p 221.

Isserstedt, J., Moffat, A.F.J. & Niemela, V.S., 1983.
    Astron. Astrophys., in press.
Kuhi, L. V., 1973. IAU Symp. 49 (eds. M.K.V. Bappu & J.
    Sahade), D. Reidel Pub. Co., Holland, p 205.
Leep, E.M., 1982. IAU Symp. 99 (eds. C.W.H. de Loore &
    A.J. Willis), D. Reidel Pub. Co., Holland, P 41.
Massey, P.M. & Conti, P.S., 1977. Astrophys.J., **218**, 431.
Massey, P.M., 1980. Astrophys.J., **236**, 526.
Massey, P.M., 1981. Astrophys.J., **246**, 153.
Massey, P.M., 1982. IAU Symp. 99 (eds. C.W.H. de Loore &
    A.J. Willis), D. Reidel Pub. Co., Holland, p 251.
Massey, P.M., 1982b IAU Symp. 99 (eds. C.W.H. de Loore &
    A.J. Willis), D. Reidel Pub. Co., Holland, p 273.
Massey, P.M. & Conti, P.S., 1981. Astrophys.J., **244**, 173.
Massey, P.M., Conti, P.S. & Niemela, V., 1981. Astrophys.J.,
    **246**, 145.
Massey, P.M. & Niemela, V., 1981, Astrophys. J., **245**, 195.
McLean, I.S., 1980. Astrophys.J. **236**, L189.
Moffat, A.F.J., 1981. IAU Coll. 59 (eds. C. Chiosi & R.
    Stalio), D. Reidel Pub. Co., Holland, p 301.
Moffat, A.F.J., 1982. IAU Symp. 99 (eds. C.W.H. de Loore &
    A.J. Willis), D. Reidel Pub. Co., Holland, p 263.
Moffat, A.F.J. & Seggewiss, W., 1979, Astron.Astrophys., **77**,
    128.
Moffat, A.F.J., Firmani, C. & Lara, E. de., 1981. IAU
    Symp. 99 (eds. C.W.H. de Loore & A.J. Willis). D.
    Reidel Pub. Co., Holland, p 577.
Niemela, V.S., 1980. IAU Symp. 88 (eds. M.J. Plavec, D.M.
    Popper & R.K. Ulrich), D Reidel Pub. Co., Holland
Niemela, V.S., 1981. IAU Coll. 59 (eds. C. Chiosi & R.
    Stalio), D. Reidel Pub. Co., Holland, p 307.
Niemela, V., 1982. IAU Symp. 99 (eds. C. W. H. de Loore &
    A.J. Willis), D. Reidel Pub. Co., Holland, p 299.
Niemela, V.S., Mendez, R.H., & Moffat, A.F.J., 1983, Astrophys.
    J., **272**, 190.
Niemela, V.S., & Moffat, A.F.J., 1982. Astrophys. J. **259**, 213.
Nussbaumer, H., Schmutz, W., Smith, L.J. & Willis, A.J.,
    1981. Astron.Astrophys.Suppl., **47** , 257.
Paczynski, B., 1973. IAU Symp. 49 (eds. M.K.V. Bappu &
    J. Sahade), D. Reidel Pub. Co., Holland, p 143.
Perry, D.N. & Conti, P.S., 1982. IAU Symp. 99 (eds. C.W.
    H. de Loore & A.J. Willis), D. Reidel Pub. Co., Holland,
    p 109.
Sanders, W.T., Cassinelli, J.P. & van der Hucht, K.A., 1982.
    IAU Symp. 99 (eds. C.W.H. de Loore & A.J. Willis),
    D. Reidel Pub. Co., Holland, p 589.
Smith, L.F., 1968. Mon.Not.R.Ast.Soc., **138**, 109.
Smith, L.F., 1973. IAU Symp. 49 (eds. M.K.V. Bappu & J.
    Sahade), D. Reidel Pub. Co., Holland, p 228.

Smith, L.J. & Willis, A.J., 1982. Mon.Not.R.Ast.Soc.,
    **201**, 451.
Smith, L.J. & Willis, A.J., 1983. Astron.Astrophys.Suppl.,
    in press.
Treves,  A.,  Chiapetti, L.,  Tanzi, E. G.,  Tarenghi, M.,
    Gursky, H., Dupree, A.K.,  Hartmann, L.W.,  Raymond, J.,
    Davis, R.J.,  Black, J.,  Matilsky,T.A.,  Vanden Bout,
    P.,Sanner, F., Pollard, G., Sanford, P.W., Joseph, R.D., &
    Meikle, W.P.S., 1980. Astrophys.J., **242**, 1114.
Underhill, A.B., 1980. Astrophys.J., **239**, 220.
Underhill, A.B., 1981. Astrophys.J., **244**, 963.
Vanbeveren, D. & Conti, P.S., 1980. Astron.Astrophys., **88**, 230.
Vanbeveren, D. & de Greve, J.P., 1979. Astron.Astrophys.,**77**,
    295.
Vanbeveren, D. & de Loore, C.W.H., 1980. Astron.Astrophys.,
    **86**, 21.
Vanbeveren, D., van Rensbergen, W. & de Loore, C.W.H.,
    1982. Astron.Astrophys., **115**, 69.
van den Heuvel, E.P.J., 1976, IAU Symp. 73 (eds. P.P. Eggleton,
    S.M. Mitton & J.A.J. Whelan), D. Reidel Pub. Co.,
    Holland, p 35.
Van der Hucht, K.A., Conti, P.S., Lundstron, I. &
    Stenholm, B., 1981. Spac.Sci.Rev., **28**, 227.
Van der Klis, M., Hammerschlag-Hensberge, G., Bonnet-Bidaud,
    M., Ilovaisky, S.A., Mouchet, M., Glencross, W.M., Willis,
    A.J., van Paradijs, J.A., Zuiderwijk, E.J., & Chevalier,
    C., 1982. Astron.Astrophys. **106**, 339.
Willis, A.J., 1982. Proc. 1st Rutherford Appleton Lab.
    Workshop (ed. P. Gondhalekar) RAL-82-075, p 1.
Willis, A.J., 1982b. IAU Symp. 99 (eds. C.W.H. de Loore &
    A.J. Willis), D. Reidel Pub. Co., Holland, p 87.
Wolf, C.J.E. & Rayet, G., 1867. Comptes Rendus, **65**, 292.
Wray, J.D. & Corso, G.J., 1972. Astrophys.J., **172**, 577.
Wright, A.E. & Barlow, M.J., 1975. Mon.Not.R.Ast.Soc.,
    **170**, 41.

# PHOTOMETRY OF ACTIVE ALGOLS

Edward C. Olson

Astronomy Department
University of Illinois,
U.S.A.

ABSTRACT

    While light curves of several  short period Algol binaries
have shown transient instabilities from the  ultraviolet to the
near infrared, only in RW Tau and  most  notably  in U Cep have
these disturbances been large enough to analyse quantitatively.
From a study of the spectral distribution of light excesses and
losses  in  U Cep, we derive properties of "equatorial bulges,"
hot spot, and  stream  during  periods  of  activity presumably
associated with  transient,  enhanced,  mass  flows.   We also
discuss the nature of large light losses  that appeared outside
eclipse  and in primary eclipse egress during active intervals.
We show  that small surges in cool star brightness preceded and
accompanied intervals of activity.  Apart from such surges, the
cool  star brightness has varied over the last eight years in a
way somewhat suggestive of a magnetic activity cycle.  Finally,
since 1972  four  abrupt  orbital period changes have occurred,
and the evidence now available suggests that these changes have
nothing to do with eruptive mass flows in U Cep.

*P. P. Eggleton and J. E. Pringle (eds.), Interacting Binaries, 127–154.*
© *1985 by D. Reidel Publishing Company.*

## 1.   INTRODUCTION

In contrast to active binaries such as the "Serpentids"
(Plavec, 1980), Algols are in a state of slow mass transfer.
Can multicolour continuum photometry of the latter tell us
anything about interactions between the stellar components? In
particular, can we recognise "hot spots," "disks," and
"streams" expected when matter flows from a contact to a
detached component? Can such flows, which are often irregular
in nature, be associated with well-known period changes? Can
we find evidence that the cool contact star is actually the
source of the mass flow?

There is no doubt that emission line spectroscopy,
particularly in the satellite ultraviolet, is the most
sensitive indicator of circumstellar flows in close binary
systems (for example Plavec 1982). Photometry is a useful
adjunct, partly because extended observing time on the required
small to moderate sized telescopes has been available to cover
some of the rapid variations associated with irregular mass
flows. Light curve instabilities occur simultaneously at
wavelengths from the ultraviolet to the near infrared,
suggesting that their sources may be optically fairly thick in
the continuum. In some cases, the full precision of
photoelectric photometry can be brought to bear on the
interpretation of these variations.

We use five-colour intermediate band (uvby, Kron I)
photometry, since these colours are free of band-width effects,
and can be converted accurately to relative monochromatic
fluxes for analysis. Such photometry has shown transient
activity, defined as phase-and-time variable light gains or
losses occurring in a few orbital cycles, in several
short-period Algos, most notably by far in U Cep.


## 2.   ALGOL SURVEY

We have carried out a three-year survey of primary
eclipses of 16 short-period Algol systems, to search for
transient photometric disturbances. Portions of about 12
eclipses each were observed for the systems RZ Cas, U Cep,
U CrB, U Sge and RW Tau, while the remaining 11 systems were
observed an average of 4 times each (Olson 1982b). These 11
(XZ And, KO Aql, SW Cyg, W Del, AI Dra, TT Lyr, RW Mon, RV Oph,
ST Per, X Tri and TX UMa) had relatively stable light curves,
though some showed small anomalies such as curvature or tilt
during eclispse totality. Partially eclipsing systems RZ Cas
(A2 + G6) and U CrB (B6 + G0) often showed light curve
fluctuations as large as 0.1 mag to 0.15 mag in the

ultraviolet, with smaller variations at longer wavelengths (see Figs. 6,7 and 10 in Olson, 1982b). Were these systems totally eclipsing, these instabilities would probably be much larger, and the more favourable eclipse geometry would aid in their interpretation. As noted above, these variations seemed to arise in optically thick sources.

Variations were larger in RW Tau (B8 + K0), as shown in Figure 1. Extra light, stronger during eclipse ingress than egress, was present on several nights; it varied in strength appreciably in as little as six days. These light excesses reached their maximum values at phases from mid-eclipse $|\Delta\phi|\gtrsim$ 0.02, and were essentially constant at larger phase displacements. This behaviour (and a study of the spectrum of light excess noted below) suggests that this light comes from an asymmetrical "bulge" that developes around the equator of the hot gainer, presumably as a result of mass flow. Because of the structure of such bulges in RW Tau and U Cep (discussed below), we do not use the term "disk." The sense of the light asymmetry shows that most of the circumstellar material was above the following hemisphere of the hot star, near the point of impact of the mass-transferring stream and the gainer's photosphere. Kaitchuck, Honeycutt & Mufson (1980) observed associated $H\alpha$ emission, once during the same eclipse in which photometric activity was seen.

Of 12 primary eclipses of U Sge (B8 + G2) observed so far (one has been added since Olson 1982a), four showed normal light curves in good agreement with earlier **uvby** observations by McNamara & Feltz (1976). The remaining 8 light curves were shifted brighter or fainter by up to several tenths of a magnitude. The simplest explanation invokes a radius change in one (or both) of the stars. There is no hint of comparable behaviour in any of the other 15 Algols. U Sge has the most massive gainer (and mass sum) of these 16 systems. Verification of radius changes in U Sge requires more extensive coverage of primary eclipse and regions outside eclipse.

## 3.   EIGHT YEARS OF U CEPHEI:  DESCRIPTIVE SUMMARY

U Cep (B7 + G5-8) is by far the most active of the short-period ( $\lesssim$ 10 days) Algols, and is bright and easily observed. It is potentially one of the best systems for unravelling the physics of mass transfer. Since early warnings of activity in late 1974 (Batten et al. 1975, and Plavec & Polidan 1975), we have obtained photometry in some 60 primary eclipses and at other orbital phases outside eclipse. Ultraviolet data from four eclipses in 1980, shown in Figure 2, illustrate the nature and size of moderate disturbances in this

system, as well as their transient behaviour. Other examples
have been published by Crawford (1979). Relative to the nearly
undisturbed eclipses of February 25 and April 25, note the
following features of the disturbed curves:

(1)     There is excess light at internal contacts $t_2$ and $t_3$; as
        in RW Tau, the larger excess occurs at second contact.
        The equatorial bulge responsible for this extra light is
        essentially hidden at mid-eclipse (see Fig. 5 for a
        sketch of the geometry near second contact).

(2)     Maximum light excess occurs before $t_2$, near phase
        0.97. Here the equatorial bulge is fully exposed, and
        there may be a contribution from a hot spot where the
        stream strikes the photosphere of the gainer.

(3)     From the light maximum near phase 0.97, excess light
        drops toward earlier phase and eventually becomes a
        light loss. Even in the undisturbed state, the ingress
        shoulder is depressed relative to egress; this
        depression probably is due to projection of the stream
        against the B-star photosphere (the depression is not
        evident in Fig 2, but is clear in the observations of
        Fig. 11).

(4)     Excess light after mid-eclipse peaks just after third
        contact (as the "egress bulge" comes fully into view);
        shortly thereafter, an accelerating light loss depresses
        the active curve below the undisturbed one.

     This kind of activity was highly transient. U Cep did not
remain in a steady "high" state for more than one orbital
period (2.5 days). Activity often developed or decayed in $\lesssim 10$
days. Particularly in late 1974, activity seemed to be
modulated in a time $\simeq 20$ days.

     If these photometric instabilities were caused by eruptive
mass flows, then the release of accretion energy should raise
the luminosity. This did not happen for the observed optical
radiation, close to the orbital plane. Nearly coincident with
disturbances in primary eclipse, the light outside eclipse
fell, dropping to as little as half its normal brightness in
the ultraviolet. Figure 3 shows this behaviour, and includes
observations from an active run in October, 1975. Light losses
during primary eclipse egress joined smoothly to losses outside
eclipse (see upper panel, Fig. 3, where losses from October 26
to 31 can be compared to those which developed from October 24
to 29 near phase 0.2, and to larger ones which developed from
October 25 to 30 near phase 0.6). A lower envelope of these
irregular light curves seems to be defined by observations on

October 28, 29 and 30.   Excess bulge light at second contact
peaked on about October 26, while  at third contact it declined
gradually from October 16 to November 5 (Olson 1980a,b).   There
is  thus  a  hint  that  light  losses  outside  eclipse  lagged
slightly behind excesses seen during primary eclipse.

Light  fluctuations described above, both in  and  outside
primary eclipse,  decreased  monotonically from the ultraviolet
to the near infrared ($\sim$350 to 840 millimagnitudes).

Finally, anticipating a more detailed discussion below, we
note that:  (1)  the  outer  hemisphere  of  the  cool  loser
brightened somewhat during  times  of photometric activity, and
(2)  several  abrupt orbital period  changes  since  1972  were
apparently uncorrelated with photometric activity.


4.    U CEPHEI:   ABNORMAL ANATOMY

a)    Primary Eclipse

Can physical sense  be  made  of  the transient brightness
variations  in  U  Cep, and they  be  related  to  effects
expected when mass transfer occurs via Roche lobe overflow?  It
is rather remarkable that  most  excess  light  during  primary
eclipse  originated  in  optically thick equatorial bulges that
radiated as stellar atmospheres,  while the losses seemed to be
produced when part of the normal B star photosphere was covered
or replaced by cooler optically  thick material, also radiating
in simple stellar atmosphere fashion.  We want to justify these
statements  by  analaysing  the spectrum of  light  variations.
Specifically, we deal with $L_\lambda^c$ ($\phi$), the light excess (or loss if
negative) at orbital phase $\phi$, expressed as a monochromatic flux
normalized in a convenient  way.  This flux excess may contain
contributions from several sources (e.g., bulge  and hot spot),
so it is more useful to look  at changes $\Delta L_\lambda^c$  between selected
phases  in  order to isolate specific sources.  Fortunately the
observational precision is just adequate to do this in a fairly
convincing way.  In  error  estimates we include errors in: (1)
the observed active light curves,  (2)  the  undisturbed  light
curves  (taken  in primary eclipse as the mean of egress curves
when activity was  a  minimum, and reflected around mid-eclipse
to ingress when necessary), and  (3)  the  colour-relative flux
calibrations (usually small).

Batten (1974)  commented  on  brightness variations of the
G star in U Cep.  Such  variations  can  be  seen in published
light  curves  (for  example Olson 1976, Fig. 1).  The G star
tends to be slightly brighter when photometric activity distorts
the light curve.  Figure 4 shows flux distributions of the cool

star in its "quiet" and "active" states, normalized to the
qui   yellow flux. Data were taken from the flat portions of
eclipse totality, to prevent contamination by bulge light.
Circles and crosses are model atmosphere fluxes (Olson 1981).
Evidently, this variation can be explained as a small
temperature increase of the entire photosphere (alternatively,
a slightly larger temperature increase of a smaller area would
also work). There is no convincing evidence for a radius
change in this star. Later, we attempt to correlate brightness
changes of the G star with episodes of mass transfer over the
past 8 years.

Ingress eclipse geometry is shown in Figure 5, and is
based on a new photometric solution (Olson 1983) which, for our
purposes, is nearly the same as the solution given by Hall &
Walter (1974). Consider first changes in excess light $\Delta L_\lambda^c$
between second contact and mid-eclipse, and specifically for
the phase steps labelled "1", "2" and "3". If the phase steps
are taken in the sense away from eclipse, then $\Delta L_\lambda^c$ is positive
for each step. The upper panel of Figure 6 shows the spectral
distribution of added light, normalized to the undisturbed
yellow flux of the B star, in two eclipses in October 1975.
Every disturbed eclipse so far shows the same pattern: close
to the B star limb, the spectrum of the extra light is typical
of a hot optically thick source, while beyond about 1.3 B-star
radii, the light decreases and becomes characteristic of a gas
of finite optical thickness (that is, the Balmer continuum goes
into emission). Filled circles are fits for phase range "3" to
model atmosphere fluxes at $T_{eff}$ = 14,000 K, log g = 4.0
(Kurucz, 1979). The actual temperature may be somewhat lower
for two reasons: (1) some Balmer emission may contaminate the
ultraviolet flux, reducing the size of the absorption Balmer
jump and pushing the estimated temperature up somewhat; (2) the
spectrum may be that of near-limb stellar intensity rather than
flux; Olson (1980a) showed that temperatures are then ~12, 000
to 13,000 K. In either case, the maximum total vertical
geometrical thickness of the bulge slice in range "3" is of the
order of, or somewhat larger than, half the B-star diameter.
The vertical thickness is smaller for range "2", and so we
sketch the bulge roughly as shown in Figure 5. This sort of
structure looks nothing like a classical accretion disk.
Notice also that essentially none of the circumstellar material
revealed by these observations projects very closely to the
edge of the B star's Roche lobe.

The lower panel of Figure 6 shows flux distributions for
the corresponding egress phase ranges, where again filled
circles represent atmospheric fluxes at $T_{eff}$ = 14,000 K, log
g = 4.0. Radiation from the optically thick parts of egress
bulges was always less intense than for their ingress

counterparts, so most of the visible mass in these asymmetrical features was concentrated above the following limb of the B star. On occasion, however, the optically thinner outer regions were more prominant during egress (as, for example, on April 5, 1980, as shown in Fig. 2). Strong emission lines, particularly of Hα, noted during active periods by many investigators (Batten et al. 1974; Plavec & Polidan 1975; Crawford 1981; Kaitchuck 1982) presumably originated in or near these bulge fringes. Piirola (1980) observed intermittent polarization from 1972 to 1978, and concluded that both optically thick and optically thin circumstellar material was present around the B star during active periods; he suggested that the optically thin material may have had a nearly spherical distribution.

One of the most striking of the early spectroscopic results was Crawford's (1981) deduction, from Hα emission profiles, of "disordered" velocities (= "turbulence") on the order of 100 km/sec. This result has been confirmed and expanded by Plavec (1983), who obtained IUE spectra during primary eclipses of U Cep. Plavec suggests that the emitting region is probably clumpy and subjected to shocks.

There is good evidence for the non-steady (clumpy?) nature of optically thick equatorial bulges in U Cep. With the kind permission of Piirola (1979) and McNamara & Feliz (1979), we are able to add data from many additional distributed eclipses in late 1975. We judge the strength of bulge light by noting the excesses at internal contacts, converted for uniformity to Johnson B magnitudes. These excesses are shown in Figure 7; similar plots are given by Olson (1980a,b). Fluctuations at ingress and egress were often uncorrelated, with large changes sometimes occurring in one or two orbital cycles. It is not even clear if these data fully resolve the fluctuations, but a large amount of "noise," possibly modulated by changes in the mass transfer rate, was present in the bulge light.

Equatorial bulges were also intermittently present in RW Tau. We show typical spectral distributions of light excesses at second and third contacts of the total primary eclipse in Figure 8. Truly optically thick radiation was never seen from such bulges. The distributions in Figure 8 can be roughly explained as flux from a plasma at $T \simeq 10,000$ K, density $\simeq 5 \times 10^{-12}$ gm cm$^{-3}$, and ultraviolet optical thickness $\simeq 2$. Vertical geometrical thicknesses were similar to those occurring in U Cep, and the total mass of circumstellar (bulge) material visible at both internal contacts $\simeq 3 \times 10^{-12}$ solar masses. RW Tau shares with U Cep the sense of bulge asymmetry and rapid time variability of bulge light. The mass of circumstellar matter in RW Tau was clearly very much less in

U Cep, but otherwise the photometric disturbances were similar. The cool loser in RW Tau is only about one-fifth as massive as its counterpart in U Cep. Kaitchuck & Honeycutt (1982) also found evidence for strong non-thermal broadening of emission lines in RW Tau. The similarities between RW Tau and U Cep are thus rather strong.

Shortly after third contact of distorted eclipses of U Cep, light losses appeared; these are obvious in Figures 2 and 3. The spectral distribution of these losses, similar in all disturbed eclipses, is shown in Figure 9. Here we plot the change in excess light $\Delta L_\lambda^c$ (now negative) between phase .03 and 0.08 for October 26, 1975. Changing the phase range does not greatly alter the relative distribution, which always showed a monotonically increasing light loss with decreasing wavelength. In every case, losses can be explained by replacing part of the normal photosphere, whose undisturbed dereddened colours correspond to effective temperature 13,500 K, with photospheric radiation of lower temperature, $T(c)$. Longward of the ultraviolet, the predicted relative spectrum is insensitive to the value of $T(c)$, so the behaviour of the radiation across the Balmer jump provides the only temperature criterion. We show predicted losses for $T(c) = 10,000$ K and 12,000 K, derived from the Kurucz (1979) flux grid; we have ignored possible centre-limb effects. Between about 25 and 70 percent of the B-star photosphere exposed between these phases was displaced by such cooler material. Recalling that $L_\lambda^c$ is normalized to the undisturbed yellow flux of the hot star, it follows that about 10 percent of the normal visual light of the B star was lost in the phase range covered in Figure 9. The simplest explanation is probably that the equatorial bulge seen at third contact actually was at $\simeq 12,000$ K, and that after orbital phase $\sim 0.03$ it was seen in projection against the B star. Thus, the bulge may have extended a considerable distance around the circumference of the B star. It may also have been related to the large light losses out of eclipse, shown in Figure 3.

Let us now return to eclipse ingress to look briefly at changes in excess light at second contact and earlier phases. Phase range "4" in Figure 5 carries us to about the peak of excess light in disturbed eclipses. We expect a contribution from the rest of the equatorial bulge, and from a hot spot where the stream strikes the B star. The spectrum of $\Delta L_\lambda^c$ "hardens" in this range, as shown in Figure 10. The lower dashed curve is an estimate of the bulge contribution ($\simeq 0.4 \times$ sum of excess light in phase ranges 2 and 3). The difference between these curves is a rough measure of the hot spot contribution. On this night (October 26, 1975) the hot spot radiation was about half of the total bulge light in the ultraviolet, and a smaller fraction at longer wavelengths. The spectrum of this hot spot radiation can (just) be explained as a hot optically thick source of temperature $T(h) \gtrsim 25,000$ K. If $T(h)$ was between 25,000 and 100,000 K, then the bolometric luminosity of the hot spot was $\simeq 0.05$ to $\approx 1.2$ times that of the normal B star. The implied mass transfer rate ranges from $2 \times 10^{-7}$ to $5 \times 10^{-6}$ solar masses per year. Since it is not possible to determine $T(h)$, this calculation gives only the crudest estimate of a plausible range of transfer rates that could explain just the hot spot luminosity.

A final consideration in primary eclipse is early ingress. Even when U Cep was inactive, the ingress shoulder before phase $\simeq 0.96$ was depressed relative to egress, as shown in Figure 11. Here undisturbed egress observations are reflected to ingress. The observations of September 14, 1980 are typical of essentially clean eclipses. For disturbed eclipses such as July 26, 1976 and March 11, 1980, excess light also decreased from phase $\sim 0.97$ toward earlier phase. Figure 12 shows $\Delta L_\lambda^c$ for phase range "5" (0.965 to 0.935) for the three eclipses plotted in Figure 11. These spectra are similar to the egress one shown in Figure 9, and their most obvious explanation is similar: that part of the normal photospheric light of the B star was intercepted by cooler material. Analogous losses are present in a number of binary systems (for example RZ Sct, Wilcken, McNamara & Hansen 1976), and are probably caused by the mass-transferring stream, which is cooler than the gainer. Electron scattering is usually assumed as the extinction process, but this cannot be the case in U Cep. Symbols in Figure 12 show light losses if the stream radiates as a stellar atmosphere with temperatures $T(str) = 8,000$, 10,000 and 12,000 K, and log $g = 3.5$. Stream temperatures, and therefore projected stream areas, are obviously uncertain; for example, an error of 1,000 K in temperature changes the projected stream area by 20 to 40 percent. Nor is it certain during active periods that some of the light losses seen outside eclipse (Fig. 3) did not affect phase range "5". Some support for the purely stream interpretation comes from six eclipses for which adequate data exist, given the following table:

| Dates | $\overline{\Delta L_v^c}$ | T(str) (K) | Stream/Exposed B-Star Area |
|-------|--------------------------|------------|----------------------------|
| Sept. 2,27, 1978; Sept. 14, 1980 (low activity) | -0.10 ±0.03 (sd) | 11700 ±600 (sd) | 0.43 ±0.11 (sd) |
| Oct. 11, 1975; July 26, 1976; March 11, 1980 (moderate to high activity) | -0.27 ±0.04 | 9700 ±1500 | 0.44 ±0.19 |

These data suggest that in the transition from low to high activity, the stream light loss increased because the stream temperature decreased, but the stream area and vertical thickness remained roughly unchanged. Via the eclipse geometry, we find a mean vertical stream thickness 0.40± 0.06 (sd of mean) of the B-star diameter. This thickness refers to those parts of the stream which were optically thick along the line of sight.

We compare this observed estimate of stream thickness to the theory of Lubow & Shu (1975, 1976). To do this, we need to integrate the Lubow-Shu stream densities along the line of sight (Olson 1982b). Because of the exponential relation between density and height, stream thickness is very insensitive to mass-transfer rate. Observed stream thicknesses are $\simeq$ 2 to 2.5 times theoretical thicknesses, for transfer rates $\sim$ few x $10^{-6}$ solar masses per year. Expressed differently, to reach the minimum observed stream thickness ($\simeq$0.35 B-star diameters) would require rates $\simeq 10^{-2}$ solar masses per year. But given the observational uncertainties, it is not clear that any serious discrepancy really exists between observations and theory.

b)    Outside Primary Eclipse

Whenever activity was obvious in primary eclipses of U Cep large light losses, such as those of Figure 3, developed outside eclipse. Conversely, such losses outside eclipse were an infallible warning of disturbances in eclipse. Our data are most numerous in fall, 1975. Some activity was present in late September. Ten consecutive clear nights at Kitt Peak National Observatory in late October and later observation at Prairie

Observatory show that the out-of-eclipse loss was increasing from the 22nd through the 27th, was at maximum from the 28th to at least the 31st, and was declining by November 1. An ultraviolet loss $\simeq$0.7mag occurred near phase 0.6 on October 30. Kondo, McCluskey & Wu (1978) obtained ultraviolet observations from 155 to 330 nanometre, with the Astronomical Netherlands Satellite, which showed a "dip" of $\sim$1 mag near phase 0.6 in early September, 1975. Only a small fraction of our observations outside eclipse seem unaffected by some anomalous losses. Possibly, such losses contributed to the rather low effective temperature (11,000 K) estimated for the B star by Dobias & Plavec (1982).

While there are variations with phase of H$\alpha$ and $\beta$ narrow-band indices (Olson 1976,1978), they are not correlated with ultraviolet losses in Figure 3. Thus, these losses were probably "photospheric" in origin, if one includes in this term the equatorial bulges seen in primary eclipse. The spectrum of the (negative) light excess $L_\lambda^c$ for phase 0.62 on October 30, relative to the undisturbed light for the same phase on May 4, 1976, is shown in Figure 13. For the third time we see the same kind of wavelength distribution, produced when cooler matter replaces part of the normal photosphere. The lower spectrum in Figure 13 assumes that the extra bulge light of primary eclipse was also visible out of eclipse. The large loss at this phase requires that roughly 0.7 to 0.8 of the projected hemisphere was covered by a new "photosphere" near $T(c) \simeq$ 10,000 K, or $\simeq$ 3,500 K cooler than the normal photosphere. Batten et al. (1975) actually observed a change of spectral type from B7 to B9 or A0 during activity in late 1974, supporting this conclusion.

The spectral fingerprint of cooler optically thick matter is evident, in whatever way we choose to analyze these observations. Losses on October 25 also correspond roughly to $T(c) \simeq$ 10,000 K, but with a smaller cool area. On October 29 near phase 0.2, $T(c) \simeq$ 11,000 to 12,000 K with the cool region covering $\gtrsim$ 0.7 of the B-star area. Local "bumps," as near phase 0.27 on October 29, were also optically thick variations produced by either temperature or area fluctuations (or both). Perhaps the clearest indication of the nature of the large light losses comes from the spectrum of light changes between October 25 and 30. Undisturbed light levels subtract out of this analysis, as does any other constant extra light; errors in $\Delta L_\lambda^c$ then depend only on photometric and flux calibration errors. Changes for phases 0.62, 0.68 and 0.71 are consistent with an increase in size of a cool region of $T(c) \simeq$ 10,000 K. Figure 14 shows results for phase 0.62. This is the best theoretical-observational agreement obtained for any active feature in U Cep.

Thus, very large light losses occurred in the radiation of the hot star in directions near the orbital plane. It is tempting to associate these losses with the thick bulges that were seen during primary eclipse. Perhaps the photospheric bulge really was cooler than the normal photosphere, particularly in directions viewed near phase 0.6. The bizarre nature of light variations out of eclipse (Fig. 3) may not be inconsistent with the non-steady nature of the bulges, mentioned in paragraph 4a. However, as noted in paragraph 3, the maximum loss outside eclipse seemed to lag several days behind the peak bulge light in eclipse. The entire situation is physically unclear.

The presence of large equatorial bulges increases the effective radiating surface of the hot star. Possibly, in spite of a light decrease near the orbital plane, the total luminosity in all directions actually increased (Olson 1980a). Uncertainties of geometry and temperature preclude any reliable estimate of this luminosity change.

We noted in paragraph 4a that out-of-eclipse losses extended back into primary eclipse egress. No such losses were ever observed in RW Tau (Fig. 1); we have no observations outside primary eclipse. Perhaps such losses require truly optically thick bulges, which were not observed in RW Tau. Or, possibly losses somehow require a large rotational velocity such as is present in U Cep ( $\gtrsim$ 300 km/sec, Batten 1974). In RW Tau, the B-star rotational velocity is $\simeq$ 80 km/sec (Olson 1975).

c)    G-Star Variations

In paragraph 4a we showed evidence that small brightness changes of the G star were due to temperature changes of its photosphere. These variations may be the single clue that allows us to associate the cool star with the activity discussed above. We plot in Figure 15 changes in visual magnitude (Stromgren y transformed to Johnson V) against orbital cycle count. Here, E is calculated from epoch 2,438,291.5 and period 2.49306 day, and the zero magnitude change is arbitrarily set at variable minus comparison V = 1.36. Only measurements are included from which the two nearby companions were excluded, and for which accurate transformations to the standard system were available. There are two rows of vertical tick marks above the magnitudes; the lower notes primary eclipses that were notably disturbed, while the upper row shows eclipses with only small disturbances. In the past 9 years, most observed activity occurred in late 1974 and 1975, with a few isolated disturbances thereafter. Activity coincided with a brightening of the outer hemisphere of the G star (in a few cases in 1976 and 1977, observations were not available to measure the G star accurately), as shown on four separate occasions in Figure 15. In one case (early 1980), brightening of the G star preceded a disturbed eclipse. If this correlation is valid, it offers some evidence that the photometric activity we have described really was due to flows of matter from the G star. An energy source for the brightness surges may be hydrogen recombination energy in matter flowing from the interior to surface of the G star. A uniform brightening of the G star by 0.2 mag corresponds to a flow rate $\simeq 4 \times 10^{-6}$ solar masses/year.

In addition to "spikes" associated with mass flows, there seems to be a slow (quasi periodic?) variation of the G star, of amplitude $\simeq$ 0.05 mag. Another decade or more of observations would be required to verify a periodic variation, but even now one wonders if this variation is caused by a magnetic activity cycle; certainly the requisite large rotational velocity is present. The spectral distribution of the variation between 1977-78 and late 1980 is qualitatively similar to that of Figure 4, implying that the change was indeed in the photospheric light of the G star. Shore (1980) has commented briefly on the similarity of radio bursts in $\beta$ Per and RS CVn flares. Recently, White & Marshall (1983) have surveyed nine Algol systems for X-rays, using Einstein data. They find that Algols are about an order of magnitude less luminous in X-rays than are RS CVn components of the same rotational period. The Algol X-ray luminosities are consistent with those of single or widely separated stars extrapolated to Algol rotational velocities. Thus, continued search for photometric cycles among cool Algol components may be worthwhile.

In RW Tau, episodes of disturbed eclipses were observed in the 1979 and 1980 observing seasons, followed by clean eclipses observed in 1981. The K subgiant was $\simeq$ 0.05 to 0.1 mag brighter during activity than it was in 1981, suggesting a correlation similar to that found in U Cep (Olson 1981).

d)   Orbital Period Changes

Unlike most Algols, the period of U Cep shows a gradual increase (Batten 1974). Hall (1975) recognized superimposed fine structure consisting of period changes of both signs, and interpreted these changes with the theory of Biermann & Hall (1973). Crawford & Olson (1979) noted that times of minima were spuriously late (by up to 0.01 day) for disturbed eclipses, due to the asymmetry introduced into the light curve by equatorial bulges and hot spot. The most recent discussion

was given by Olson et al. (1981), and is brought up to date
by Figure 16. Here we plot observed minus calculated (O-C)
times of minima (days) based on the ephemeris JD (hel.)
min I = 2438291.5020 + 2.4930410E. Only photoelectric
determinations are plotted. Prior to E = 1500, only (presumed)
undisturbed eclipses are shown. Filled circles represent known
undisturbed eclipses, based on the simple criterion (Crawford &
Olson 1979) that the apparent length of totality in the visual
be $\geq$ 0.075 day. Open circles are eclipses that fail to satisfy
this requirement. Obviously, only undisturbed eclipses can be
used to estimate the period and period changes.

In the 11 years covered in Figure 16, we see four abrupt
period changes of alternating sign, separated by intervals of
essentially constant period up to five years in length. The
period changes were $\Delta P/P \simeq$ -2.5 x $10^{-5}$ , +1.0 x $10^{-5}$ ,
-0.8 x $10^{-5}$ ; the final period change in 1982 cannot yet be
determined. These period changes seem to be unrelated to any
observed episodes of mass transfer. During the interval of
highest activity, from late 1974 through 1975, the period
remained constant, as it did for three additional years.
Primary eclipses of U Cep can be seen from a fixed point on
earth in two windows separated by 6 months, so constant
monitoring from a fixed location is impossible. It is
conceivable that very short intense bursts, unlikely to be
observed, may have produced period changes. The time of period
decrease in early 1979, however, was covered by International
Ultraviolet Explorer spectra obtained by Kondo, McCluskey &
Harvel (1981) and noted in Figure 16. These spectra show no
unusual features. Such hypothetical short bursts would have to
be truly short ($\lesssim$ 2 weeks?), and leave no circumstellar
residue, to have gone undetected in these IUE spectra.
Perhaps, then, mass exchange is not involved in abrupt period
changes.

Recently, Matese & Whitmire (1983) proposed a model in
which abrupt period changes occur because of fluctuations in
the quadrupole moment of the cool component, due to radius
changes, changes in the apsidal motion constant, or both. The
best test of this model will be a detailed study of period
variations in U Sge, where radius changes may have been
detected.

5.    DISCUSSION

Disturbances large enough to reveal some of the effects of
mass flows in close binaries occurred only in U Cep and RW
Tau. Light excesses in primary eclipses revealed asymmetrical
equatorial bulges in both systems, and in U Cep these features
were mostly optically thick. While hot spots may have been

present in both systems, bulges were the most prominent contributors to extra light observed in primary eclipse. The maximum vertical bulge thickness at peak flow rates may have exceeded half the hot-star diameter. At such times U Cep, at least, resembled some "Serpentids", whose hot components seem to be nearly "buried" in a thick accretion disk (in Serpentids, however, the stream does not strike the photosphere of the gainer). Perhaps some insight can be gained by looking at the structure of these bulges.

Figure 7 shows most clearly that bulge light in U Cep can change significantly in one orbital cycle. However, there were times when these changes seemed to be slower as, for example, in late October 1975. At such times, bulges may have been close to hydrostatic equilibrium in the vertical direction. We crudely modelled this vertical structure by using a polytropic equation of state, assuming that the vertical acceleration of gravity is linear with the height, and using an upper boundary condition given by model atmospheres. This procedure works only if the material is truly optically thick, and the boundary layer is geometrically thin. Density and temperature variations with vertical coordinates z are

$$\rho(z) \simeq \rho(H) + AH^{2n}\left[1 - \left(\frac{z}{H}\right)^{2n}\right] \quad ; \quad T(z) \simeq T(H)\left[\frac{\rho(z)}{\rho(H)}\right]^{1/n},$$

where H is the geometrical height of the bulge above the orbital plane, and A is determined by the upper boundary condition. For the most disturbed eclipses in U Cep, and for polytropic index $n \simeq 3/2$ to 3, we find in the orbital plane: $\rho(0) \simeq 10^{-8}$ to $10^{-9}$ gms cm$^{-3}$, $T(0) \simeq 10^5$ K, $P_{gas}(0) \simeq (1$ to 5) $\times 10^5$ dyne cm$^{-2}$, and $P_{rad}(0) \simeq 5 \times 10^5$ dyne cm$^{-2}$. The total mass of the bulge $\lesssim 10^{-6}$ solar masses. In other words - and this is the only point emphasized - radiation pressure may play an important role in the structure of the bulges.

As Ulrich & Burger (1976) have noted, the kinetic energy of the hypersonic stream is converted mostly into radiation after it penetrates below the photosphere of the gainer. With radiation as the key, it is tempting to link part of the accretion energy with formation of the bulge, assuming that some of this energy can be distributed around the subsurface equatorial region of the gainer. Olson (1980b) sketched roughly how this might happen: the stream, striking at angle $\simeq 45$ deg. to the surface, accelerates an equatorial ring of material; if the transfer rate is high enough and the stream penetrates deeply enough, then part of the "photon excess" can survive sub-surface passage around the star and contribute to raising an extensive equatorial bulge. This process could be particularly efficient in U Cep, whose gainer rotates on the average at several times the synchronous rate. A large equatorial acceleration would be required to accomplish the same thing in RW Tau, however. If something like this process actually occurs in close binaries, it would help the gainer to survive episodes of intense matter transfer such as may occur early in the evolution of these systems.

Rather than providing decisive results, our analysis of observations of Algol systems has mainly left questions. The cause of abrupt orbital period changes is still obscure. For many binary systems, the size of these changes $|\Delta P/P| \simeq$ (1 to 3) $\times 10^{-5}$ is apparently independent of system parameters, with the qualification that in many systems observations are too few to determine such changes with great confidence. In U Cep, visible evidence of mass flows does not correlate with abrupt period changes. Most photometric activity followed within two years of a period increase in 1974. Will similar activity soon follow the (possibly larger) period increase of 1982? Would such activity be a reaction to an internal change in the cool star of the sort proposed by Matese & Whitmire? Further observations of primary eclipses (or even of points out of eclipse, particularly in the ultraviolet) over the next year or so will answer this question.

The other system possibly relevant to the question of abrupt period changes is U Sge, where stellar radius changes may have been detected. Our three latest times of minima hint that a period decrease may have occurred sometime before 1978. An intensive observing programme of many years duration will be required to uncover a possible correlation between orbital period and radius changes in U Sge.

Continued observation of at least U Cep and U Sge is partly justified by the indication (Fig. 15) of an activity cycle in the cool component of U Cep. Our observations of U Sge in primary eclipse totality cover only five years, but even here there is a suggestion of a gradual photospheric brightness change $\lesssim$ 0.05mag. RW Tau should perhaps be added as a candidate for such variation.

In summary, it is becoming clear that some Algols are more active and interesting than once thought. As Plavec (1983) has emphasized, they may be a bridge to other active systems like the "Serpentids". It will also be interesting to explore possible similarities, suggested by points discussed above, to RS CVn systems.

## ACKNOWLEDGEMENTS

This research has been supported by grants from the National Science Foundation and the Research Board of the University of Illinois. I am grateful to the Directors of Kitt Peak National Observatory for generous amounts of observing time, and to Dr Kenneth Yoss who not only kept Prairie Observatory running smoothly, but also made possible the flexible observing scheduling necessary for eclipse chasing. Dr Richard C. Crawford has been a constant source of ideas and help for many years, at UCLA and later at Illinois. Dr Mirek Plavec has given encouragement and many helpful comments. Anyone who works on U Cep must be grateful to Dr Alan Batten and Dr Douglas Hall for their enlightening efforts to understand this binary. I am grateful to Dr Ron Webbink, Dr Ron Kaitchuck and Dr Yoji Kondo and his colleagues for much helpful advice. Dr Kaitchuck communicated to me several recent times of minima of U Cep. Dr Italo Mazzitelli suggested to me the possible importance of radiation pressure in support of equatorial bulges in U Cep. I also thank Dr D.H. McNamara, Dr K.A. Feltz, Jr. and Dr V Piirola for use of some of their observations of U Cep. Finally, I record my thanks to those who assisted with observations: at Prairie Observatory, Dr Richard Crawford, Dr Drake Demming, Dr William Hartkopf, Dr Deidre Hunter and Mr Richard Karman; at Mount Laguna Observatory, Mr Tom Bryant, Mrs JoAnn Eder, Mr Dean Espitallier, Mr Clayton Heller, Mr Jeffrey Hickey, Ms Victoria Payler and Mr Patrick Smiley.

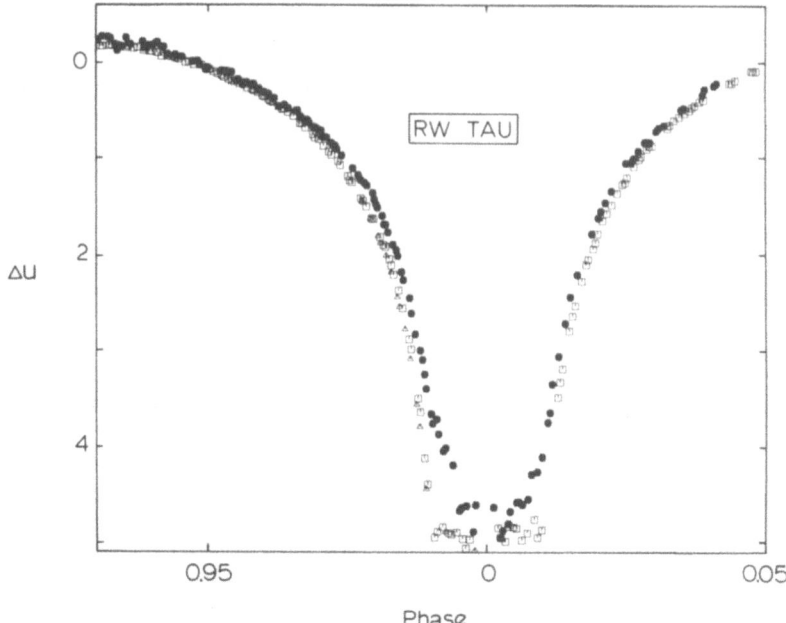

Figure 1. Ultraviolet photometry of RW Tau, showing two undisturbed and one disturbed eclipse.

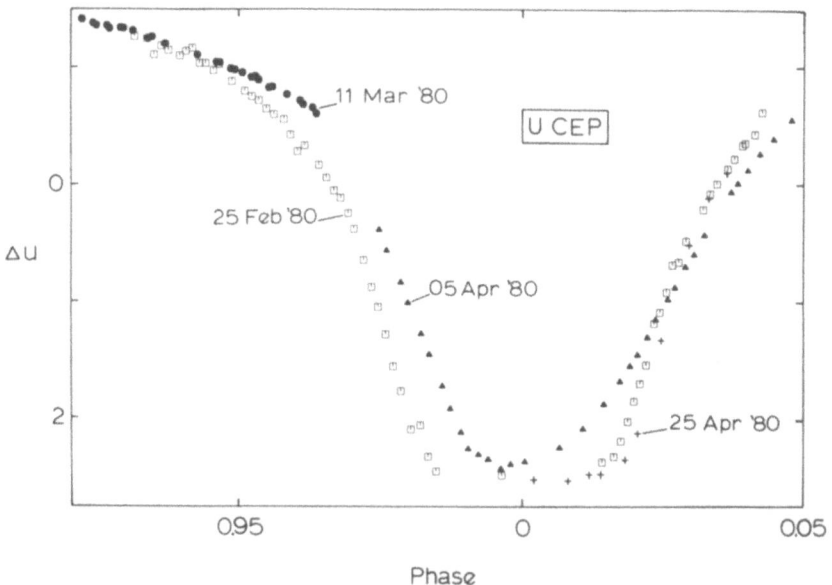

Figure 2. Examples of nearly undisturbed and moderately active
ultraviolet light curves of U Cep.

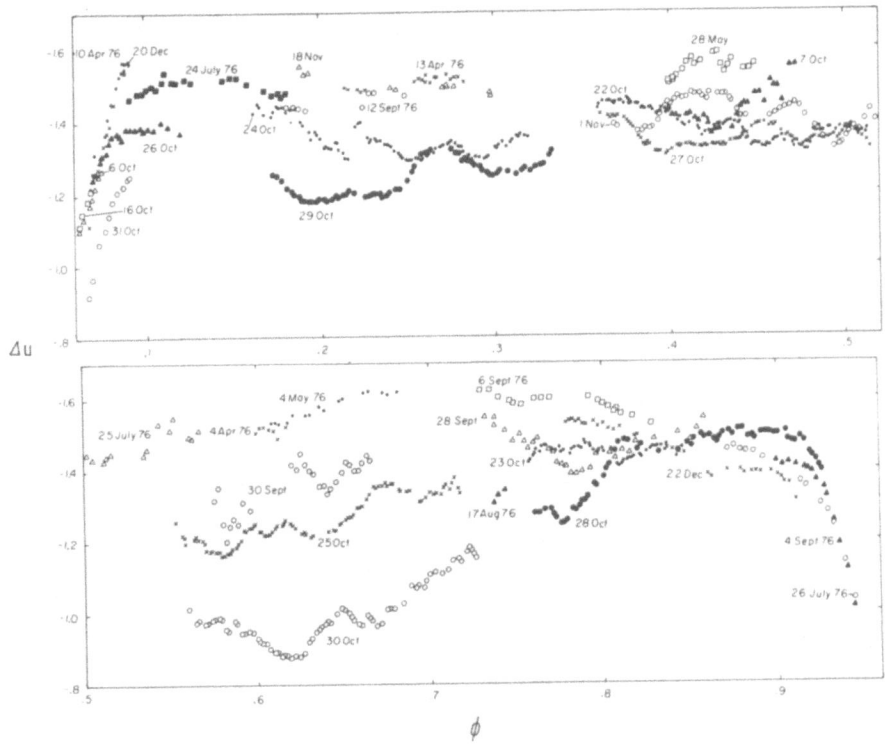

Figure 3.  Ultraviolet light variations of U Cep outside eclipse,
showing various degrees of activity.  Dates without years are
1975.  Courtesy of the University of Chicago Press.

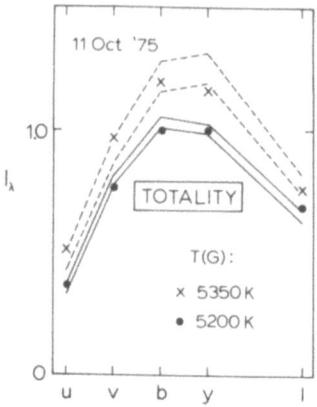

Figure 4. Flux distributions of the cool star in U Cep. Lower
curve: at normal minimum brightness; upper curve: during a
brightness surge associated with photometric activity. Symbols
give best-fitting model atmosphere fluxes, for log g = 3.5. As in
all such spectra, the estimated error spread is shown.

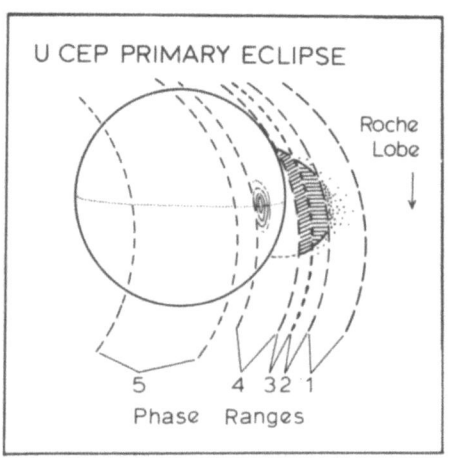

Figure 5. Eclipse geometry for U Cep. Cross hatched areas represent optically thick parts of the "ingress equatorial bulge." Dots
schematically show the location of radiating plasma of finite
optical thickness. Phase range 4 includes the hot spot, also
shown schematically. The egress bulge is omitted for clarity.

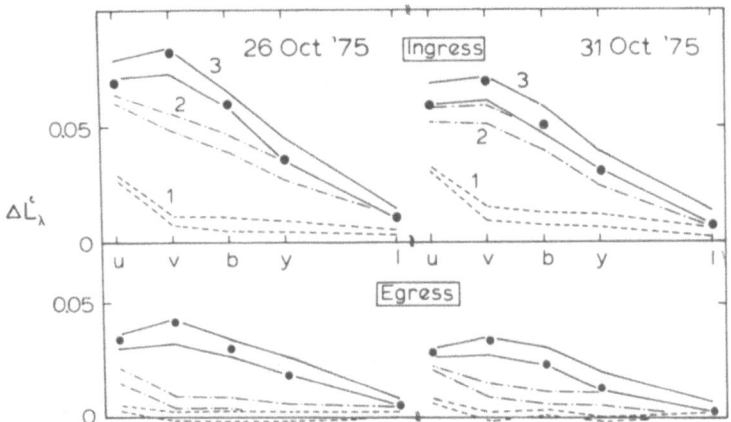

Figure 6. Upper panel: spectra of changes in light excess in ingress phase ranges "3", "2", and "1", shown in Figure 5, for two active eclipses of U Cep in October 1975. Lower panel: spectra for analogous egress phase ranges. Filled circles are model atmosphere fluxes for $T_{eff}$ = 14,000 K, log g = 4.0.

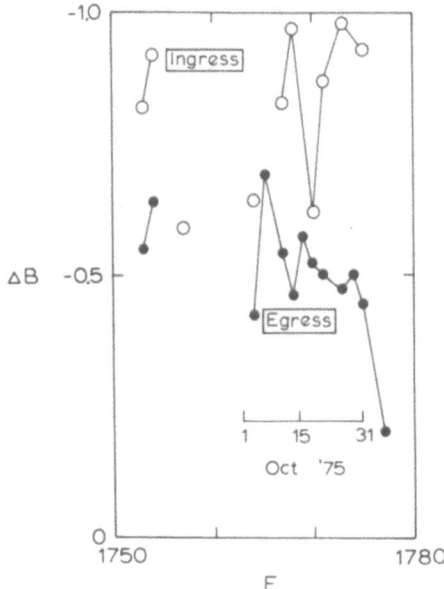

Figure 7. Light excesses expressed as Johnson B magnitudes at second and third contacts of primary eclipse (noted as ingress and egress, respectively), for a series of disturbed eclipses of U Cep in fall, 1975. Data are from McNamara & Feltz (1979), Olson & Piirola (1979). Lines join points for which variations are probably well-resolved.

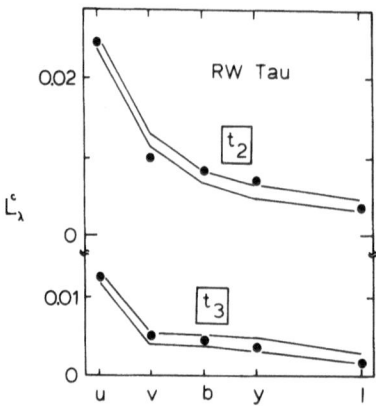

Figure 8.   Spectra of light excesses at second and third contacts
("ingress" and "egress bulge" light, respectively) of primary
eclipse in RW Tau.

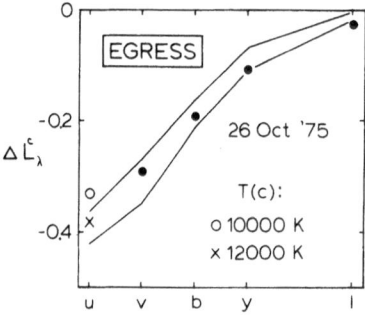

Figure 9.   Spectrum of the light excess change (which is negative
and therefore represents a light loss) between egress phases 0.03
and 0.08 in U Cep.   Symbols give the predicted spectrum when part
of the photosphere of the hot star is replaced by cooler atmospheric material.   Longward of the ultraviolet, relative fits are
insensitive to temperature, and are shown with filled circles.

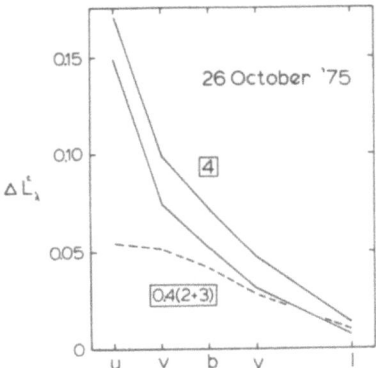

Figure 10. Upper curve: spectrum of the light excess change in U Cep for phase range "4" (Figure 5), which includes the hot spot. The lower dashed curve is a crude estimate of the contribution of equatorial bulge light, so the difference between these curves approximates the hot spot contribution.

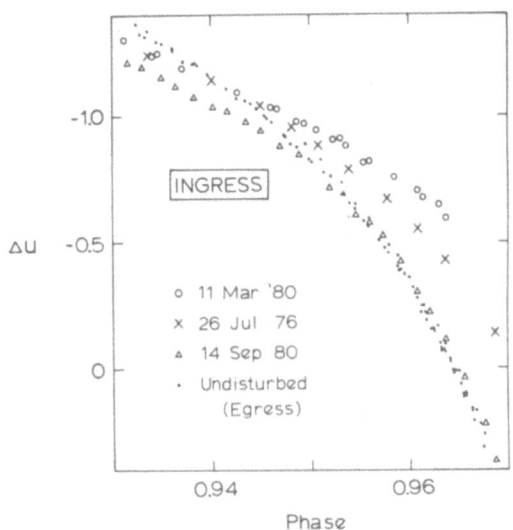

Figure 11. Ultraviolet observations of early ingress of primary eclipses of U Cep, where the stream may distort light curves. Undisturbed observations are reflected from egress and are from the least-disturbed eclipses.

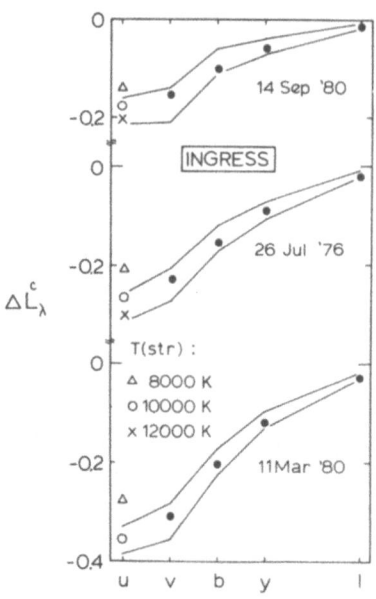

Figure 12.  Spectra of light excess changes (losses) from phase
0.965 to 0.935 for eclipses whose ultraviolet observations are
shown in Figure 11.  Models with streams projected against the hot
star are also shown.

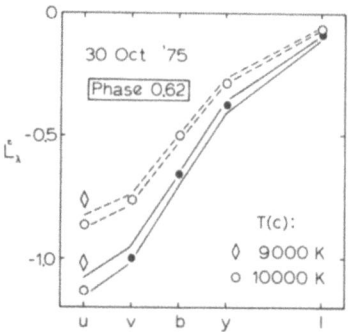

Figure 13.  Spectrum of the light loss at orbital phase 0.62 in
U Cep on October 30, 1975, relative to undisturbed observations at
the same phase on May 4, 1976.  The lower curve assumes that the
contemporary extra light seen in primary eclipse is also visible
out of eclipse.  Symbols are predicted fluxes, if large cool regions
cover much of the normal photosphere of the hot star.

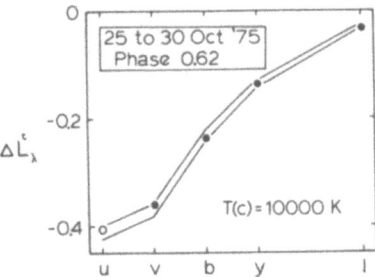

Figure 14.   Spectrum of the light change in U Cep at phase 0.62
between October 25 and 30, 1975, and best-fitting models.

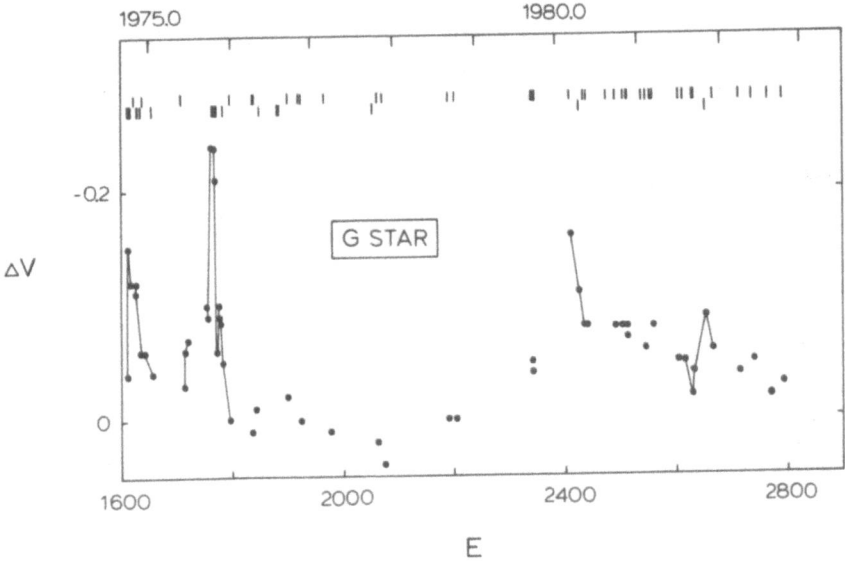

Figure 15.   Variations of the outer hemisphere of the cool star
U Cep, as observed during primary eclipse totalities, over about
eight years.   The lower series of tick marks notes photometrical
disturbed eclipses, while the upper series shows relatively clea
eclipses.   Here, " V" means "variable minus comparison - 1.36m.'
The visual magnitudes are Strömgren y's transformed to Johnson '

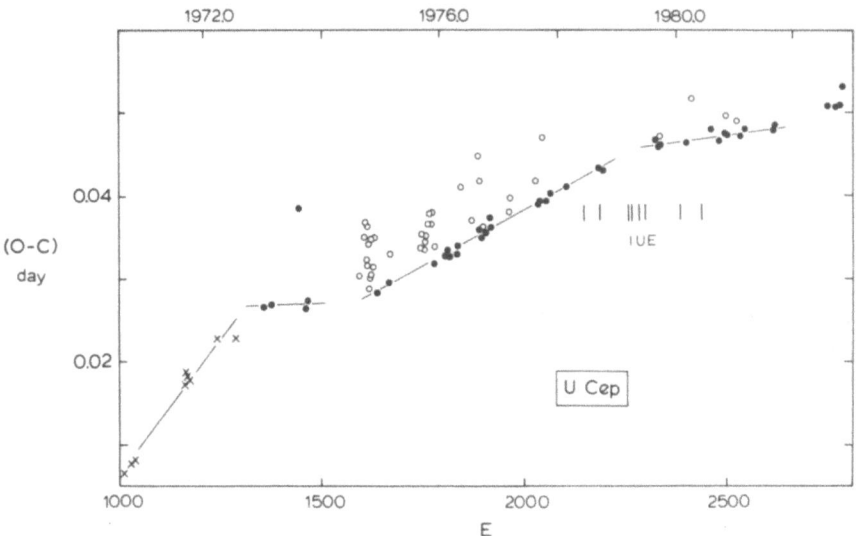

Figure 16.  Variation in observed minus calculated times of minima of U Cep.  Open circles show eclipses in which photometric distortion spuriously delayed the apparent time of minimum.  See text for other details.

REFERENCES

Batten, A.H., 1974. Publ.Dom.Astrophy.Obs. Victoria **14**, 191.
Batten, A.H., Fischer, W.A., Baldwin, B.W. & Scarfe, C.D.,
    1975. Nature **253**, 174.
Biermann, P. & Hall, D.S., 1973. Astr.Astrophy., **27**, 249.
Crawford, R.C., 1979. Pub.Astron.Soc. Pacific., **91**, 111.
Crawford, R.C., 1981. Ph.D. thesis, UCLA.
Crawford, R.C. & Olson, E.C., 1979. Publ.Astron.Soc. Pacific,
    **91**, 413.
Dobias, J.J. & Plavec, M.J., 1982. In "Advances in Ultraviolet
    Astronomy," ed. Y. Kondo, J.L. Mead & R.D. Chapman (NASA
    Conf. Publ. 2238), 538.
Hall, D.S., 1975. Acta Astr., **25**, 1.
Kaitchuck, R.H., 1982. Private communication.
Kaitchuck, R.H., Honeycutt, R.K. & Mufson, S.L., 1980. In
    "Close Binary Stars: Observations and Interpretation,"
    ed. M.J. Plavec, D.M. Popper and R.K. Ulrich, (Dordrecht:
    Reidel), 233.
Kaitchuck, R.H. & Honeycutt, R.K., 1982. Astrophys.J.,
    **258**, 224.
Kondo, Y., McCluskey, G.E. & Harvel, C.A., 1981. Astrophys.J.,
    **247**, 202.
Kondo, Y., McCluskey, G.E. & Wu, C.-C., 1978. Astrophys.J.,
    **222**, 635.
Kurucz, R.L., 1979. Astrophys.J.Suppl., **40**, 1.
Lubow, S.H. & Shu, F.H., 1975. Astrophys.J., **198**, 383.
Lubow, S.H. & Shu, F.H., 1975. Astrophys.J. (Letters)
    **107**, L53.
Metese, J.J. & Whitmire, D.P., 1983. Astr.Astroph. (letters)
    **117**, L7
McNamara, D.H. & Feltz, K.A., Jr., 1976. Publ.Astron.Soc.
    Pacific., **88**, 688.
McNamara, D.H. & Feltz, K.A., Jr., 1979. Private
    communication.
Olson, E.C., 1975. Unpublished.
Olson, E.C., 1976. Astrophys.J., **204**, 141.
Olson, E.C., 1978. Astrophys.J., **220**, 251.
Olson, E.C., 1980a. Astrophys.J., **237**, 496.
Olson, E.C., 1980b. Astrophys.J., **241**, 257.
Olson, E.C., 1981. Astrophys.J., **250**, 704.
Olson, E.C., 1982a. Publ.Astron.Soc.Pacif., **94**, 70.
Olson, E.C., 1982b. Astrophys.J., **259**, 702.
Olson, E.C., 1983. In preparation.
Olson, E.C., Crawford, R.C., Hall, D.S., Louth, H., Markworth,
    N.L. & Piirola, V., 1981. Publ.Astron.Soc.Pacif.,
    **93**, 464.
Piirola, V., 1979. Private communication.
Piirola, V., 1980. Astr.Astrophys., **90**, 48.

Plavec, M.J., 1980.  In "Close Binary Stars:  Observations and
    Interpretation," ed. M.J. Plavec, D.M. Popper, and R.W.
    Ulrich (Dordrecht:Reidel), 251.
Plavec, M.J., 1982.  In "Binary and Multiple Stars as Tracers
    of Stellar Evolution," ed. Z. Kopal and J. Rahe
    (Dordrecht:Reidel), 159.
Plavec, M.J., 1983.  Preprint.
Plavec, M.J. & Polidan, R.S., 1975.  Nature, **253**, 173.
Shore, S.N., 1980.  In "Close Binary Stars: Observations and
    Interpretations," ed. M.J. Plavec, D.M. Popper and
    R.W. Ulrich (Dordrecht:Reidel), 395.
Ulrich, R.K. & Burger, H.L., 1976.  Astrophys.J., **206**, 509.
White, N.E. & Marshall, F.E., 1983.  Astrophys.J., **268**, L117.
Wilcken, S.K., McNamara, D.H. & Hansen, H.K., 1976.  Publ.
    Astron.Soc.Pacif., **88**, 262.

# FROM ALGOL TO BETA LYRAE

Mirek J. Plavec

Department of Astronomy
University of California
Los Angeles
U.S.A.

## ABSTRACT

We believe that we understand reasonably well the structure and evolutionary status of Algol and similar semi-detached systems, which will for brevity be called Algols. β Lyrae is much more complicated, and some of the problems of revealing its structure are discussed. The best model available at the present considers β Lyrae as an Algol-type semi-detached system in a phase of fairly rapid mass transfer, i.e. younger than typical Algols. The most important conclusion drawn in this article is that β Lyrae is not so unique as it is usually believed. Some characteristics that are so puzzling in β Lyrae actually appear already in such classical Algols like U Cephei and RW Tauri. The mass-accreting components in these systems are surrounded by a hot, turbulent layer which probably expands and which is the seat of emission lines of fairly high ionization discovered in the far ultraviolet. In β Lyrae and the so-called W Serpentis stars, the circumstellar hot turbulent shell is much more extensive and probably also denser, and may be forming a kind of a "superchromosphere". Also, a trend appears among the W Serpentis stars for the gainers to be surrounded by thick disks, and this trend, too, appears more forcefully in β Lyrae, where the disk probably completely surrounds and hides the accreting star proper.

The original title of my contribution promised new models for β Lyrae and related objects. However, I will be talking more about raw observations and their interpretation, rather than about all-embracing models. I think it will be useful to

*P. P. Eggleton and J. E. Pringle (eds.), Interacting Binaries, 155–178.*
© *1985 by D. Reidel Publishing Company.*

attract your attention to some outstanding problems which must
be answered before any complete models can be considered.
Moreover, who knows where we would get if I presented models?
This place seems to be conducive to idolatry of models.
Physics Today in its 1983 May issue (Vol. 36, No. 5, p. 17)
reports that ".... Just five months earlier, the inflationary
universe had been a major focus of effort at a three-week
worship in Cambridge, England." Thus, if you do not want to
spend three weeks worshipping a model of a close binary star,
we had better look at the observations. There really do not
exist models of interacting binaries worth much worshipping.
Every day we realize more and more that the now-almost-
canonical Roche model of a semidetached binary with a
mass-transferring stream either impinging on the gainer or
creating a flat equatorial disk is probably a large
oversimplification. The extent of this oversimplifiction will
emerge gradually as we review the various interacting systems
with nondegenerate components that probably can be placed
somewhere between Algol and β Lyrae as to their complex
structure and evolutionary stage.

John Goodricke, a young Englishman, discovered the
eclipsing nature of Algol in November 1782, and the variability
of β Lyrae in September 1784. Thus, within less than two
years, he enriched astronomy enormously by pointing out two
important archetypes of eclipsing binaries. Since then,
several generations of illustrious and industrious
astronomers have attempted to understand their character and
evolutionary scheme. For a long time, the two stars were
considered to be on an essentially equal footing, each believed
to represent a large category of eclipsing pairs. Algol, with
its flat light curve between eclipses, was the prototype of a
system with spherical, i.e., rotationally and tidally
undistorted, component stars. The β Lyrae category included
all the various eclipsing systems in which component distortion
leads to a continuous light variation, and in which the two
eclipses are of unequal depth. A third category, the W UMa
stars, was introduced to represent systems also with a
continuously variable light but with nearly equally deep
eclipses; but in this case, a little more physics crept in with
the postulate of a fairly narrow range of orbital periods and
spectral types.

Gradually, however, the investigations of Algol and
β Lyrae began to take very different paths. For Algol, all
the modelling essentially developed and of course greatly
refined the original model proposed by Goodricke. His "dark
eclipsing body" was replaced by a subgiant dark only relatively
to the bright primary component. The existence and shape of
the subgiant were found to be products of a geometry and an

evolution very specific to close binary systems. Although it is not the best and simplest of the systems (with its partial eclipses and a third star light contamination), Algol has maintained its role as the prototype of a populous and well-defined class of semidetached interacting binaries.

The story of $\beta$ Lyrae has been different. As the years went and data accumulated, it was becoming more and more isolated from the multitude of eclipsing binaries. The "$\beta$ Lyrae type of the light curve" was found to be a superficial and almost meaningless criterion (Kopal 1955; Plavec 1980a). The $\beta$ Lyrae system itself became to be labelled "enigmatic," "bizarre," or the like, after the word "peculiar" degenerated, in stellar astrophysics, to something meaning "rather ordinary." No matter which ornamental adjective you prefer, $\beta$ Lyrae was in essence considered as unique, unrelated to anything else among binary systems. This trend culminated some ten years ago, when several investigators suggested that $\beta$ Lyrae may harbor a black hole. Since then the trend has been reversed and it appears reasonably justified to speak about $\beta$ Lyrae with the same breath as about Algol, since $\beta$ Lyrae may well be just an extreme case of an Algol-like semidetached system, rare but not unique.

Today, I would like to approach the problem of $\beta$ Lyrae from a new angle, namely by searching for systems that would represent a transition between a very simple, classical Algol system, and $\beta$ Lyrae. I think that the <u>International</u> <u>Ultraviolet Explorer (IUE)</u> satellite has contributed significantly to the progress in the studies of interacting binary stars, and I would like to mention briefly those of the recent developments that are directly relevant to the problem I am pursuing here.

## LOOKING AT ORDINARY ALGOLS WITH IUE

It is reasonably certain that our basic theoretical understanding of the evolution of the Algol-type semidetached binaries is correct. The now-less-massive and later-type component initially was the more massive star, and lost a great deal of its mass. It is still losing mass, since it continues to fill its critical Roche lobe, but in most cases we are witnessing the last phases of the process, with the mass loss rate already greatly abated.

It is worthwhile to read objections to this scheme by Kopal (1978). Kopal argues, among other things, that the <u>gainer</u> in a typical Algol system does not differ, in its <u>external</u> characteristics, from a primary component in a

detached, noninteracting system. Indeed, our Figure 1 shows
that, for example, the flux distribution of the primary
component in U Sagittae over the wide range of wavelengths
between 1200 to 8200 Å does not differ from that of an ordinary
single star, since it can be matched very well by a Kurucz
(1979) model atmosphere. In a simple case like that, the IUE
observations either confirm the spectral classification derived
from the optical region, or enable us to define it more
precisely. The latter case applies to U Cephei. While in
U Sagittae the absorption lines are reasonably sharp and make
spectral classification on the MK system rather easy, in
U Cephei the absorption lines are broadened by very fast
rotation. As a consequence, classification based on them has
always been rather uncertain and varied between B6 V and A0 V,
although part of this uncertainty may be real (Batten 1974;
Olson 1980 a,b). Our Figure 2 shows that, at least for the
phase and epoch of our observations, an effective temperature
as high as 13,000 K is ruled out, and the correct value is much
closer to 11,000 K (Figure 3).

However, high-dispersion IUE spectra of these Algol
primaries show that their spectra are not entirely normal.
Figure 4 is taken from the forthcoming Ph.D. thesis written at
UCLA by Jan Dobias. It shows the ultraviolet resonance doublet
of C II at 1336 Å in the spectrum of the primary component (the
gainer) in U Sagittae. Lines of other elements were used to
determine the atmospheric parameters including the rotational
and turbulent velocities. The only adjustable parameter left
is the abundance of carbon. The figure shows that the observed
line profile cannot be fitted by the solar abundance of carbon;
one has to reduce it to about 9% of the solar value. The same
result applies to another classical Algol, U Coronae Borealis
(Figure 5). This underabundance of carbon is to be expected if
the large-scale mass transfer theory for Algols is correct. As
mass flows away from the loser, deeper and deeper layers are
exposed, until eventually the star is stripped down to the
region where the chemical composition of the gas has already
been affected by the CNO process. Initially, carbon is more
abundant than nitrogen, but the reaction involving $N^{14}$ is so
slow that equilibrium can be reached only when most of the $C^{12}$
has been converted to $N^{14}$, and the abundance ratio is strongly
reversed in favour of nitrogen. While the underabundance of
carbon should be, in fact is, quite striking, the nitrogen will
be relatively mildly enriched. Indeed, the synthesis of the
N I lines near 1493 Å by Jan Dobias shows that the profiles are
better matched if an overabundance of nitrogen over its solar
value is assumed (Figure 6).

Please note that the layers of the denuded loser are
actually observed on the surface of the gainer. Carbon and

nitrogen do not display good spectral lines in the optical region, so that the late-type loser is less suitable for abundance studies. They have been attempted, though (for a summary, see Lambert 1982) and although the results do not yet yield a consistent picture, they can be considered as confirming the large-scale mass loss from the gainers in Algols.

Our results for U Sagittae and U Coronae Borealis show that some material from the loser must have been landing on the gainer even at an advanced stage of the mass loss process. There still exists no consistent answer to the question, though, as to how much has on the whole been intercepted by the gainer and how much has escaped into space. Theoretical as well as observational evidence has been growing in favour of nonconservative evolution. For example, from the radial velocities of the circumstellar absorption lines in U Cephei, Kondo, McCluskey & Stencel (1979) concluded that a stream is leaving the system; a similar conclusion was reached for other systems by Polidan & Peters (1980). I will have more to say on this problem later.

Although the IUE observations of absorption line profiles and velocity shifts reveal that the "classical" Algols are not completely dormant systems, they do confirm that the component stars are essentially normal; none of the Algols examined would be classified as anything like "enigmatic". Thus, $\beta$ Lyrae stands out as an unusual system.

## WHAT MAKES BETA LYRAE A PECULIAR BINARY SYSTEM?

Even this question has no straightforward answer. All investigators would agree with me that $\beta$ Lyrae is enigmatic, but different generations would give different reasons. At the time of the famous "frontal attack on the mystery of $\beta$ Lyrae" by Kuipal, Kuiper, Struve and others (in the Ap.J., 93, 1941) it was believed that the object consists of two probably most massive stars known. This opinion was based on a firm belief in the general validity of the mass-luminosity relation. Rather surprisingly, one characteristic of $\beta$ Lyrae is simple and well-established: this is the radial velocity curve of the B9 II component. It is simple, smooth, well-defined, and the amplitude of 184 km/s makes it much larger than the observational errors (Sahade et al. 1959; Batten & Fletcher 1975). The resulting mass function is $f(m) = 8.5 M_\odot$, and since the rather deep eclipses ensure that the orbital inclination $i \geq 85$ deg., we can write:

$$M_U^3 \, (M_B + M_U)^{-2} = 8.6 \, M_\odot$$

(1)

where B denotes the observed giant while U stands for the largely unknown secondary component. The following table shows the various pairs of masses we get if we vary the mass ratio $q = M_U/M_B$ :

Possible masses in $\beta$ Lyrae

"Primary more massive"                    "Secondary more massive"

| q | $M_U$ | $M_B$ | $M_t$ | q | $M_U$ | $M_B$ | $M_t$ |
|-----|-------|-------|-------|-----|-------|-------|-------|
| 1.0 | 34.4 | 34.4 | 68.8 | 1.0 | 34.4 | 34.4 | 68.8 |
| 0.9 | 38.3 | 42.6 | 80.9 | 2.0 | 19.4 | 9.7 | 29.1 |
| 0.8 | 43.5 | 54.4 | 97.9 | 3.0 | 15.3 | 5.1 | 18.4 |
| 0.7 | 50.7 | 72.4 | 123.2 | 4.0 | 13.4 | 3.4 | 16.8 |
| 0.6 | 61.1 | 101.9 | 163.0 | 5.0 | 12.4 | 2.5 | 14.9 |
| 0.5 | 77.4 | 154.8 | 232.1 | 6.0 | 11.7 | 2.0 | 13.7 |
| 0.4 | 105.3 | 263.3 | 368.6 | 8.0 | 10.9 | 1.4 | 12.3 |

The table shows that if the B9 star is to be more the massive one, unusually large masses for both components result. But that appeared to be the natural solution, since the eclipse of the B9 star is deeper and the absorption lines of the U component are invisible. A favourite solution was to assume $M_B = 1.5 \, M_U$ , giving $M_B = 80.5 \, M_\odot$, $M_U = 53.7 \, M_\odot$. With such a large mass, the B9 star was naturally believed to be a supergiant of large size and high luminosity. Hence, one could read at that time that "$\beta$ Lyrae shines through our Galaxy like a beacon." Nowadays we have ample evidence that it is just a bright giant with an absolute visual magnitude fainter than -4 mag (Abt et al. 1962; Plavec, Dobias & Weiland 1984).

We also prepared to accept serious deviations from the mass-luminosity relation, in particular, in stars affected by mass loss. The recent comprehensive article by Sahade (1980) explains how and why the idea was abandoned, that the more luminous B9 star must also be the more massive one. But how much more massive is the U star than the B star still remains very uncertain. Only indirect and rather approximate arguments are available. Wilson (1974, 1982) allows a wide range of mass ratios $M_U/M_B$ between about 3 and 8, although he prefers a

value near 6. In spite of this uncertainty, we will probably
not be far off if we asssume that the mass of the secondary
component lies near 12 $M_\odot$ and the mass of the primary near
2 $M_\odot$.

Are these masses unusual and unreasonable? No! Let us
adopt the fundamental model of an Algol semi-detached binary
system for $\beta$ Lyrae, with the B9 II star as the loser. It is
true that "classical" Algols like U Cephei or U Sagittae have
losers of a much later spectral type, as a rule between late F
and early K. But those systems are much less massive. There
exists a small subgroup among the Algols consisting of pairs in
which both components are of a substantially earlier spectral
type: V356 Sagittarii, $\lambda$ Tauri, RZ Scuti. In all three, the
loser is a genuine giant near A0, not unlike that of $\beta$ Lyrae;
and the primary components, or gainers, are early B-type stars
either on the main sequence or possibly above it (RZ Scuti).
In fact, in V356 Sgr we have a system rather reminiscent of $\beta$
Lyrae: the early B-type primary has a mass of about 13 $M_\odot$,
while the A1 II giant secondary has about 5 $M_\odot$. The
secondary, in spite of its much smaller mass, is a genuine
giant and actually dominates the flux from the system in the
optical region, contributing about 52% of the V light (Wilson &
Caldwell 1978). Both components satisfy, at least roughly, the
mass-luminosity relation, and as we go to shorter wavelengths,
the B star quickly gains superiority. This is not the case in
$\beta$ Lyrae.

Thus, nowadays the real problem in $\beta$ Lyrae is the
radiation of the secondary star, and in general its very
nature. Qualitatively, it seems that we understand why we are
receiving less light from the secondary than we should. This
is most likely not due to its actual underluminosity, but
rather to an unusual spacial distribution of its emittance.
The shape of the light curve implies that at least one of the
components must be much more flattened than a normal star can
be, and the consensus seems to be that the secondary is
actually a star embedded - or, better, completely hidden - in a
thick disk. Wilson (1974, 1982; Wilson & Lapasset 1981) has
done a lot of work on modelling of this disk, and the agreement
of his models with the observed properties is very encouraging.
The disk is pictured as rather massive (containing perhaps
half a solar mass), and optically as well as geometrically
quite thick. Its thickness perpendicular to the orbital plane
is, in Wilson's model, some 30% of the separation between the
components, or about 16 $R_\odot$. The radial extent of the disk is
probably roughly equal to the mean radius of the star's
critical Roche lobe, which is about 30 $R_\odot$. The separation
between the centres of the two components is, for a mass ratio
of 6, about 55 $R_\odot$. While we may be satisfied with this

qualitative picture, we must make every effort to verify it and derive the characteristics of the two components by all available means. I wish to present a preliminary report on the work being done at the University of California in Los Angeles.


## FLUX DISTRIBUTION IN BETA LYRAE

Our approach to the problem starts with a study of the flux distribution in β Lyrae. We can now cover the spectral interval between 1200 Å and 8200 Å almost without a gap. Two IUE scans (from the SWP and LWR cameras) cover the region 1200 Å to 3200 Å, and IDS scans taken at Lick Observatory reach from 3200 Å to 8200 Å with about the same spectral resolution as the low-dispersion IUE spectra. A narrow gap in the vicinity of 3200 Å must in fact be assumed, since neither instrument gives a reliable response there. Lack of allocated observing time does not permit us to cover all phases. So, instead of solving a complete light curve, we have concentrated on the observations of the the two eclipses and we are trying to obtain the flux distribution of either component by suitably subtracting the scans. Since the eclipses are probably not complete, one must adopt a model of the system.

Without any detailed analysis or modelling, Figure 7 shows what kind of results we may anticipate. It shows the flux distributions in full light (actually at phase 0.13P), in primary eclipse, and in secondary eclipse. The three curves run nearly parallel in the optical region, which implies that the components have not too dissimilar colour temperatures there; the secondary component is about 3,000 K cooler. The eclipse of the B9 II star, which is definitely deeper optically, becomes the shallower of the two in the far ultraviolet. Thus, we conclude that shortward of about 1500 to 1600 Å, the secondary star has a higher colour temperature; again, the difference is only at most a few thousand degrees K. This result confirms the analysis of the UV light curves by Kondo, McCluskey & Eaton (1976).

The crossover of the eclipse depths is not due to emission lines, as is sometimes believed; it is a genuine effect in the continuum. It implies that at least one component has an anomalous flux distribution. It should not be surprising that the secondary cannot be fitted by an ordinary stellar model atmosphere. However, at least part of the effect may be due to a flux deficiency of the primary star in the far ultraviolet, perhaps due to an unusually large line blocking. Absorption lines of the primary star are observable in between the emission lines all the way to the short wavelength limit.

If the secondary dominates (albeit not by too much) in the far ultraviolet, why don't we observe absorption lines formed in its atmosphere? First of all, let us not be so very definitive. That part of the spectrum has become accessible only recently, the few available spectra are not always properly exposed, and our search for the absorption lines moving with the secondary has not been concluded yet. Nevertheless, the result will probably be negative. Why? We may attempt partial answers. The surface layers of the disk are most likely in rapid rotation, which will wash out weak lines. Many strong lines may be obliterated by emissions. Moreover, very broad lines require careful techniques and excellent signal to noise ratios for detection. In any case, the answer is not yet final.

Returning to the flux distribution in the continuum, one should be aware that both components are surrounded by extensive gas clouds. Continuous hydrogen radiation is evident at the Balmer limit, where it reduces the depth of the Balmer jump. But a much more striking evidence of the presence of the circumstellar material is the strange bulge observed between wavelengths (approximately) 1700 Å and 2400 Å. It is best seen in Figure 8, which compares the ultraviolet flux distribution in full light and in secondary eclipse. One notices that there is practically no loss of light during the eclipse over the above interval of wavelengths. Primary eclipse is not much different in this respect: it also becomes extremely shallow. This phenomenon, too, was detected alredy from an ultraviolet photometry at the wavelength of 1910 Å by Kondo, McCluskey & Eaton (1976). However, only spectrophotometric scans show how wide a range of wavelengths is affected. Naturally, the same large areas of the components are eclipsed at these wavelengths as at any others. But clearly the dominating contribution to the observed flux within the region of the bulge comes from circumstellar material distributed over a volume so large that the bodily eclipses do not reduce this part of the total flux significantly. I will return to the problem of what is shining there in a later section.

## ULTRAVIOLET EMISSION LINES

The ultraviolet IUE spectrum, in particular its part shortward of 2400 Å, abounds in emission lines. These come from the following ions: C II, C IV, N V, Si II (weak), Si III, Si IV, Mg II, Al II, Al III, Fe III and Ni III, to name at least those obviously present. The He II line at 1640 Å is conspicuous by its absence. This unusually rich emission-line spectrum, when discovered with the Copernicus satellite (Hack et al. 1975) was believed to be unique; and

probably there was no other such spectrum within the reach of
that satellite. But the IUE satellite permits us to observe,
albeit at a much lower dispersion, the spectra of many fainter
interacting binaries. Among them, Plavec & Koch (1978)
identified a group which Plavec (1980b) later called the W
Serpentis stars. Initially, this group included RX Cas, SX
Cas, V367 Cyg, W Cru, β Lyr and W Ser. (While β Lyrae is no
doubt the most prominent member, it would have been confusing
to introduce just another "β Lyrae-type" of eclipsing
binaries.) Their common property is a conspicuous
emission-line spectrum in the far ultraviolet. Figure 9 shows
that actually the emission spectrum of W Serpentis is richer
and stronger than that of β Lyrae. Noticeable is a virtual
absence of the emission lines of Al II and Al III from the
spectrum of β Lyrae, while these lines are quite conspicuous
in the spectrum of W Serpentis. One must be cautious, however,
in drawing any conclusions from this fact. High-dispersion
spectra of β Lyrae show those aluminum lines as quite strong
emission, flanked on the short-wavelength side by a rather deep
absorption. In this P-Cygni profile, the lines do not differ
materially from the other emissions in the far-ultraviolet
spectrum of β Lyrae. Probably the degradation in the
low-dispersion spectrum happens to work in such a way as to
cancel the emission by the absorption. Yet this circumstance
does not change the conclusion that the emission spectrum of β
Lyrae is not unique, since all the W Serpentis stars show
similar spectra.

How can this discovery help us in resolving the mystery of
β Lyrae? First of all, we must show that the similarity of
the far ultraviolet spectra is not just a superficial property.
This could easily be the case if the emission lines were formed
in the chromosphere (or, to speak more precisely, in the
transition region between chromosphere and corona) of the
late-type components. Such a chromospheric spectrum in G- or
K-type giants bears considerable resemblance to what we observe
in the W Serpentis stars. Yet there exist powerful arguments
showing that the emissions in the Serpentids do not come from
the chromospheres of the late-type components. In the first
place, there is no suitable late-type component in β Lyrae or
in V367 Cygni. Again, where a suitable late-type component is
available, as in SX Cas, RX Cas, and so on, the surface
emission flux per unit surface area of the cool star would be
several orders of magnitude ($10^3$ to $10^5$ times) larger than
even in giants with active chromospheres, such as the RS CVn
stars. More importantly, the emission lines show weakening
(albeit not too conspicuous) at the time of the eclipse of the
gainer. Thus, the emitting region is associated with the
gainer, but occupies a much larger volume and forms perhaps a

kind of a "superchromosphere" whose dimensions are comparable
with the dimensions of the whole system.

Thus, the W Serpentis stars do appear to form a family of
objects with more than superficial resemblance. But the
physical similarity may still go only as far as the existence
of the superchromosphere, while the character and evolutionary
state of the stellar components may be very different from one
system to another. Thus, in order to find those systems
genuinely related to $\beta$ Lyrae, one must examine each individual
case very carefully. This is not easy to do, since the W
Serpentis systems show such a complex spectroscopic and
photometric behaviour that their modelling is no easier than
for $\beta$ Lyrae. In fact, at least some of them may have even
more circumstellar material than $\beta$ Lyrae, and their mass
transfer rates may be higher, too. Nevertheless, progress is
being made in unraveling their properties, and the results do
seem to be relevant to the case of $\beta$ Lyrae. Thus, an analysis
of SX Cassiopeiae (Plavec, Weiland & Koch 1972) indicates the
existence of a thick disk surrounding the gainer, although that
disk does not fully obscure the accreting star.

The similarities also go as far as to show, in some
Serpentids, the same kind of "$\lambda$ 2000 Å bulge" seen in
$\beta$ Lyrae. It is prominent in SX Cas, W Cru and W Ser. A
comparison must ultimately help decide what is the nature of
the bulge. For $\beta$ Lyrae, it was suggested that the excess flux
between about 1700 and 2400 Å is due to an accumulation of
emisssion lines of Fe III (Viotti 1976). High-dispersion IUE
spectra do indeed show strong emission lines of several
multiplets of Fe III, and many fainter lines of that ion do
indeed fall within the region of the bulge. Our line
identification shows that the onset of the excess flux near
1700 Å may be associated with the occurrence of many emission
lines of Ni III. Yet I am still not fully convinced that the
whole bulge can be explained in terms of an accumulation of
individual emission lines. The bulge in W Serpentis is broader
and extends as far as 2600 Å or even beyond. Fe III lines do
not cover the whole range; one would have to include probably
Fe II, but these lines have not yet been seen in emission in
the Serpentids. If part of the bulge is due to an elevated
continuum, the question comes about the source. If one is to
apply Occam's razor and postulate as few components as
possible, then a circumstellar hydrogen cloud offers itself,
since its presence is indicated by an excess radiation at the
Balmer jump. But, in order to get a peak near 2100 Å in the
flux from a hydrogen cloud, that cloud would have to be very
optically thick and at an electron temperature near 14,000 K.
The former requirement does not agree with the fact that at the
Balmer jump, continuous emission is observed.

BETWEEN ALGOL AND W SERPENTIS

    With the progress in the W Serpentis stars being of necessity so slow, perhaps the most important discovery was the detection of the "W Serpentis" - type emission lines in ordinary gainers such as V356 Sagittarii, U Cephei and RW Tauri. Being intrinsically much fainter than the corresponding lines in the Serpentids, these lines can be observed only when the continuous radiation of the (early-type) gainer is substantially weakened (at advanced phases of the primary eclipse) or completely absent (during its total eclise). The reasoning that led to their discovery was as follows: systems such as W Serpentis, $\beta$ Lyrae, SX Cassiopeiae and other Serpentids display emission lines in the Balmer series (and, as in $\beta$ Lyr or W Ser, also in the He I lines), and this emission is so strong that it can be observed at all phases. Some more ordinary Algols, notably U Cephei and RW Tauri, are known to show the Balmer emission only during eclipses; hence, these emissions are much weaker relatively to the continuum of the component stars. One can anticipate that the far-ultraviolet emission lines may behave in the same fashion, and they indeed do. Figure 10 shows the emissions in the spectrum of V356 Sgr, and Figure 11 the totality spectrum of U Cephei. The relative intensity of the C IV and N V resonance doublets (each blended in the low-dispersion spectrum) in U Cep signals some CNO processing, since for solar-like abundances the intensity ratio should be near 10:1, while in U Cep it is only 4:1. While this deviation is not far from being marginal, the case for carbon underabundance and nitrogen overabundance is striking in V356 Sgr. There are several lines of argument suggesting that V356 Sgr is at the very end of its mass-transfer phase (Wilson & Caldwell 1978), while U Cephei is still active.

    Now U Cephei and V356 Sagittarii are bright enough to be observable outside the primry eclipse in the high-dispersion mode of the IUE spectrograph. We have found (Plavec 1983) that the line-emitting region can be observed against the photosphere of the B9 V gainer at phases outside its total eclipse. The lines seen in emission at totality are now observed in absorption, and because of the high dispersion, their profiles can be recognized. One sees, as Figure 12 shows for the case of the Fe III line at 1895 Å, that their profiles are different from those of the photospheric lines, which are greatly broadened by the rapid rotation of the photosphere (which amounts to about 315 km/s, as determined from the optical absorption lines). It is also good to remember that absorption lines such as the N V doublet at 1240 Å, the C IV doublet at 1550 Å, etc., should not be seen in the photosphere of a star with an effective temperature of only 11,250 K. Thus, they are formed in a shell surrounding the gainer. This

shell (I am using this term without implying any  special shape
and  thickness of the layer, which are largely unknown) must be
at an  electron  temperature  near 100,000 K; judging from the
profiles of the absorption lines, the hot region must be highly
turbulent, since the lines are too broad for their shapes to be
explained by thermal motions only. One can presume some amount
of rotational broadening, but turbulence  obviously  dominates.
From  the  emission  line  intensities one also infers that the
emitting ions have been not only collisionally excited but also
collisionally  ionized.  The  total  power  emitted  in  the
far-ultraviolet  emission  lines  in U Cephei  is  only a  few
percent of the  solar  luminosity  and  represents  only a few
percent  of  the  energy  probably  released  as  the
mass-transferring stream descends to the surface of the gainer.
Thus,  we  conclude that  U  Cephei  is  surrounded  by  a hot
turbulent  region  which  most  likely  obtains  the  necessary
kinetic,  ionization  and excitation energy from the process of
accretion. Now the  same  kind  of  a  hot turbulent layer was
found  in  other Algols (such as U CrB, TX UMa, AU Mon) by
Polidan & Peters (1982) and was in fact already anticipated for
U Cephei and U Sagittae  by  Kondo,  McCluskey & Harvel (1981).
Since  the  emission lines in the W Serpentis  stars  are  very
similar to those  seen  in  U  Cephei, only much stronger (by a
factor of about 100 in SX Cas),  we observe  most  likely  the
same  fundamental phenomenon in all these systems.  It would be
hard to  deny  that  here  seems  to  be a direct link between,
speaking symbolically, Algol and β Lyrae.  It appears that the
hot  turbulent  layer  in  β  Lyrae is unusually  extended  and
probably  surrounds  the  whole  system,  although  some
concentration to the gainer is noticeable.


## ONE LAST LOOK AT BETA LYRAE

Fortunately for us, β Lyrae  is  so bright that it can be
observed  at  high  dispersion  at  all phases,  including  the
eclipses,  and  the profiles of its  far  ultraviolet  emission
lines can be studied detail. Figure 13 shows the emission line
profiles of the  Si  IV  resonance  doublet  lines near 1400 Å.
Their P Cygni character is quite obvious, and  this  holds  for
all  the far ultraviolet emission lines we observe in this star
(and also in another, noneclipsing, bright W Serpentis star, KX
Andromedae =  HD 218393).  It is worth noticing that no hint of
the P Cygni  character  of  the  emissions  is  visible  on
low-dispersion spectra: compare Figure 13 with Figure 9.  The
P Cygni profiles tell  us  that  the  "superchromosphere"  is a
dynamic  structure and that gas is flowing out of  the  system.
The terminal  velocity  is  of the order of 500 km/s, much less
than in hot luminous stars  where the stellar wind is driven by
radiation pressure.  But this terminal velocity  is  of the same

order as the escape velocity from the system, and there is little doubt that the material can escape either into space or at least far from the two components. This stellar wind in β Lyrae, probably induced in the process of accretion, shows us a surprising aspect of the mass transfer process. Quite possibly, at least at the present evolutionary stage of β Lyrae, more gas is being driven out of the sytem than is accreted. Perhaps, in better understanding of the phenomenon of the induced wind, lies an answer to the puzzling dilemma of mass transfer versus loss. Although the faintness of the other W Serpentis stars does not permit us to recognize the P Cygni profiles of their emission lines, it is probable that they also indicate an outflow from the system, or at least from the vicinity of the gainer.

So far the association of the induced wind with the gainer has been based only on the observed weakening of the emission lines at the time of the eclipse of the gainer. This effect is strongest in U Cephei, but is very weak in β Lyrae. Figure 14 shows the profile of the Si IV line at three crucial phases. We can see that the emission component is very little affected by the primary eclipse, and is only somewhat weakened in secondary eclipse. The line emitting region must surround the whole system in β Lyrae, and its concentration to the secondary component (the gainer) is not very pronnounced. In accord with the optical observations of Batten & Sahade (1973), we suspect that the line profiles are actually composite, and that there exists a low but broad emission profile which is probably coming from the inner parts of the system and is more affected by the eclipses, and is probably greatly broadened by rotation. The phenomenon of the wind in β Lyrae is very complex. From optical observations with a Reticon, Etzel & Meyer (1983) argue that there are two separate winds, coming from the two components. Further studies are urgently needed to clarify the structure of the two components and of the surrounding circumstellar material.

CAVEATS

The preceding section ended with a note of caution, and justly so. It is useful to look back, for example, to the great β Lyrae effort of 1941. Many valuable facts were established at that time, yet the model that emerged was considerably faulted because of the assumption that the mass-luminosity relation is sacrosanct, and also that both objects are more or less spherical stars. Could there be a similar fatal flaw in our present-day models? Of course it can be, and it may lie in just any one assumption that we consider as quite safe, for example, adopting the model of a semi-detached binary for β Lyrae. A number of arguments seem to support this model. What if we are wrong? What if, for example, stellar winds from both components play a much more important role than Roche lobe overflow? At least one fact is alarming: β Lyrae seems to be in a fast (although apparently not the very fastest possible) phase of mass transfer, while V356 Sgr is believed to be at the very end of even slow mass transfer. Yet the mass ratio in β Lyrae is believed to be 12/2, very far from near equality of the component masses, while in V356 Sgr it is 13/5, much closer to equality. Clearly, a variety of initial conditions must play a role, and it is not possible to accept general validity of a simple qualitative relation between the mass tranfer rate (or the corresponding phase of mass transfer) and the instantaneous mass ratio, as it has been done (perhaps correctly in that case) for Algol, U Sagittae, and U Cephei (Tomkin 1979). Nevertheless, there is good reason to be cautious.

## ACKNOWLEDGEMENTS

My work on the problems of Algols, W Serpentis stars, and β Lyrae has been made possible by continuous grants by the National Science Foundation and NASA. The studies have been conducted in cooperation with my graduate students Mr Jan J. Dobias and Ms Janet L. Wiland. Mr Robert L. O'Daniel assisted in the final editing of the manuscript. My participation in the Cambridge meeting has been made possible by a generous grant offered to me by the organizers. All these contributions are gratefully acknowledged.

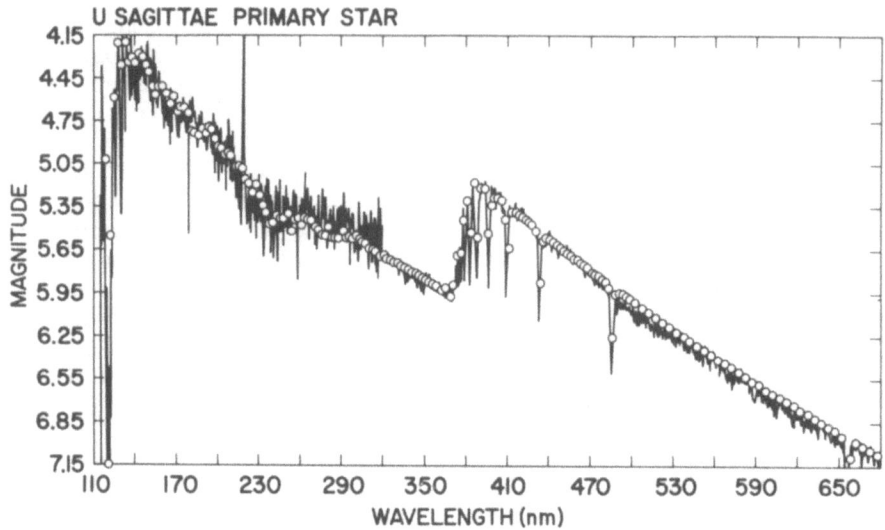

Figure 1. The accreting star in U Sagittae has a flux distribution that is well matched by Kurucz (1979) model atmosphere (circles).

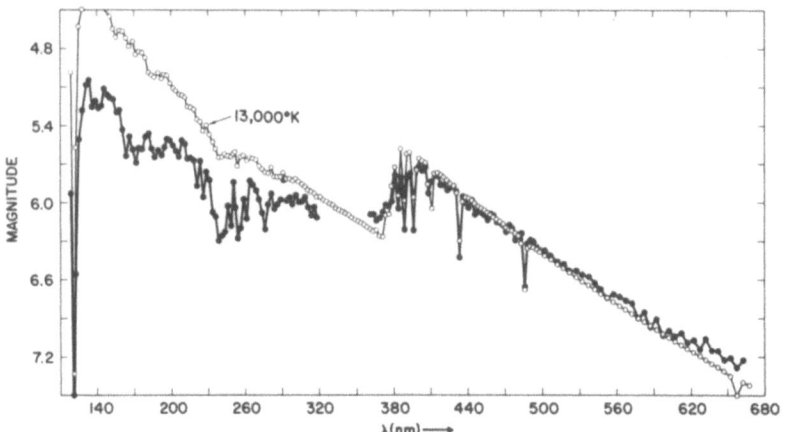

Figure 2. While the optical flux distribution in U Cephei seems to be tolerably well fitted by a Kurucz model with an effective temperature as high as 13,000 K, IUE observations show that such a high temperature is not acceptable.

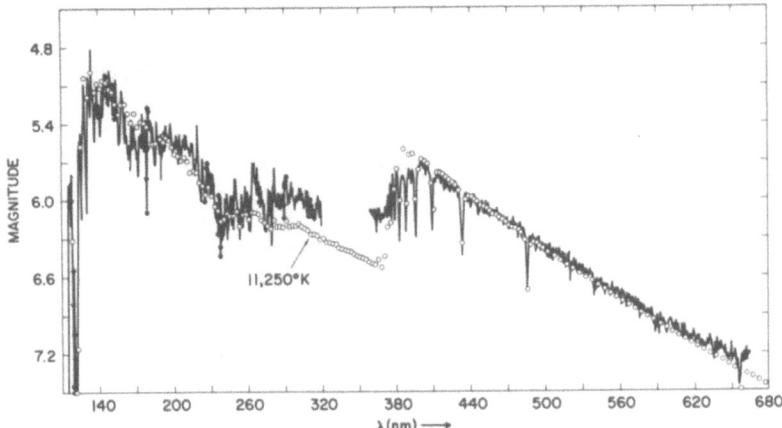

Figure 3.  An effective temperature of 11,250 K gives a much better overall match to the flux distribution in U Cephei.

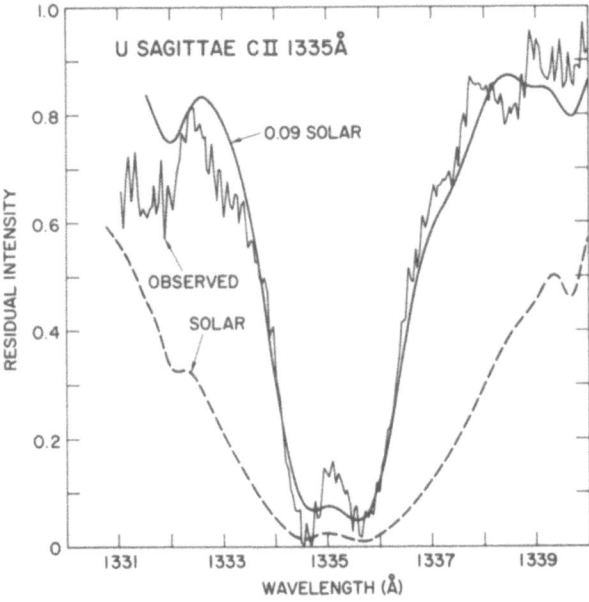

Figure 4.  If U Sagittae had the same photospheric abundance of carbon as the Sun, the profile of the C II line would be much broader.  The observed profile requires a carbon abundance reduced to 9% of the solar value (from the Ph.D. thesis of Jan Dobias, UCLA).

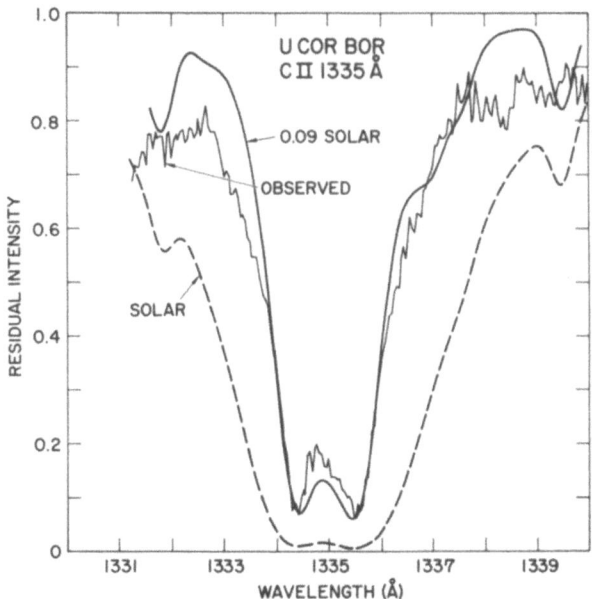

Figure 5. U Coronae Borealis, another classical Algol system, shows about the same carbon underabundance as U Sagittae.

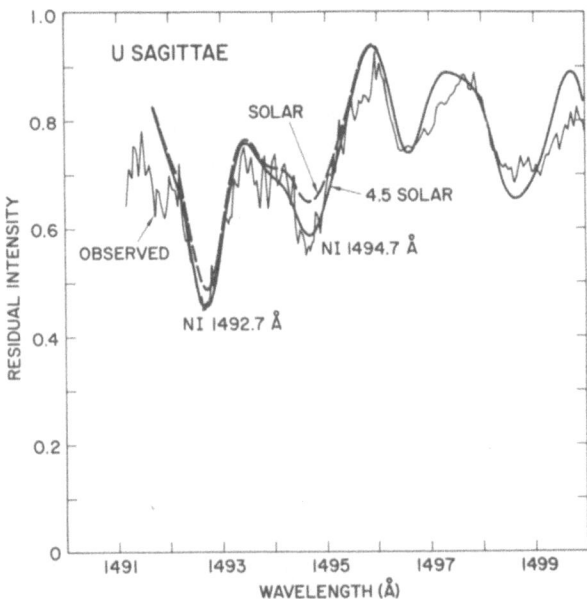

Figure 6. If the material has been CNO-processed, an underabundance of carbon implies an overabundance of nitrogen, and this is
what the ultraviolet N I lines indicate (from the Ph.D. thesis of
Jan Dobias, UCLA).

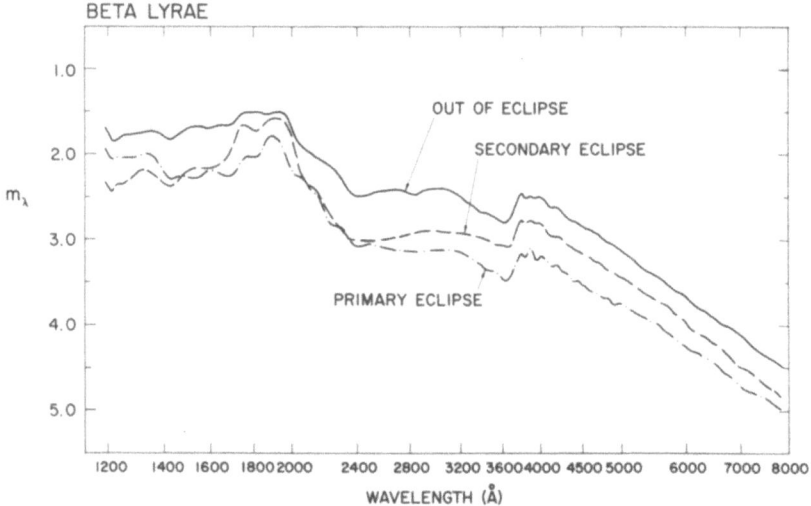

Figure 7. The depth of the eclipses in β Lyrae varies with wave-length, and in the far ultraviolet, the role of the eclipses is interchanged. Note that around λ2000 Å, both eclipses are very shallow.

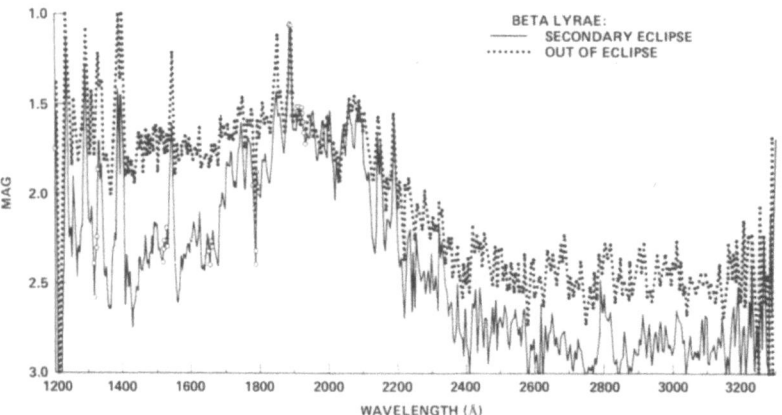

Figure 8. The existence of the "λ2000 Å" bulge is best seen when we compare the flux distributions in full light and in secondary eclipse.

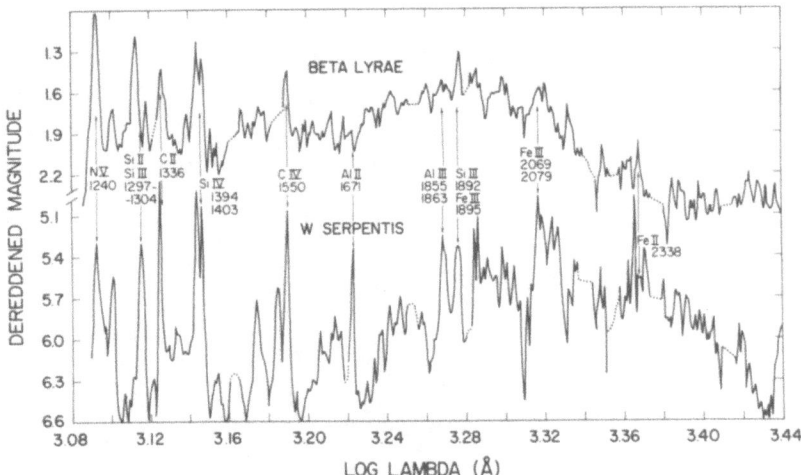

Figure 9. A comparison of the ultraviolet spectra of β Lyrae and
W Serpentis shows considerable similarity.

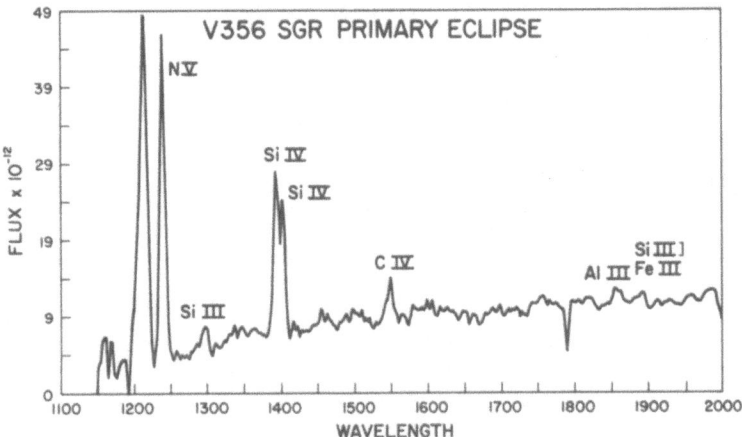

Figure 10. The emission-line spectrum observed during the total
eclipse of the gainer in V356 Sagittarii strongly suggests CNO-
processing.

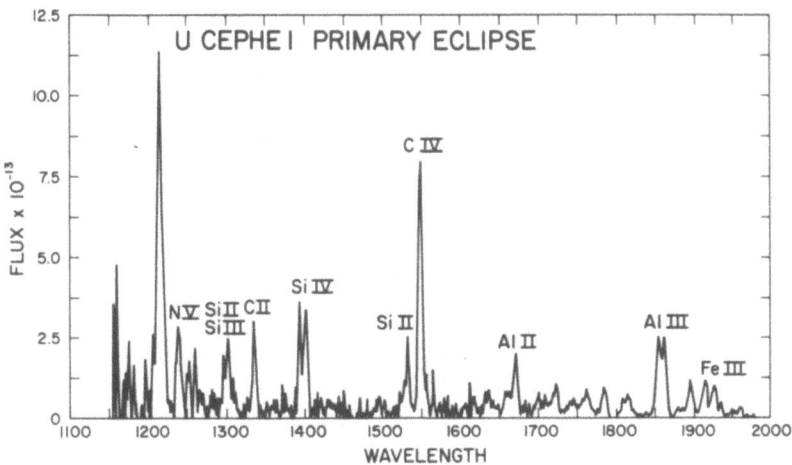

Figure 11. When the primary star (gainer) in U Cephei is totally eclipsed, a rich emission-line spectrum appears in the far ultraviolet.

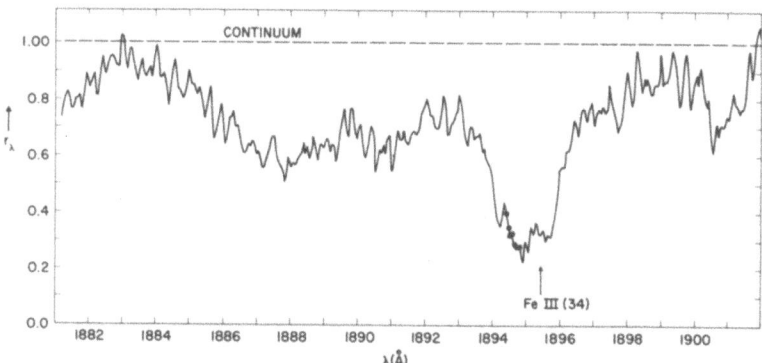

Figure 12. The high-dispersion far-ultraviolet spectrum of U Cephei shows absorption lines (like Fe III 1895 Å) which are not formed in the photosphere but rather in a circumstellar hot turbulent region.

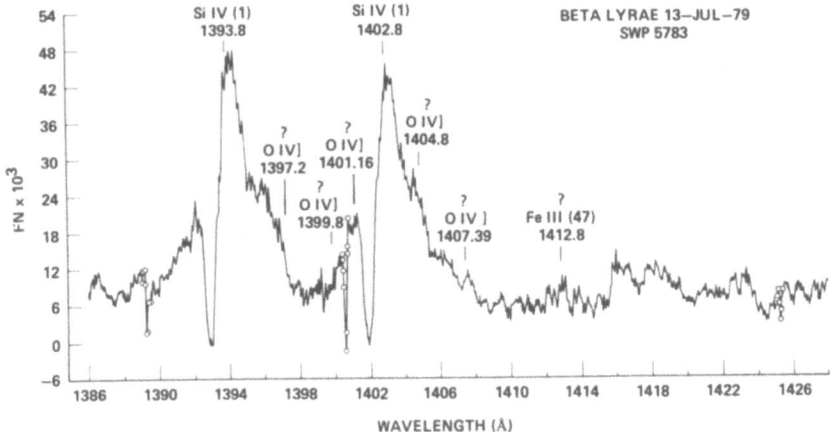

Figure 13.   The emission lines in the far ultraviolet spectrum of
β Lyrae all display P Cygni profiles when observed at high dispersion.

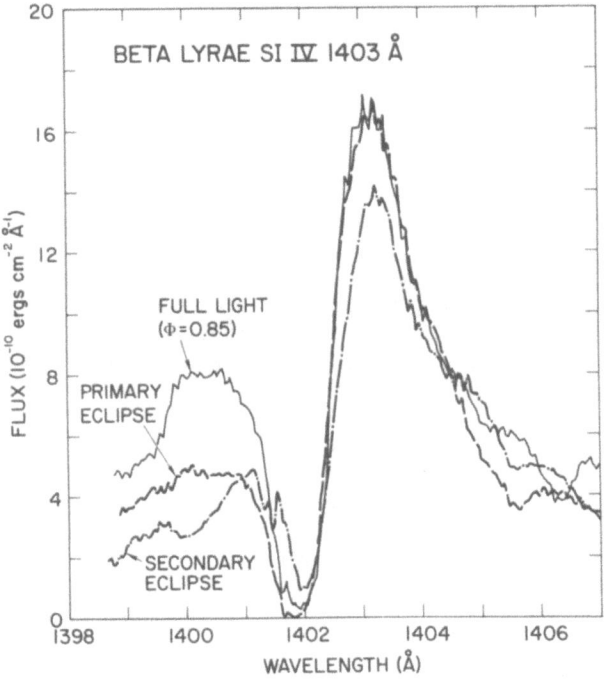

Figure 14.   The emission line of Si IV in β Lyrae as observed in
the eclipses and at full light (from the Ph.D. thesis of Janet
Weiland, UCLA).

REFERENCES

Abt, H.A., Jeffers, H.M., Gibson, J. & Sandage, A.R., 1962.
    Ap.J., **135**, 429.
Batten, A.H., 1974.  Publ.Dom.Astrophys.Obs., **14**, 191.
Batten, A.H. & Fletcher, J.M., 1975.  Publ.Astr.Soc.Pacif.,
    **85**, 599.
Hack, M., Hutching,s J.B., Kondo, Y., McCluskey, G.E.,
    Plavec, M.J. & Polidan, R.S., 1975.  Ap.J., **198**, 453.
Kondo, Y., McCluskey, G.E. & Stencel, R.E., 1979.  Ap.J.,
    **233**, 906.
Kondo, Y., McCluskey, G.E. & Harvel, C.A., 1981.  Ap.J.,
    **247**, 202.
Kopal, Z., 1955.  Ann.Astrophys., **18**, 379.
Kopal, Z., 1978.  In "Dynamics of Close Binary Systems"
    (Dordrecht: Reidel), p. 420.
Kurucz, R.L., 1979.  Ap.J.Suppl., **40**, 1.
Lambert, D.L., 1982.  In "Advances in Ultraviolet Astronomy",
    ed. Y. Kondo, J.L. Mead and R.D. Chapman (NASA Conf.
    Publ. 2238), p. 114.
Olson, E.C., 1980a.  Ap.J., **237**, 496.
Olson, E.C., 1980b.  Ap.J., **241**, 257.
Plavec, M.J., 1980a.  In "Close Binary Stars: Observations
    and Interpretation", ed. M.J. Plavec, D.M. Popper and
    R.K. Ulrich, (Dordrecht: Reidel), 3.
Plavec, M.J., 1980b.  Ibid., p. 251.
Plavec, M.J., Dobias, J.J. & Weiland, J.L., 1984.  Submitted
    to Ap.J.
Polidan, R.S. & Peters, G.J., 1980.  In "Close Binary Stars:
    Observations and Interpretation", see ref. above, p. 293.
Polidan, R.S. & Peters, G.J., 1980.  In "Advances in
    Ultraviolet Astronomy", ed. Y. Kondo, J.M. Mead and R.D.
    Chapman (NASA Conf. Publ. 2238), p. 534.
Sahade, J., 1980.  Sp.Sci.Rev., **26**, 349.
Sahade, J., Huang, S.S., Struve, O. & Zebergs, V., 1959.
    Trans.Am.Phil.Soc., **49**, 1.
Tomkin, J., 1979.  Ap.J., **231**, 495.
Viotti, R., 1976.  Mon.Not.R.A.S., **177**, 617.
Wilson, R.E., 1974.  Ap.J., **189**, 319.
Wilson, R.E., 1982.  In "Binary & Multiple Stars as Tracers
    of Stellar Evolution", ed. Z. Kopal and J. Rahe
    (Dordrecht: Reidel), 261.
Wilson, R.E. & Caldwell, C.N., 1978.  Ap.J., **221**, 917.
Wilson, R.E. & Lapasset, E., 1981.  Astron.Astrophys.,
    **95**, 328.

# SYMBIOTIC STARS

Scott J. Kenyon

University of Illinois Urbana-Champaign
and
Harvard-Smithsonian Center for Astrophysics
U.S.A.

## ABSTRACT

Possible binary models for the small class of eruptive variables known as symbiotic stars are outlined. A discussion of the observable properties of such models leads to the conclusion that symbiotic stars are likely to be main sequence or white dwarf stars accreting material from a red (super)giant primary. Recent observations of symbiotic stars confirm this notion, and three systems (CI Cyg - a red giant with an accreting main sequence star companion, V1016 Cyg - a Mira variable with a hot white dwarf companion, and V443 Her - a red giant with a hot white dwarf companion) have been chosen to illustrate the impact modern observations have had on understanding the symbiotic phenomenon.

*P. P. Eggleton and J. E. Pringle (eds.), Interacting Binaries, 179–203.*
© *1985 by D. Reidel Publishing Company.*

## 1.   INTRODUCTION

In the early 1900's, A. Cannon and P. Merrill discovered a
group of stars whose optical spectra showed the strong red
continuum and absorption features characteristic of late-type
giants, as well as the weak blue continuum and bright emission
lines usually associated with planetary nebulae. These dozen
or so "symbiotic stars" (also known as "stars with combination
spectra") were found to be low amplitude, long period
variables; in some cases these fluctuations represent orbital
motion in a binary system (Belyakina 1970; Kenyon 1982).
Periodically, a symbiotic star undergoes a 2-7 magnitude
eruption in which the M-type features vanish and are replaced
by either an F-type absorption spectrum (e.g., CI Cygni;
Belyakina 1979) or an intense nebular continuum (e.g., V1016
Cygni; Andrillat, Ciatti & Swings 1982). The late-type
features return to dominate the optical spectrum once the
system returns to minimum. As of this writing, 150
symbiotics are known to exist (Allen 1982), and they appear to
belong to the old disk population (Boyarchuk 1975; Wallerstein
1981). Excellent reviews of the observed properties of
symbiotic stars can be found in Friedjung & Viotti (1982), as
well as in Allen (1979, 1984), Boyarchuk (1969, 1975) and
Swings (1970).

It is commonly accepted that most, if not all, symbiotic
stars are binaries. The M-type features can then be associated
with a late-type (usually giant)star, while a **hot component**
is responsible for the blue continuum and emission lines.
However, the physical nature of these components is poorly
known. The late-type star may be either a normal giant (RGB
star), a bright asymptotic branch giant (AGB star) or (rarely)
a supergiant (Boyarchuk 1975; Kenyon & Gallagher 1983; Kenyon
1983; Allen 1979, 1984). A recent study by Kenyon & Webbink
(1983) suggests **most** symbiotic **hot** components are either
very hot white dwarfs or accretion disks surrounding low mass
main sequence stars. Some systems may contain neutron stars or
other types of hot stars. In most instances, the hot star
luminosity is believed to be derived either directly or
indirectly from accreted matter donated by the red (super)
giant primary. It is possible that a few symbiotics are **not**
binaries, and single star models have been proposed to explain
the behaviour of selected systems. These models, not
considered here, are reviewed in some detail by Kenyon (1983)
and Persic (1983).

In this article, possible binary models for symbiotic
stars are surveyed. Since a late-type giant or supergiant is a
common denominator in all such models, I will consider a wide
range of binary stars containing a red (super) giant component.

The primary goal is to discuss the physical characteristics of each of these classes of binaries, and then to identify those systems likely to be classified as symbiotic stars by terrestrial observers. Along the way, I hope to relate symbiotic stars to other interacting binaries containing evolved late-type stars, such as VV Cephei and Algol systems. Recent observations of symbiotic stars in the ultraviolet and the infrared have greatly expanded our understanding of these systems. Three stars (CI Cygni, V1016 Cygni and V443 Herculis) have been chosen to illustrate the impact modern observations have had on symbiotics, and to present examples of the binary models described in Section 2. These systems are discussed in Section 3. Although some progress has been made on understanding the symbiotic phenomenon, much work remains to be accomplished: a few suggestions for new observational studies are made in Section 4.

## 2. BINARIES CONTAINING A RED (SUPER) GIANT

In this section, I want to examine the properties of various binaries with a red (super) giant component. Two important physical processes generally determine the observable characteristics of such a system:

1) How the giant loses mass - mass lost via a stellar wind tends to form a large, **low density** circumbinary nebula. **Dusty** nebulae are expected if the mass-losing star is a Mira variable or a red supergiant (Merill & Ridgway 1979). Mass lost via tidal interaction tends to form a **high density** accretion disk surrounding the binary companion.

2) How the luminosity generated by the mass gaining star (intrinsic plus accretion luminosity) interacts with gaseous matter in the binary. This depends primarily on the mass loss process for the primary star (wind versus tidal overflow) and the nature of the accreting star. We consider, in turn, the effects of disk and wind accretion onto a (i) main sequence star, (ii) white dwarf star or (iii) neutron star companion to the red (super) giant.

A. Red Giant + Main Sequence Star Binaries

A number of interesting binaries contain an evolved late-type star and a relatively unevolved main sequence companion; some of these systems (e.g. Algols) are reviewed elsewhere in this volume. The physical appearance of such a binary depends primarily on:

1)      the evolutionary  status  of the evolved star (subgiant,
        giant or supergiant),

2)      the nature of the  unevolved  (main sequence) star (high
        mass or low mass), and

3)      the binary period.

For sufficiently long binary periods, the  evolved  star cannot
lose  mass via tidal overflow, and the luminosity  produced  by
accretion processes is apt to be quite small  when  compared to
the intrinsic luminosity of the binary components.  The reverse
is true if the late-type star fills its tidal lobe,  since very
large  mass  loss  rates  are  likely  to  occur  (Paczynski  &
Sienkiewicz 1972;  Plavec,  Ulrich  &  Polidan  1973;  Webbink
1979a,b).

    A flow chart for  binaries containing an evolved late-type
star (RG) and a main  sequence  star  (MS)  is  illustrated  in
Figure  1.  (Note: in Fig.  1 , V stands for visual brightness
and not V  magnitude.)  For  simplicity,  I have divided these
binaries into two groups based on the mass of the main sequence
star.  The  bolometric luminosity of a massive  main  sequence
star is comparable  to  (or marginally less than) that of a red
giant or supergiant, and may  contribute an observable fraction
of the optical flux to the  binary  as a whole.  In addition, a
significant  fraction of this flux is capable of  ionizing  the
hydrogen and  helium  present  in the evolved primary's stellar
wind.  VV Cephei stars are a good  example of binaries in which
an  O  or  B-type star ionizes the stellar  wind  (or  extended
atmosphere) of its supergiant  companion.  The periods of these
systems  are  comparable to or  greater  than  those  found  in
symbiotic stars (P $\geq$ 2-3 years; cf. Sahade & Wood 1978; Cowley
1969).  Both types of  binaries  exhibit a lack of short period
systems (P $\leq$ 1 yr), and it  is  tempting to consider BQ[] stars
as  the  short  period  counterparts  of VV Cep  and  symbiotic
systems.  BQ[] stars (Wackerling 1970) are hot objects (Bep and
Beq spectral types) with infrared excesses  usually interpreted
as  circumstellar  dust emission (e.g., Swings &  Allen  1972).
Recent observations of these systems (Ciatti & Mammano 1976 and
references  therein)  show  that  an  M-type  star,  perhaps  a
supergiant or luminous giant, is responsible for part of the IR
excess.  Ciatti and Mammano suggest the late-type star is about
to become a planetary nebula.  An alternate possibility is that
BQ[] stars are  shorter period versions of symbiotic and VV Cep
systems,  in  which the  late-type  star  transfers  mass  very
rapidly to its  main  sequence  companion  star  (e.g., Webbink
1979a).

The situation is quite different if the main sequence star has a rather low mass ($\lesssim 3M_\odot$). An "Algol-type binary" occurs if the evolved star is a K subgiant, and if it has an A or B-type companion. The "RS Canum Venaticorum binaries" are similar to Algols in having a K subgiant, but the companion star is an F or G-type star, rather than an A-type star. A very interesting binary is formed if the evolved star is not a subgiant, but a giant or supergiant. In this case, the evolved star totally dominates the optical luminosity of the binary (unless accretion plays an important role, as discussed below), unlike the Algol and RS CVn systems, where the hot component may dominate the optical light of the system. To become a significant optical light source in a binary with a luminous red giant primary, a lower mass main sequence companion needs to accrete matter at roughly $10^{-5} M_\odot \, yr^{-1}$ (Bath & Pringle 1982; Kenyon & Webbink 1983). It is these systems which may produce a symbiotic optical spectrum, and I now discuss these binaries in more detail.

The flow chart for a binary containing a red (super) giant and a low mass main sequence star ($\lesssim 3M$) is shown in Figure 2. Wind accretion is expected to be the primary mass transfer mechanism in long period binaries (P $\gtrsim$ 30 yr; Webbink 1979b), and only a small fraction of this material can be accreted by the companion star. Even at the highest possible mass loss rates from the (super) giant ($10^{-4} M_\odot yr^{-1}$, Zuckerman 1980), the optical luminosity resulting from accretion ($\lesssim 1-30 \, L_\odot$) is insignificant when compared to the optical luminosity of the mass losing star ($\gtrsim 100-1000 \, L_\odot$). The main sequence companion might only be visible at UV wavelengths, where the light output from the giant is very small. The supergiant TV Gem might be such a system: IUE observations revealed an abnormally strong UV continuum which might be photospheric emission from a late B or early A giant or subgiant, or perhaps radiation from an optically thick accretion disk (Michalitsianos, Hobbs & Kafatos 1979). Note that periods of enhanced mass loss (due perhaps to an instability in the late-type star) could result in optical outbursts, as indicated in Figure 2.

While wind accretion onto a low mass main sequence star produces a rather "unexciting" interacting binary, the reverse is true of disk accretion. As noted by Webbink (1979a) high accretion rates (approaching or exceeding $10^{-3} \, M_\odot yr^{-1}$) are expected of a lobe-filling giant, especially if it is the more massive binary component (as is likely to be the case here, since the main sequence star is unevolved). At these rates, the disk dominates the giant component in the blue, and is a strong permitted emission-line source (Bath 1977; Bath & Pringle 1982; Kenyon & Webbink 1983). The strength of the emission lines is a direct function of the accretion rate. At low rates ($\simeq 10^{-6} M_\odot yr^{-1}$) the emission lines are rather weak; an example of such a system may be the recurrent nova T CrB in its quiescent state (cf. Webbink 1979a). At higher accretion rates ($1\text{-}5 \times 10^{-5} M_\odot yr^{-1}$), the blue continuum and the emission lines are very strong, comparable to those observed in the symbiotic star CI Cygni (cf. Bath & Pringle 1982; Kenyon & Webbink 1983; Kenyon 1983). At very high accretion rates ($\simeq 10^{-3} M_\odot yr^{-1}$), but below the Eddington critical luminosity, the optical continuum resembles that of an A-F supergiant, which typifies the outburst state of a symbiotic star (e.g., Swings 1970; Belyakina 1979). Once the Eddington limit is reached (or exceeded), radiation pressure disrupts the disk and generates an optically thick, outflowing wind (Bath 1977). The P Cygni profiles observed in the outbursts of Z And and other symbiotic stars may be a result of such a wind triggered by super-Eddington mass transfer rates (Bath 1977). An episodic mass transfer burst can lead to a transformation from a low state ($\lesssim 1\text{-}10 \times 10^{-6} M_\odot yr^{-1}$) to a high state ($\lesssim 10^{-3} M_\odot yr^{-1}$); Webbink (1976) and Bath & Pringle (1982) have considered such a process to explain the outbursts of T CrB and CI Cyg, respectively. In both these studies, the observed evolution of the outburst required the mass-gaining star to be a low mass main sequence star.

## B.   Red Giant + White Dwarf Binaries

The observable properties of a red giant binary containing a white dwarf (WD) are summarized in Figures 3 and 4. These once again depend primarily on the mass accretion rate and the physical extent of the giant, but have an additional sensitivity to the evolutionary state of the white dwarf. An important difference between white dwarf and main sequence accretors is that the incoming material can undergo nuclear processing on the surface of a white dwarf, and provide an additional energy source which may exceed that released by accretion. If the white dwarf should accrete matter at a high rate ($\gtrsim 3 \times 10^{-7} M_\odot yr^{-1}$) its outer envelope expands to supergiant dimensions, and hydrogen burns in a thin shell above the degenerate core (e.g., Paczynski & Zytkow 1978). This expansion results in either a contact binary or a double red giant (cf. Figure 3); such systems may not be easily detected as interacting binaries.

At lower accretion rates, the hydrogen may burn stably ($M \lesssim 1\text{-}3 \times 10^{-7} M_\odot yr^{-1}$; Paczynski & Zytkow 1978; Iben 1982) or unstably ($M \lesssim 10^{-7} M_\odot yr^{-1}$). [These limits are somewhat sensitive to the underlying white dwarf mass; cf. Iben 1982 and references therein]. Stable hydrogen burning produces an extremely luminous EUV source ($L \lesssim 10,000 L_\odot$), which probably ionizes the primary's outer envelope (cf. Belyakina 1970), and might be detectable as a strong soft X-ray source. When the accretion rate is below $10^{-7}$ $M_\odot yr^{-1}$, the total luminosity depends on the structure (and the mass) of the white dwarf, as well as the accretion rate. The accretion luminosity is likely to dominate the light output if the white dwarf is "cold" ($L \lesssim 10$ $L_\odot$) as in cataclysmic binaries. The accreted hydrogen accumulates on the white dwarf's surface until it reaches a "critical mass" and triggers a thermonuclear runaway (see, for example, MacDonald 1980). A "hot" white dwarf generates additional luminosity in a helium burning shell (cf. Iben 1982), and this may be sufficient to dominate the accretion luminosity. A hydrogen envelope accumulates in this case as well, although the shell flash may be weaker (Iben 1982). Paczynski & Rudak (1980) identified the two types of shell flashes possible in the hydrogen envelope of an accreting white dwarf, namely "strong flashes" and "weak flashes". Strong flashes occur when the envelope is degenerate (a "cold" white dwarf); this produces a dynamic expansion which is observed as a classical nova outburst. Weaker flashes are produced in a non-degenerate envelope (a "hot" white dwarf); these result in a weak optical outburst with little or no ejection of the accreted envelope. The strength of the flash depends on the white dwarf mass as well: for a given temperature structure and accretion rate, stronger flashes occur on more massive white dwarfs. Very massive white dwarfs ($M \gtrsim 1.3$ $M_\odot$) tend to have strong flashes irrespective of the accretion rate, while low mass white dwarfs ($M \lesssim 0.6-0.7$ $M_\odot$) nearly always have weak shell flashes (Iben 1982; MacDonald 1980 and references therein). Kenyon & Truran (1983) have identified strong H-shell flashes with the outbursts of the nova-like symbiotic stars AG Peg, RT Ser and RR Tel, while the weak H-shell flashes appear to explain the outbursts of the planetary nebula-like symbiotic stars V1016 Cyg, V1329 Cyg and HM Sge. Note that these systems distinguish themselves from the more classical symbiotic stars (such as Z And and CI Cyg) by the protracted length of their **single** outbursts.

Although the total luminosity of the accreting white dwarf depends on the accretion rate and the hydrogen burning rate, the appearance of the optical spectrum depends on the manner in which matter is accreted as well. If the red giant loses matter tidally, a **high density, optically thick** disk forms about the white dwarf. Radiation from the disk and the white dwarf's surface can photoionize optically thin "coronal" material above and below the disk. The material is likely to have $n_e > 10^8 cm^{-3}$: permitted lines (e.g., H I, He I, He II, N III, O III and C IV) are more likely to be emitted than forbidden lines (e.g., [O III] and [Ne III]), since these latter lines will be collisionally quenched. This hypothetical binary will look like a symbiotic star if $L_{wd}$ (due to the combined effects of nuclear burning and accretion) is large ($L_{wd} \gtrsim$ 10-100 $L_\odot$; Kenyon & Webbink 1983; Tutukov & Yungel'son 1982). Lower luminosity white dwarfs do not emit sufficient numbers of ionizing photons to produce observable emission lines. These latter systems would, perhaps, be observed in the optical as normal M-type stars, but ultraviolet observations would reveal the M-type star to have an abnormally strong continuum and perhaps some weak emission lines (e.g., CIV $\lambda\lambda$ 1548, 1550).

The situation is quite different if the red giant loses matter in a stellar wind. Only a small fraction ($\lesssim 1\%$) of this material can be captured by the white dwarf: the remainder forms a relatively **low density** circumbinary nebula. The collision rates are far too small to quench the forbidden lines ($n_e \lesssim 10^6$ cm$^{-3}$ ), and these lines (e.g., [O III] and [Ne III]) are expected to be extremely strong. As noted above, hydrogen and helium burning may make the white dwarf a strong EUV source, and such a system is likely to be identified as a symbiotic star. If the white dwarf is cool and faint, it cannot photoionize an observable portion of the nebula, and thus cannot produce a symbiotic star. An example of such an object is Mira (Warner 1972), in which a cool white dwarf accretes a small amount of material from Mira's stellar wind.

The observational appearance of a red giant + white dwarf binary will also depend on the evolutionary status of the giant. Normal red giants lose matter at fairly low rates ($\lesssim 10^{-7} M_\odot$ yr$^{-1}$; Reimers 1975), and substantial amounts of dust do not generally form in their outflowing winds (Merrill & Ridgway 1979). Mira variables, which probably arise from intermediate and low mass stars in the double shell burning AGB phase, lose mass much more rapidly, and dust does form in the outflowing wind (Merrill & Ridgway 1979 and references therein). Interestingly, this division of giants into those with large dust shells and those without is present in symbiotics, as well. Webster & Allen (1975) separated symbiotic stars into S-types (systems with "stellar" IR continua and T(colour) $\simeq$ 3000K) and D-types (objects with "dusty" IR continua and T(colour) $\simeq$ 1000K). The S-type systems contain relatively non-variable giants ($\Delta K \lesssim 0.2$ mag), while the D-types usually contain long period Mira-like variables ($\Delta K \simeq 1$ mag). The Mira-like stars in the D-type systems appear to be completely enshrouded in dust (although the hot components may not be!), with estimates of visual extinction ranging from 12-20 magnitudes (e.g., Thronson & Harvey 1981; Allen 1983b). In most late-type giants, the dust is rather cool (200 K; Merrill & Ridgway 1979), but UV photons from a hot binary companion can cause the dust to radiate at a higher temperature. This appears to be the case in some symbiotic stars (e.g., H1-36; Allen 1983), as they seem to contain hot dust (T $\lesssim$ 1000K).

## C.  Red Giant + Neutron Star Binaries

Although neutron stars are undoubtedly rare among binary systems containing luminous red stars, they can be observed at large distances since they tend to be strong X-ray sources, even at modest accretion rates. Their observational appearance at non-X-ray wavelengths tends to resemble the white dwarfs discussed earlier. As long as the accretion rate remains low, X-rays from the neutron star can ionize the expanding wind from a red giant or form a corona around an accretion disk (e.g., Galeev, Rosner & Vaiana 1979). If the accretion rate should become large ($\gtrsim 10^{-7}$ $M_\odot yr^{-1}$), the outer envelope of the neutron star expands to red giant dimensions, and perhaps forms a contact binary. Such a system is unlikely to be an X-ray source, and would probably be undetectable as a peculiar system.

An example of a neutron star accreting material from an evolved red giant is the strong X-ray source GX1+4 (Davidsen, Malina & Bowyer 1977 and references therein). The X-ray spectrum of this source is very hard and can be fit by a Planck law, $F \propto E^3 /[\exp(E/kT)-1]$, with kT = 27 keV (Doty, Hoffman & Lewin 1981). Intense emission lines are observed in optical spectra of GX1+4 (Davidsen, Malina & Bowyer), and a few TiO bands are also apparent. The combination of a strong X-ray source and a symbiotic optical spectrum is unique among symbiotics; calculations by the author suggest the observed optical spectrum can be reproduced by an M-type giant, an accreting neutron star and a photoionized nebula.

## 3.  EXAMPLES OF THE SYMBIOTIC PHENOMENON

The study of symbiotic stars has become quite popular in recent years, spurred on by recent advances in ultraviolet, optical, infrared and radio instrumentation. Low resolution 2 $\mu$m spectroscopy has been quite useful in analysing the nature of the late-type star (e.g., Allen 1980), and near-infrared spectroscopy may become even more important (e.g., Andrillat 1982). Strong photospheric absorption bands are visible in near-IR (CN, TiO and VO) and IR (CO and $H_2O$) spectra of late-type giants, and the stength of these bands is sensitive to the temperature and luminosity in the stellar photosphere (Sharpless 1956; Baldwin, Persson & Frogel 1973). These bands are just beginning to be used as probes of the conditions in the outer atmosphere of the late-type star in symbiotics (Andrillat 1982; Allen 1980; Kenyon & Gallagher 1983). These

studies are, as yet, inconclusive for symbiotic stars as a
group; preliminary results imply the late-type stars in these
systems range from normal giants not filling their tidal lobes
to bright AGB stars that do fill their tidal lobes (Kenyon &
Gallagher 1983). A combination of luminosity indicators is
probably necessary to determine accurately the evolutionary
status of red giants in most symbiotic systems. The existence
of giants in symbiotics (and in the longer period Algols, as
well) that do **not** fill their tidal lobes may be very
important for understanding mass transfer processes in binary
stars.

The bulk of the observations of symbiotic stars have been
obtained in the optical; usually these data are best
interpreted when nearly simultaneous ultraviolet and infrared
data are available. A great deal of information concerning
symbiotic nebulae can be obtained from "simultaneous" optical
and ultraviolet data (since the emission lines in some systems
vary quite rapidly; Swings 1970, Nussbaumer 1982), but this has
not been accomplished for more than a few systems (e.g.
Oliversen & Anderson 1983). The inaccessible EUV spectrum of
the hot source can be probed using the strong emission lines of
hydrogen and helium. Iijima (1981) estimated temperatures of
roughly 100,000 K for a few systems, and this temperature seems
a reasonable estimate for other symbiotics as well (Kenyon
1983).

Broad-band or narrow-band photometric programmes may prove
to yield the most reliable results concerning symbiotic binary
periods. Recent analyses by Meinunger (1979, 1981) and Kenyon
(1982, 1983) have established binary periods for AG Dra (554
days), SY Mus (627 days), AG Peg (827 days) and AX Per (682
days), as well as a pulsational period for RR Tel (374 days).
Similar studies of many systems are needed, since very few
symbiotic binary periods are well-established. Optical
spectroscopic studies are still required to determine the
physical parameters of symbiotic binaries (e.g., masses,
luminosities, etc.). To date, reasonably good spectroscopic
orbits are available only for two symbiotics: AR Pav
(Thackeray & Hutchings 1974) and AG Peg (Hutchings, Cowley &
Redman 1975 and references therein). In AR Pav, the mass ratio
is roughly 1 (Andrews 1974), while the giant is 3-4 times more
massive than the hot component in AG Peg.

The first satellite ultraviolet observations of symbiotics
were obtained for AG Peg (Gallagher et al. 1979) and SY Mus
(Thompson et al. 1979) and revealed these systems to have
strong UV continua. The launch of the IUE satellite made many
more systems accessible, and has confirmed that most symbiotics
are binary stars. As noted by Slovak (1982), the UV continua

of symbiotic stars (when corrected for interstellar reddening) are either: (i) stellar (resembling the Raleigh-Jeans tail of a blackbody) or (ii) flat ($F_\lambda \simeq$ constant over $\lambda\lambda$1200-3200). Providing circumstellar dust does not substantially redden the system (e.g., as in V1016 Cyg discussed below), the division into stellar and flat continua is physically significant. Symbiotics displaying **flat** UV continua appear to be main sequence stars accreting material at rates of $\lesssim$ a few x $10^{-5}$ $M_\odot yr^{-1}$, while the remaining systems are understood best as hot stellar sources with $T(eff) \lesssim$ 25-100,000K (Kenyon & Webbink 1983).

The preceding discussion only touches on the new observational data that have revolutionized the study of symbiotic stars (see also Allen 1984). In order to provide a more detailed illustration of this revolution, I would like to describe how new observational techniques have affected our understanding of three particular symbiotic stars (CI Cygni, V1016 Cygni and V443 Herculis). The available data suggests that each of these systems serves as a good type-example for one of the binary models discussed in Section 2 above. Observations of other symbiotics are related in the Proceedings of IAU Colloquium No.70 (Friedjung & Votti 1982).

1) CI Cygni: A Red Giant with an Accreting Main Sequence Star Companion

CI Cygni is one of the original symbiotic stars, having been discovered by Cannon in the early 1920's and then identified as a symbiotic star by Merrill & Humason (1932). Swings & Struve (1940) identified many strong emission lines, including H I, He I, He II, [O III], [Ne V], [Fe VII] and perhaps [Fe X], while Merrill (1950) classified the late-type absorption spectrum as gM4. CI Cygni is one of two symbiotics generally acknowledged to undergo total eclipses, the other being AR Pavonis (e.g. Hoffleit 1968). These eclipses occur every 855.25 days, and continue to follow Whitney's ephemeris (Aller 1954) faithfully:

$$Min = JD\ 2411902 + 855.25.E.$$

Recent observations of CI Cyg (Figure 5) have concentrated on understanding the nature of the hot component and its occasional outbursts. Broad-band UBV photometry has demonstrated that the **hot component** is totally eclipsed at primary minimum (Belyakina 1979 and references therein), and this has been confirmed by subsequent optical and ultraviolet spectroscopic observations (e.g., Stencel et al. 1982; Mikolajewska & Mikolajewski 1982; Oliversen & Anderson 1983). These latter observers found that many strong emission lines (e.g., H I, He II and O III]) were eclipsed by the giant, while others (e.g., C IV, [O III] and [Ne III]) appear **not** to be eclipsed. The available data imply the nebular region in CI Cyg is highly stratified, with He II and O III] concentrated toward the centre of the hot component, and [O III] and [Ne III] preferentially emitted in a much larger region. The behaviour of the velocity displacements of the eclipsed "absorption" lines visible in outburst suggests that these lines are formed in an optically thick accretion disk (Kenyon et. al. 1982). An analysis of the line fluxes from H I, He I and He II emission lines observed in quiescence implies the disk in CI Cyg has a luminosity of 300-1000 $L_\odot$ (e.g., Kenyon 1983).

Lambert et al. (1970) and Stencel et al. (1979) first suggested the UV continuum of CI Cyg arises from a disk surrounding a compact star, and speculated the central star might be a main sequence star or a white dwarf (since CI Cyg is not an X-ray source, it is unlikely the central accreting object can be a neutron star). This conclusion was supported by the high nebular density ($n_e \geqslant 10^9$ $cm^{-3}$) as well as the overall shape of the continuous flux distribution. Kenyon (1983 and references therein) and Kenyon & Webbink (1983) extensively analysed the UV flux distribution, and concluded the hot component in CI Cyg is a main sequence star accreting matter at $3 \times 10^{-5} M_\odot yr^{-1}$ in quiescence. Bath & Pringle (1982) independently modelled the behaviour of the visual light curve of CI Cyg in and out of outburst, and reached a similar conclusion. The total accretion luminosity of this model is in good agreement with that implied by studies of the optical emission lines quoted above.

The red giant component in CI Cyg (and symbiotic stars in general) has not been as extensively observed as the hot component. Broad-band IR photometry (e.g., Swings & Allen 1972; Kenyon & Gallagher 1983 and references therein) reveals the giant to be a typical M-type star, with no evidence for large amplitude variability or a circumstellar dust shell. Kenyon et al. (1982) noted the eclipse duration requires the giant to fill its Roche lobe in outburst **and in quiescence.** Kenyon & Gallagher (1983) obtained low resolution spectroscopy of the 2.3 $\mu$m CO absorption band, and found it to resemble absorption bands observed in some M-type supergiants and bright giants. The observations reported by Kenyon & Gallagher require the giant component in CI Cyg to be an evolved AGB-type star, and to fill its Roche lobe (this latter requirement is essential, given current theory, if the high accretion rate deduced by Kenyon & Webbink is to be achieved by the giant).

2)    V1016 Cygni:    A Mira Variable with a Hot White Dwarf
      Companion

V1016 Cygni was discovered by Merrill & Burwell (1950) as
a very strong Hα source, and by Nassau & Cameron (1954) as a
long-period variable.    Thus V1016 Cyg appears to have been a
symbiotic star **before** its  1964 outburst,  when V rose from 14
to  10.  The optical spectrum of V1016 Cyg (Figure 6) after
out-burst  differs  considerably  from  many  symbiotic  stars.
Strong M-type  absorption features are not usually visible, but
have been detected  at  (perhaps) regular intervals (Taranova &
Yudin 1983; Andrillat, Ciatti & Swings 1982).

Recent observations of V1016 Cyg have been obtained at all
wavelength regimes.  The system is  a  fairly weak X-ray source
($F[0.2keV] = 7.5 \pm 3 \times 10^{-14}$erg  cm$^{-2}$ s$^{-1}$ ; Allen 1981), and the
UV  continuum is  also fairly weak (Nussbaumer & Schild 1981).
The ratio $L_x/L_{UV}$ implies a hot component temperature of roughly
150,000 K attenuated by  a  foreground reddening, E(B-V) = 0.2
(Kenyon & Webbink 1983).  An analysis of  the UV emission lines
by Nussbaumer & Schild (1981) yields similar values  for  the
temperature  (160-200,000  K)  and  reddening  (0.28).  For
reasonable distance estimates,  the radius of the hot component
is $\simeq 0.2R_{\odot}$ (Nussbaumer & Schild 1981), and this is more or less
what is  expected if  the  hot  star  in  the  midst  of  a
thermonuclear runaway (Iben 1982; Kenyon & Truran 1983).

V1016 Cyg  is  a  strong  radio source (Seaquist & Gregory
1973) with  a  spectral  index  close  to  that  expected  for
free-free  emission  in  a constant velocity,  outflowing  wind
($F_{\nu} \propto \nu^{0.6}$; Wright & Barlow 1975).  The  nebula  is  clearly
bipolar at high  resolution (Hjellming & Bignell 1982; see Solf
1983 for an excellent optical demonstration of bipolarity), and
Kwok (1982a) has  interpreted  these  data  in  terms  of  an
expanding wind from the  hot  component.  Ahern et al. (1977)
had  previously estimated an outflow rate of 1.5  x  $10^{-6}$ (d/1
kpc)$^2$ M$_{\odot}$ yr$^{-1}$ ; Solf (1983) estimates a total nebular mass of 2
x 10$^{-4}$ M$_{\odot}$ based  on  high resolution optical spectra.  When
combined  with  the  outflow  rate  derived by Ahern et al.
(1977), this nebular mass suggests the  material ejected by the
hot  component  has  swept up material in  the  wind  from  the
Mira-like component.

This  symbiotic  is an extremely dusty system,  as  strong
excesses are  visible  in  the near infrared (Swings & Allen 1972)
and at 10-20 μm (Knacke 1972; Harvey 1974).  Although the dust
emission is quite substantial, M-type  absorption features have
been identified in the optical (TiO, VO; Andrillat, Ciatti &
Swings  1982; Taranova & Yudin 1983 and references therein) and
in  the  infrared  (CO,  $H_2O$;  Puetter  et  al.  1978).  Harvey

(1974) found the star to be periodically variable in the
near-IR, and this was confirmed by Yudin (1982) and Taranova &
Yudin (1983 and references therein). Kenyon & Webbink (1983)
combined these data to estimate a period of 472 days. The IR
variability and the detection of photospheric CO and $H_2 O$
absorption bands imply the late-type component in V1016 Cyg is
a Mira variable.

3) V443 Herculis: A Red Giant with a Hot White Dwarf
    Companion

    This high velocity symbiotic star was discovered by Tifft
& Greenstein (1958): its proper motion and large negative
radial velocity imply the giant in this system is a Pop II
star. Belyakina (1974) established V443 Her as a variable in
V, B-V and U-B, but did not report periodic variability. Aside
from these studies, V443 Her has been overlooked by many
observers. This is somewhat surprising, since it is one of the
brightest systems in the optical and in the infrared
(V = 11.5, K = 5.4). In spite of this lack of attention, V443
Her is a good example of a non-dusty symbiotic with a hot white
dwarf component.

    The untraviolet spctrum of V443 Her (Figure 7) was first
investigated by Kenyon (1983) and Kenyon & Webbink (1983). The
continuum is produced by a hot star at a temperature of roughly
70,000 K, with a bolometric luminosity comparable to that of
the cool giant companion. This temperature is in fair
aggreement with that estimated from the intensities of the H I,
He I and He II emission lines (T = 100,000 K; Kenyon 1983).
Oliversen & Anderson (1982) briefly discuss the optical
spectrum of V443 Her: they detect emission from HI, He II and
[O III]. The Hα profile of this system is fairly symmetrical
with an occasional "shoulder" on the blue edge of the peak.

    The IR continuum of V443 Her shows no evidence for
circumstellar dust emission as in V1016 Cyg (Swings & Allen
1972), and Taranova & Yudin (1982) found no evidence for
Mira-like variability. The 2.3 μm CO absorption band is fairly
strong, and similar in strength to that of normal field giants
(Kenyon & Gallagher 1983). The IR observations are of
insufficient resolution to determine if the giant in V443 Her
is indeed a Pop II star. If the giant is metal poor, then the
CO absorption band is probably much stronger than expected for
a normal, metal poor red giant.


4.  DIRECTIONS FOR FUTURE RESEARCH

    In the previous sections, I have tried to outline possible

binary models for symbiotic stars, and have identified those models which, at present, appear to explain the available observational data best. The three type examples, CI Cyg, V1016 Cyg and V443 Her, will, I hope, serve as guideposts in planning new observations of symbiotics and in developing theoretical models. It should be emphasized that direct evidence for binarity exists for only a handful of systems (10-15); indirect evidence (e.g., fitting the continuous energy distribution) suggests that nearly all well-studied symbiotics are binaries. Some systems may indeed be single stars, but, as yet, single star models do not explain quantitatively the observational data of any well-studied symbiotic system.

It should be obvious that modern X-ray, ultraviolet and infrared detectors have revolutionized the field of interacting binaries in general and symbiotic stars in particular. These techniques will continue to play a major role in observational astronomy, and I would like to conclude this review with a few comments concerning new observations that appear to be needed to understand the symbiotics and perhaps other interacting systems.

1.  X-ray Observations

Only a handful of symbiotic stars have been detected at X-ray wavelengths, and all but one have been very soft sources. When combined with IUE data, X-ray data provide useful constraints on the temperatures ; luminosities of symbiotic hot components. Future X-ray observations need to be complemented with EUV observations to probe (i) the boundary layer in accreting systems and (ii) the area of the Planck peak in symbiotics with hot stellar sources.

2.  Optical and Ultraviolet Observations of the Nebula

The optical and ultraviolet spectrum of a typical symbiotic star is packed with emission lines of many ionized species. Simultaneous optical and ultraviolet observations are essential to (i) determine the physical conditions throughout the nebula, (ii) determine the nature of the hot component and (iii) correlate variations in the UV continuum (radiated by the hot component) with variations in the emission lines.

3.  Infrared Observations

The cool components of symbiotic stars are best viewed in the infrared, far from the contaminating light of the hot component. Additional observations of CN and CO absorption bands are needed to measure the luminosities of the cool stars, and to determine to what extent they fill their tidal surfaces.

High resolution observations can, in principle, determine
elemental abundances (e.g., $^{12}C/^{13}C$) in the atmospheres of the
late-type stars in these systems. These would be quite helpful
in ascertaining the nature of the late-type giant.

## 4. Radio Observations

Aside from a few surveys, very little quantitative radio
work has been accomplished for symbiotic stars. The new
generation of radio interferometers should allow more detailed
mapping of individual sources. As pointed out by A. Wright, a
few systems might actually be resolved in the near future.
This would allow direct tests of models for systems such as
V1016 Cyg and HM Sge.

## ACKNOWLEDGEMENTS

The author would like to thank organizers of this meeting
for providing a stimulating environment in which to "interact",
and also J.S.Gallagher, A.V. Tutukov, R.F. Webbink, M.
Friedjung and A.E. Wright for their comments and criticisms of
my presentation. This work was supported by the NSF through
grants AST 80-18198 and AST 80-18859 to the University of
Illinois.

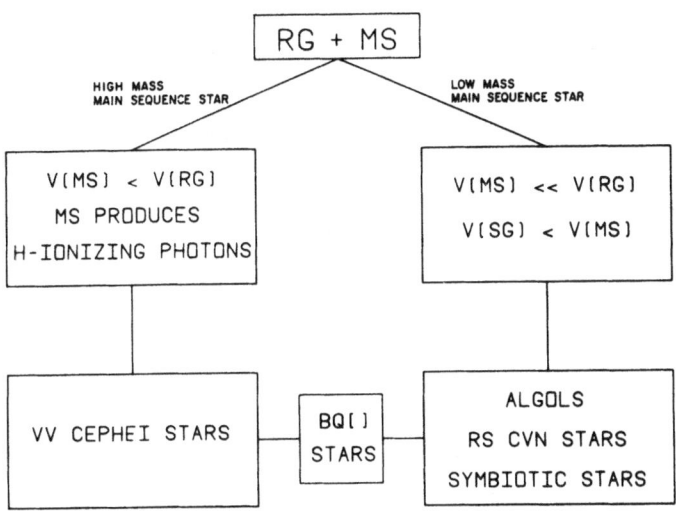

Figure 1. Flow chart for binary stars consisting of a red (supergiant (RG) and a main sequence star (MS). At high masses, the
main sequence star can ionize the red giant wind: this is observed
as a·'/VCephei star. Many types of binaries, including Algols, RS
CVn's and symbiotic stars, result if the main sequence star is of
low mass. In these latter cases, the optical luminosity of the
main sequence star (V) may be smaller or larger than that of its
cool companion.

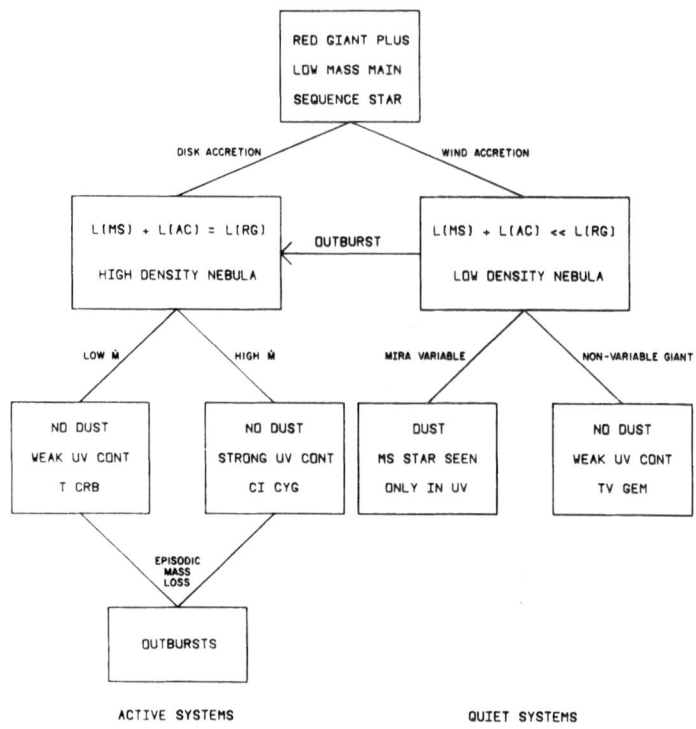

Figure 2. Flow chart for binary stars consisting of a red (super
giant) and a low mass main sequence star. If the giant loses mass
via a wind (quiet systems) only a small fraction can be accreted
by the binary companion. The companion might then be visible as
a weak UV source. If matter is lost via tidal interaction (active
systems), most of the material can be accreted by the companion,
and it should be observed as a strong UV source (with a relatively
large bolometric luminosity, L). In both modes of accretion, outbursts are a result of episodes of high mass loss from the giant.

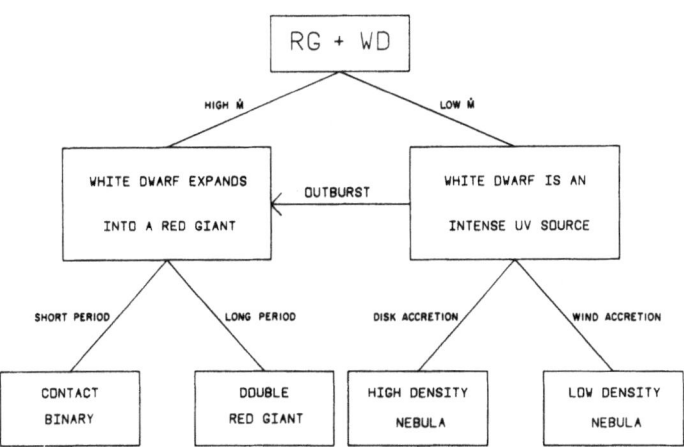

Figure 3.  Flow chart for binary stars consisting of a red (super) giant and a white dwarf (WD). At high mass transfer rates, the white dwarf expands to become a red giant. At lower transfer rates, the white dwarf is an intense UV source, and may be capable of ionizing a surrounding gaseous nebula. This would be observed as a symbiotic star/planetary nebulae.

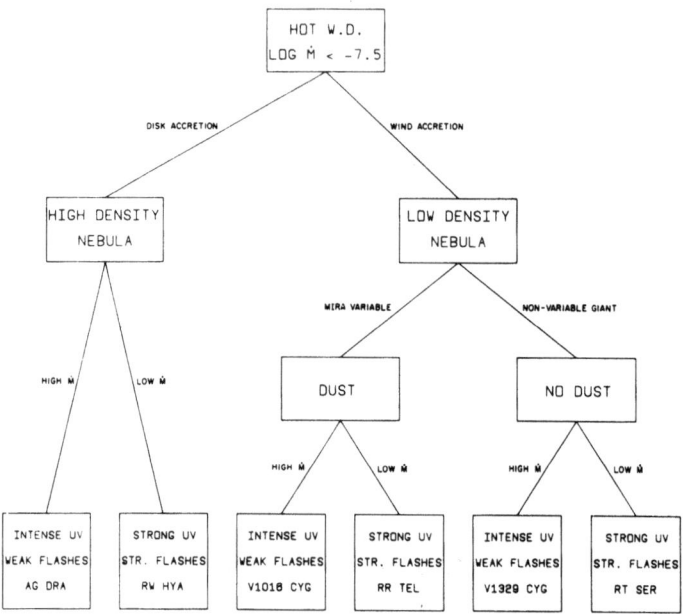

Figure 4. Flow chart for binary stars of a red (super)giant and a white dwarf accreting material at a relatively low rate. Wind-driven mass loss leads to the formation of a low-density nebula, and dust may form if the mass-losing star is a Mira variable. Tidal mass loss leads to the formation of an accretion disk which is surrounded by a high density corona. If the white dwarf is intrinsically luminous, hydrogen shell flashes in the accreted matter will be weak; strong flashes occur if the white dwarf is intrinsically faint.

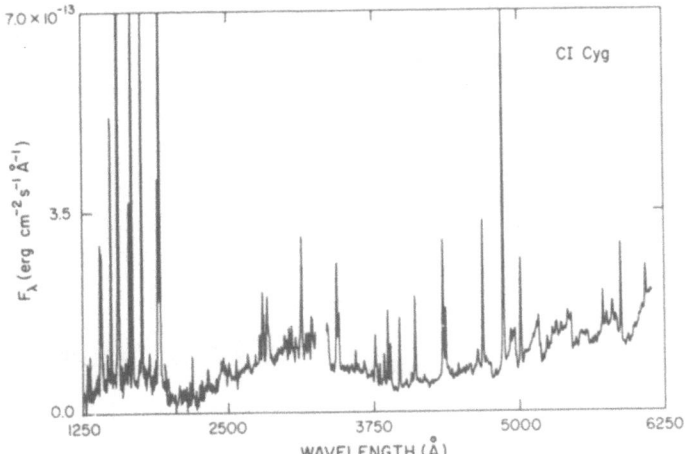

Figure 5. Combined optical and ultraviolet spectrum for CI Cygni.
This spectrum shows the features typically observed in symbiotics:
TiO and Ca I absorption features plus strong emission lines from
H I, He II and C IV. The flat UV continuum is characteristic of
an accreting main sequence star.

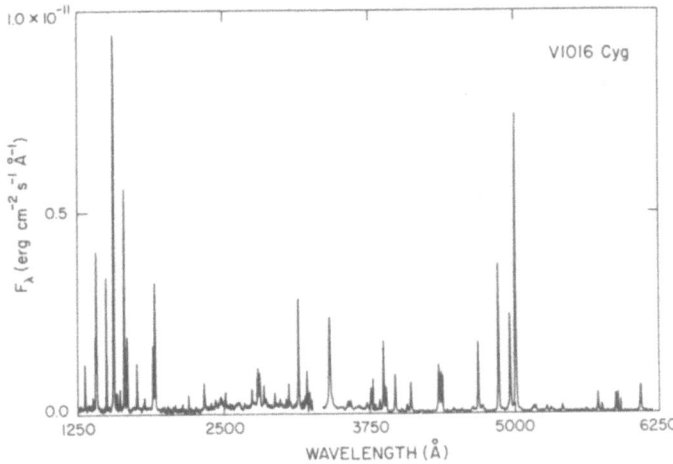

Figure 6. Combined optical and ultraviolet spectrum of V1016 Cyg.
This spectrum shows the weak continuum and very strong emission
lines that are usually observed in D-type symbiotic stars. The
UV-optical continuum is primarily nebular, with some contribution
from a very hot star (T = 160-200,000 K) at short wavelengths.

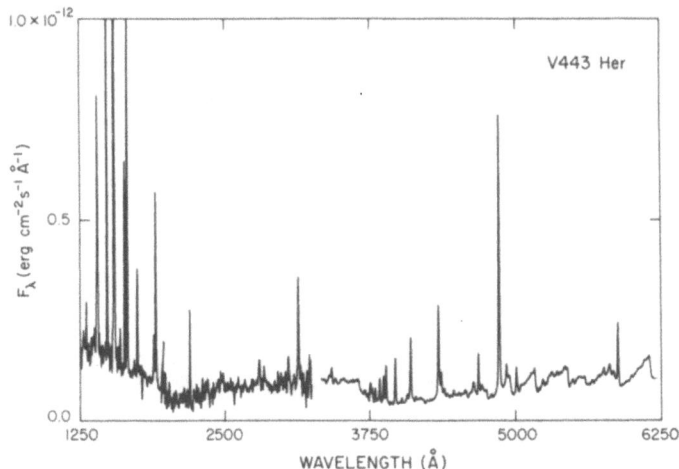

Figure 7. Combined optical and ultraviolet spectrum of V443 Her. The late-type absorption features and strong emission lines characteristic of symbiotics are visible at long wavelengths; the Rayleigh-Jeans tail of a hot blackbody (70,000 K) is visible in the ultraviolet.

## REFERENCES

Ahern, F.J., FitzGerald, M.P., Marsh, K.A. & Purton, C.R.,
    1977. Astr.Ap., **58**, 35.
Allen, D.A., 1979. In "Changing Trends in Variable Star
    Research", IAU Colloquium No. 46, eds, F. Bateson, J. Smak
    & I. Urch (Hamilton, N.Z.: U. Waikato Press), p. 125.
Allen, D.A., 1980. Mon.Not.Roy.Astr.Soc., **192**, 521.
Allen, d.A., 1981. Mon.Not.Roy.Astr.Soc., **197**, 739 [Erratum:
    Mon.Not.Roy.Astr.Soc., **201**, 1199].
Allen, D.A., 1982. In "The Nature of Symbiotic Stars", IAU
    Colloquium No. 70, eds. M. Friedjung & R. Viotti
    (Dordrecht: Reidel), p. 27.
Allen, D.A., 1983. Mon.Not.Roy.Astr.Soc., **204**, 113.
Allen, D.A., 1984. Proceedings of IAU Colloquium No. 80,
    Ap.Space Sci., **99**.
Aller, L.H., 1954. Pub.DAO Victoria, **9**, 321.
Andrews, P.J., 1974. Mon.Not.Roy.Astr.Soc., **167**, 635.
Andrillat, Y. 1982. In "The Nature of Symbiotic Stars",
    IAU Colloquium No. 70, eds, M. Friedjung & R. Viotti
    (Dordrecht: Reidel), p. 47.
Andrillat, Y., Ciatti, & Swings, J.P., 1982. Ap.Space Sci.,
    **83**, 423.
Baldwin, J.R., Frogel, J.A. & Persson, S.E., 1973. Ap.J.,
    **184**, 427.
Bath, G.T., 1977. Mon.Not.Roy.Astr.Soc., **178**, 203.
Bath, G.T. & Pringle, J.E., 1982. Mon.Not.Roy.Astr.Soc.,
    **201**, 345.
Belyakina, T.S., 1970. Astrofizika, **6**, 49.
Belyakina, T.S., 1974. Izv.Krym.Astrofiz.Obs., **50**, 103.
Belyakina, T.S., 1979. Izv.Krym.Astrofiz.Obs., **59**, 133.
Boyarchuk, A.A., 1969. In "Non-Periodic Phenomena in Variable
    Stars", ed. L. Detre (Budapest: Academic Press), p. 395.
Boyarchuk, A.A., 1975. In "Variable Stars and Stellar
    Evolution", IAU Symposium No. 67, eds. V.E. Sherwood &
    L. Plaut (Dordrecht: Reidel), p. 377.
Ciatti, F. & Mammano, A., 1976. Mem.Soc.Roy.Sci. Liege, 6
    serie, tome IX, 379.
Cowley, A.P., 1969. Pub.Astr.Soc.Pac., **81**, 297.
Davidsen, A., Malinna, R. & Bowyer, S., 1977. Ap.J., **211**, 866.
Doty, J.A., Hoffman, J.A. & Lewin, W.H.G., 1981. Ap.J.,
    **243**, 257.
Friedjung, M. & Viotti, R., 1982. "The Nature of Symbiotic
    Stars", IAU Colloquium No. 70 (Dordrecht: Reidel).
Galeev, A.A., Rosner, R. & Vaiana, G.S., 1979. Ap.J., **229**, 318.
Gallagher, J.S., Holm, A.v., Anderson, C.M. & Webbink, R.F.,
    1979. Ap.J., **229**, 994.
Harvey, P.M., 1974. Ap.J., **188**, 95.
Hjellming, R.M. & Bignell, R.C., 1982. Science, **216**, 1279.
Hoffleit, D., 1968. Irish Astr.J., **8**, 149.

Hutchings, J.B., Cowley, A.P. & Redman, R.O., 1975. Ap.J., **201**, 404.

Iben, I. Jr., 1982. Ap.J., **259**, 244.

Iijima, I., 1981. In "Photometric and Spectroscopic Binary Systems", NATO Adv. Study Inst., eds. E.B. Carling and Z. Kopal, (Drodrecht: Reidel), p. 517.

Kenyon, S.J., 1982. Pub.Astr.Soc.Pac., **94**, 165.

Kenyon, S.J., 1983. Ph.D. thesis, U. Illinois.

Kenyon, S.J. & Gallagher, J.S., 1983. Astr.J., **88**, 666.

Kenyon, S.J. & Truran, J.W., 1983. Ap.J., **273**, No. 1.

Kenyon, S.J. & Webbink, R.F., 1983. Ap.J., **279**, No. 1.

Kenyon, S.J., Webbink, R.F., Gallagher, J.S. & Truran, J.W., 1982. Astr.Ap., **106**, 109.

Knacke, R.F., 1972. Ap.Lett., **11**, 201.

Kwok, S., 1982. In "The Nature of Symbiotic Stars", IAU Colloquium No. 70, eds. M. Friedjung & R. Viotti (Dordrecht: Reidel), p. 17.

Lambert, D.L., Slovak, M.H., Shields, G.A. & Ferland, G.J., 1979. In "The Universe at Ultraviolet Wavelengths", ed. R.D. Chapman (NASA: CP-2171), p. 461.

MacDonald, J., 1980. Ph.D. thesis, Cambridge University.

Meinunger, L., 1979. Inf.Bull.Var.Stars, No. 1611.

Meinunger, L., 1981. Inf.Bull.Var.Stars, No. 2016.

Merrill, K.M. & Ridgway, S.T., 1979. Annu.Rev.Astr.Ap., **17**, 9.

Merrill, P.W., 1950. Ap.J., **111**, 484.

Merrill, P.W. & Burwell, C.G., 1950. Ap.J., **112**, 72.

Merrill, P.W. & Humason, M.L., 1932. Pub.Astr.Soc. **44**, 56.

Michalitsianos, A.G., Hobbs, R.W. & Kafatos, M., 1979. In "The Universe at Ultraviolet Wavelengths", ed. R.D. Chapman (NASA: CP-2171), p. 367.

Mikolajewska, J. & Mikolajewski, M., 1982. In "The Nature of Symbiotic Stars", IAU Colloquium No. 70, eds. M. Friedjung & R. Viotti (Dordrecht: Reidel), p. 147.

Nassau, J.J. & Cameron, D.M., 1954. Ap.J., **119**, 75.

Nussbaumer, H., 1982. In "The Nature of Symbiotic Stars", IAU Colloquium No. 70. eds. M. Friedjung & R. Viotti (Dordrecht: Reidel) p. 85.

Nussbaumer, H. & Schild, H., 1981. Astr.Ap., **101**, 118.

Oliversen, N.G. & Anderson, C.M., 1982. In "The Nature of Symbiotic Stars", IAU Colloquium No.70, eds. M. Friedjung & R. Viotti (Dordrecht: Reidel), p. 71.

Oliversen, H.G. & Anderson, C.M., 1983. Ap.J., **268**, 250.

Paczynski, B. & Rudak, B., 1980. Astr.Ap., **82**, 349.

Paczynski, B. & Sienkiewicz, R., 1972. Acta Astr., **22**, 73.

Paczynski, b. & Zytkow, A.N., 1978. Ap.J., **222**, 604.

Persic, M., 1983. M.S. thesis, Trieste Univ.

Plavec, M., Ulrich, R.K. & Polidan, R.s., 1973. Pub.Astr.Soc. Pac., **85**, 769.

Puetter, R.C., Russell, R.W., Soifer, B.T. & Willner, S.P., 1978. Ap.J. (Letters), **233**, L93.

Reimers, D., 1975. In "Problems in Stellar Atmospheres and Envelopes", eds. B. Baschek, W.K. Hegel & G. Traving (Berlin-Heidelberg: springer), p. 229.

Sahade, J. & Wood, F.B., 1978. "Interacting Binary Stars", (New York: Pergamon).

Sequist, E.R. & gregory, P.C., 1973. Nature Phys. Sci., **245**, 85.

Sharpless, S., 1956. Ap.J., **124**, 342.

Slovak, M.H., 1982. Ph.D. thesis, U. Texas.

Slovak, M.H. & Lambert, d.L., 1982. In "The Nature of Symbiotic Stars", IAU Colloquium No. 70, eds. M. Friedjung & R. Viotti (Dordrecht: Reidel), p. 103.

Solf, J., 1983. Ap.J. (Letters), **266**, L113.

Stencel, R.E., Michalitsianos, A.M., Kafatos, M. & Boyarchuk, A.A., 1979. In "The Universe at Ultraviolet Wavelengths", ed. R.D. Chapman (NASA: CP-2171), p. 459.

Stencel, R.E., Michalitsianos, A.M., Kafatos, M. & Boyarchuk, A.A., 1982. Ap.J. (Letters), **253**, L77.

Swings, J.P. & Allen, D.A., 1972. Pub.Astr.Soc.Pac., **84**, 523.

Swings, P., 1970. In "Spectroscopic Astrophysics", ed. G. Herbig (Berkeley: U. Calif.) p. 189.

Swings, P. & Struve, O., 1940. Ap.J., **92**, 295.

Taranova, O.G. & Yudin, B.F., 1982. Soviet Astr.J., **25**, 710.

Taranova, O.G. & Yudin, B.F., 1983. Astr.Ap., **117**, 209.

Thackeray, A.D. & Hutchings, J.B., 1974. Mon.Not.Roy.Astr.Soc., **167**, 319.

Thackeray, A.D. & Webster, B.L., 1974. Mon.Not.Roy.Astr.Soc., **168**, 101.

Thompson, G.T., Nandy, K., Jamar, C., Monfils, A., Houziaux, L., Carnochan, D.J. & Wilson, R., 1978. "Catalogue of Stellar UV Fluxes", (Edinburgh: Science Research Council).

Thronson, H.A. & Harvey, P.M., Ap.J., **248**, 584.

Tifft, W.G. & Greenstein, J.L., 1958b. Ap.J., **127**, 160.

Tutukov, A.V. & Yungel'son, L.R., 1982. In "The Nature of Symbiotic Stars", IAU Colloquium No. 70, eds. M. Friedjung and R. Viotti (Dordrecht: Reidel), p. 283.

Wackerling, L.R., 1970. Mem.Roy.Astr.Soc., **73**, 153.

Wallerstein, G., 1981. Obs., **101**, 172.

Warner, B., 1972. Mon.Not.Roy.Astr.Soc., **159**, 95.

Webster, B.L. & Allen, D.A., 1975. Mon.Not.Roy.Astr.Soc., **171**, 171.

Webbink, R.F., 1976. Nature, **262**, 271.

Webbink, R.F., 1979a. In "Changing Trends in Variable Star Research", IAU Colloquium No. 46, eds. F.M. Bateson, J. Smak & I.H. Urch (Hamilton, N.Z.: U. Waikato), p. 102.

Webbink, R.F., 1979b. In "Dwarfs and Variable Degenerate Stars", IAU Colloquium No. 53, eds. H.M. Van Horn & M.P. Savedoff (Rochester: U. Rochester), p. 426.

Wright A.E. & Barlow, M.J., 1975. Mon.Not.Roy.Astr.Soc., **170**, 41.

Zuckerman, B., 1980. Annu.Rev.Astr.Ap., **18**, 263.

# OPTICAL CHARACTERISTICS OF X-RAY BINARIES

Sergio A. Ilovaisky

Observatoire de Besancon
41bis, Av. de l'Observatoire
25044 Besançon Cedex
France

## 1. INTRODUCTION

The number of optical identifications of X-ray sources presumed to be compact objects in binary systems has been steadily increasing during the last few years and has now reached about 100 (Bradt & McClintock 1983), mainly as a result of refined X-ray positions obtained with the Ariel 5, SAS-3 and HEAO-1 satellites. In these sources there are good reasons to believe that the X-ray emission results from accretion onto a collapsed object, matter being transferred from a normal stellar companion. The increased number of such identifications allows a classification into three observational optical categories: massive systems comprising an early-type supergiant or main sequence star and a neutron star (or a 'black hole'), low-mass systems where the companion is a low-mass, low-luminosity dwarf, and cataclysmic systems where the companion is also a low-mass, low luminosity star but the accreting object is a white dwarf. I will be concerned here only with the first two.

Several reviews on the optical properties of binary X-ray sources have appered in the recent past (Hutchings 1982, Ilovaisky 1982, 1983, Charles 1982, van Paradijs 1983). For reviews dealing more with X-ray properties, see Rappaport & Joss (1983), Rappaport & van den Heuvel (1983) and McClintock & Rappaport (1983). Cordova & Mason (1983) have reviewed the properties of cataclysmic variables.

In view of the recent extensive review of optical observations of compact X-ray sources by van Paradijs (1983),

*P. P. Eggleton and J. E. Pringle (eds.), Interacting Binaries, 205–248.*
© *1985 by D. Reidel Publishing Company.*

in this article I will only briefly summarize the outstanding
optical properties of X-ray binaries while dealing in
considerably more detail with recent progress made on several
remarkable sources.

## 2.   OPTICAL CHARACTERISTICS OF MASSIVE X-RAY BINARIES (MXRB)

### 2.0  General Properties

In these systems the companion star is a massive, early
type star transferring mass to the compact object by stellar
wind or incipient Roche-lobe overflow. The X-ray emission is
very often pulsed implying that accretion takes place onto a
rotating, magnetized neutron star (oblique rotator). Periodic
X-ray eclipses and Doppler shift analysis of X-ray pulsations
together with optical radial velocity curves have yielded basic
data on the systems such as the masses of neutron stars, for
example (Rappaport & Joss 1983, Joss & Rappaport 1984). X-ray
spectra are characteristically hard with a high-energy
cut-off (White et al. 1983) while the X-ray to optical
luminosity ratio $\overline{L_x / L_{opt}}$ is generally lower than unity due to
to the large optical luminosity of the companion. These stars
can be fitted into an already-existing astrophysical context.
Most of the MXRB fall in either of two classes:

Class I. The primaries have early spectral types and
appear to have evolved from the main sequence (luminosity
classes I, II and III). They may also be slightly undermassive
for their luminosities and temperatures (Hutchings 1982), and
fill or are close to filling their Roche lobes. The orbital
periods are short, less than ten days.

Class II. Here the primaries are Be stars, with
luminosities close to values corresponding to the main sequence
and which underfill their Roche lobe. The orbits are often
elliptical and the orbital periods long, from tens to hundreds
of days, and difficult to determine. Several of these sources
exhibit X-ray transient behaviour (the so-called "hard"
transients) probably due to episodic (equatorial?) mass loss
events.

In what follows I will discuss only the characteristics of
Class I MXRB. Table 1 presents an up-to-date (March 1984) list
of massive binaries with optical identification. See also the
chapter on massive binaries by White (1984) in this volume,
dealing in more detail with Be X-ray binaries.

## 2.1    Optical Brightness Variations

The total light from the massive systems, dominated by the companion, is found to be modulated by at most 5-10% in the course of an orbital period. This modulation, which exhibits two maxima and two minima per period (see Figure 1), has been usually interpreted in terms of an ellipsoidally-shaped primary presenting to the observer different parts of its surface at different orbital phases: centrifugal and tidal effects causing the average optical flux from the primary to depend upon the direction of observation. Normally the two minima (phases 0.0 with the X-ray source behind the primary and 0.5 with the X-ray source in front) are expected to be of different depths, $A_{0.0}$ and $A_{0.5}$, for inclination angles close to 90 deg., since gravity darkening should be more pronounced in the direction of the Lagrangian point $L_1$, thus making the minimum at 0.5 deeper (Fig. 1a). In principle, a measurement of the amplitude of this modulation, when combined with a priori knowledge of the orbital plane inclination, can yield an estimate of the mass ratio of the system assuming that we know the effective temperature of the primary and have an idea of the fraction of the Roche lobe that is actually filled by the primary.

However, several factors affect the light curve and make such estimates highly unreliable. First, in some systems, X-ray heating of the hemisphere of the primary facing the X-ray source can alter the depth of the minimum at phase 0.5 (see Van Paradijs & Zuiderwijk 1977 and Fig. 1b). In some systems, some optical light can be contributed by the accretion disk surrounding the compact object, in fact mimicking the basic ellipsoidal modulation (extra light at phases 0.25 and 0.75 and minima at 0.0 and 0.5 - the latter when the disk passes in front of the primary) and can thus effectively increase the amplitude of the light curve (Chevalier et al. 1981, Ilovaisky et al. 1984). See the section on LMC X-4.

## 2.2  Spectroscopic Variations

The optical spectrum of massive binaries is dominated by the normal stellar companion. The absorption lines from the primary exhibit the radial velocity variations expected from orbital motion.

Some emission lines are often present (mostly He II $\lambda$ 4686 and C III-N III $\lambda\lambda$ 4630-4650) and in several cases (Hutchings et al. 1977, Crampton et al. 1978, Hutchings et al. 1978 and Mouchet et al. 1980) they move in antiphase with the absorption lines (Fig. 2).

Table A

Comparison between He II $\lambda$ 4686 emission-line and
X-ray source velocity amplitudes

| Source | $K_{4686}$ | $K_X$ |
|--------|------------|-------|
| SMC X-1 | 245 ± 16 km/s | 301 ± 2 km/s |
| 1538-52 | 302 ± 2 | 323 ± 2 |
| LMC X-4 | 475 ± 25 | 465 ± 77 |
| Cen X-3 | 330 ≤ K ≤ 415 | 415 ± 0.4 |

These emission lines arise close to the compact  star  and
their  amplitudes  are  generally smaller than the X-ray source
orbital  velocities.  (See Table A.) Most likely  these  lines
come from a  region  close  to  the  accretion disk.  In several
cases  the  emission  line  velocities  trail  the  absorption
velocity  curve (phase shift of up to 0.05) implying the region
may be  in  fact close to the "impact" point of the stream from
the primary onto the accretion disk.

The  ultraviolet  spectra  of  massive  systems,  such  as
LMC  X-4  and  SMC X-1 (Bonnet-Bidaud et al. 1981; van der
Klis 1982), Vela X-1 (Dupree et al. 1980) and Cygnus  X-1
(Treves  et  al. 1980) are  consistent  with  the  expected
early-type model atmosphere fluxes showing reasonable reddening
(Fig. 3a)  and  exhibit  (at  least  where  sufficient data is
available)  the  characteristic  double-wave  curves  seen  at
visible  wavelengths  (Fig. 3b).  The resonance absorption-line
doublets of NV,  CIV and Si IV show marked changes with orbital
phase, being particularly weak  when  the  X-ray  source  is in
front  of  the  primary (Fig. 3c).  These  changes  can  be
understood  in  terms of an anisotropic ionization structure (a
'hot bubble')  in  the  expanding  atmosphere  of  the  primary,
described  by  Hatchett  &  McCray  (1977), caused by the X-ray
emission from the compact object.

2.3  A Special Massive Binary:  LMC X-4

The  binary  X-ray  source  LMC  X-4,  the  second  to  be
discovered in an external galaxy, exhibits  a  large  degree of
variability.  Described  as  possibly variable on the basis of
early UHURU data, it  was  not  detected  with  OSO-7  nor  the
Copernicus,  but was seen again with the Ariel 5 SSI (Griffiths
& Seward 1977) and SAS-3 RMC experiments (Epstein et al. 1977).

Photometric observations by Chevalier & Ilovaisky (1977) of an early-type star in the Large Magellanic Cloud found by Sanduleak & Philip led to the discovery of a 1.4 day periodic modulation. Soon thereafter, X-ray eclipses with that period were found in the SAS-3 and Ariel 5 data (White 1978, Li et al. 1978). The first HEAO-1 observations (Skinner et al. 1980) showed large X-ray variability on a time scale of a month, explaining the early non-detections. Analysis of the complete HEAO-1 500-d data base (Lang et al. 1981) demonstrated the X-ray flux to be modulated with a 30-day period.

Analysis of the extensive set of optical photometry of the massive X-ray binary LMC X-4 obtained at ESO from 1976 through 1983 shows the presence of the X-ray 30-day cycle discovered by Lang et al. During X-ray OFF states, the amplitude of the 1.408-day orbital period double-wave light curve is 0.07 mag, while during ON states it increases to 0.20 mag. (Fig. 4a). The periodogram of the optical data exhibits peaks at the sums of the orbital and half-orbital frequencies with the 30-day frequency itself (Fig. 4b).

In order to sort out the changes which take place in the light curve as a function of the long period, the B data have been plotted in Fig. 6 into ten 30-day period bins (Fig. 4c). The low-amplitude curves are restricted to the 30-d phase interval 0.9-0.1, corresponding to the X-ray OFF state, while the large-amplitude curves are found in the interval 0.3-0.6, during the ON state. The mid-orbital phase minimum deepens in the 30-d phase interval 0.3-0.5, close to the time of maximum X-ray flux.

The observed effects are explained (Ilovaisky et al. 1984) in terms of an X-ray illuminated, tilted, counter-"precessing" accretion disk, as in Her X-1. The excess light comes both from the accretion disk itself and from the illuminated hemisphere of the primary, the latter experiencing variable X-ray shadowing due to the changing aspect of the disk. Each source contributes approximately half of the observed excess. The disk, taken to be 20 deg. thick, appears to be 'tilted' by 20 deg. with respect to the orbital plane.

Although tilted "precessing" accretion disks are required by the observational data available for both LMC X-4 and Her X-1, their true nature has been the subject of considerble discussion. (Katz 1973, Roberts 1974, Petterson 1975, 1977, Merrit & Petterson 1980, Papaloizou & Pringle 1983). The clock responsible for the apparent precession in Her X-1 has been attributed to episodic mass transfer (Boynton

et al. 1980). In this interpretation the disk precession is
not "steady" but "apparent" due to the changing disk rim
structure          imposed by quasi-periodic mass transfers when
the inner Lagrangian point $L_1$ is in the disk shadow (Crosa
et al. 1980).

Other similar long-period X-ray modulations, which could
be also due to such "precessing" disks, have been reported in
the massive binaries Cyg X-1 (Priedhorsky et al. 1982),
where the modulation also appears in the optical light curve
(Kemp et al. 1983), and 4U1907+09 (Priedhorsky & Terrell
1983). The basic scenario appears to be the same as in
Hercules X-1 and raises the same unanswered questions: How can
a "tilted" disk be formed? What causes the "apparent"
precession?

## 2.4 Massive unseen stars in LMC X-3 and LMC X-1

In recent months much excitement has been generated
concerning two X-ray binaries in the Large Magellanic Cloud
which exhibit soft X-ray spectra and large-amplitude
absorption-line radial velocity amplitudes, both features
thought to indicte the presence of massive accreting objects.
In what follows I review the state of affairs.

LMC X-3. This is a very strong and highly variable X-ray
source, possibly the brightest and most variable (by a factor
of more than 30) of the LMC sources (Griffiths & Seward 1977).
It is characterized by a soft X-ray spectrum very similar to
the 'high' state spectra of Cygnus X-1 and GX 339-4 (see White
and Marshall 1984). Its luminosity is $4 \times 10^{38}$ ergs/s in
X-rays (at maximum and $L_x/L_{opt}$ is 140. X-ray observations
with the MPC of Einstein (Weisskopf et al. 1983) show no
evidence for the rapid variability typical of Cyg X-1 and GX
339-4 when in their 'low' (or hard) state.

The optical star considered as a likely identification
(Warren & Penfold 1975) is faint (average magnitude B = 17),
blue (U-B = -0.7) variable (by 0.5 mag), and exhibits an early
B type with variable absorption-line velocities. An Einstein
HRI position has confirmed the identification with the
Warren-Penfold star (Cowley et al. 1983).

According to Cowley et al., optical spectroscopy shows
large amplitude radial velocity variations with a 1.7 day
period and with a total range of 500 km/s (Fig. 5a). The
systematic velocity of 310 km/s agrees with an LMC membership.
The spectral type is B3V with weak lines. The expected
absolute magnitude is $M_v = -1.9$, but if the continuum from the
X-ray source contributes, then probably $M_v = -1.5$. This is in

agreement with the expectations for such a star.

The X-ray data (White & Marshall 1984, Weisskopf et al. 1983) show no eclipses at the optical 1.7 day period. This constrains the orbital inclination to i < 70 deg. The derived mass function is large, f(M) = 2.3 $M_\odot$ . Various considerations yield mass estimates for the star between 4 and 8 $M_\odot$ with a most likely value of 6 $M_\odot$ . For masses larger than this the main sequence star radii become smaller than the Roche lobe. Most likely values for the radius lie within 5-6 $R_\odot$. The He I absorption lines show v sin i = 130 $\mp$ 20 km/s. If the star corotates then a lower limit of 100 km/s implies i > 50 deg. From the above considerations, the unseen star's mass (the X-ray source) lies within the limits: 7 $M_\odot$ < $M_x$ < 14 $M_\odot$ . Even if the visible primary is undermassive by a factor of 2 the X-ray source mass lower limit is 6 $M_\odot$. The most probable value is 9 $M_\odot$ (Fig. 5c).

Optical photometry obtained in December 1981 at ESO (van der Klis et al. 1983) combined with other data reveals a double-waved light curve of 0.2 mag amplitude with long-term brightness variations of up to 0.5 mag. One minimum coincides with the phase when the X-ray source is behind the primary. Both minima are of equal depth. The light curve may be dominated by ellipsoidal variations, but an additional light source, presumably an accretion disk, is likely to be present.

Several questions remain unanswered: What is the origin of the large-amplitude optical curve? What is the relation between the X-ray flux and the long-term optical brightness variation? Why is there no apparent strong heating effect in spite of the high $L_x$/$L_{opt}$ ratio? Does a thick accretion disk shield the primary from the X-rays?

LMC X-1. This somewhat fainter LMC X-ray source was localized in the star-forming region N159D by Copernicus (Rapley & Tuohy 1974) and HEAO-1 (Johnston et al. 1978), and to date no precise identification has been possible. In spite of an accurate (3") Einstein HRI X-ray position (Hutchings et al. 1983) two stars separated by 6" are still possible candidates for identification as the optical counterpart since the HRI position lies halfway between them: R148, a B5 supergiant of V =12.0 and an O star (V = 14.5), discovered by Pakull (1980) and noted by Cowley et al. (1978) as No 32. Both stars are in a nebulous region whose emission lines fill-in the strongest stellar absorption lines. The X-ray flux shows no regular variations and the luminosity is 2 x $10^{38}$ erg/s. The X-ray spectrum is soft, similar to that of LMC X-3 (White & Marshall 1984).

While spectroscopic observations (Hutchings et al. 1983) show R148 to be a velocity variable with a full range of $\sim$ 50 km/s but no discernable period, the fainter star No. 32 (Pakull's candidate) appears to be a binary with a period of about 4 days. Absorption line radial velocities (He I $\lambda$ 4471 and $\lambda$ 4026, He II $\lambda$ 4541 and $\lambda$ 4200) vary by a full amplitude of 140 km/s. The period is not yet well determined but the best choice is 3.91 days (possible 4.04 day period, too). Emission-line velocities (weak, broad CIII-NIII $\lambda\lambda$ 4640 and $\lambda$ 4686 appear to move in antiphase with a full amplitude of about 300 km/s (Fig. 5b).

According to Hutchings et al. the mass function is $f(M) = 0.12 \ M_\odot$. If the emission lines arise near to the X-ray source, as in other massive binaries (see above), the ratio of the velocity amplitudes implies $M_{opt}/M_x \sim 2$. This is probably an upper limit, as if the lines arise between the two stars, the amplitude ratio would lie between 2.5 and 5.

The optical spectrum of the primary is 07. The absence of X-ray eclipses (White & Marshall 1984) implies $i \lesssim 65$ deg. Adopting $A_v = 0.6$ mag and since $T_{eff} \sim 36000$ K, $M_{bol} = -8.2$ and $R \sim 10 \ R_\odot \pm 2 \ R_\odot$. If the primary underfills its Roche lobe by no more than 10%, this yields $R_{Roche} \sim 13 \ R_\odot$ if $R \sim 12 \ R_\odot$. The primary has a mass lying between 20 and 30 $M_\odot$ (Fig. 5d). Normal stars of type 07 have masses between 20 and 30 $M_\odot$. It thus seems undermassive for its temperature and luminosity, not unlike other MXRBs (Hutchings 1982). The most probable value is 14 $M_\odot$. This implies a mass for the secondary (the X-ray source ??) of 4 $M_\odot$.

In summary then, LMC X-3 probably contains a massive unseen X-ray emitting star with a mass around 9 $M_\odot$. LMC X-1 may contain a massive unseen (X-ray emitting ??) star with a mass around 4 $M_\odot$. Both these masses are above acceptable limits for neutron stars and these systems may be, together with Cygnus X-1 (Bolton 1972, 1975; Hutchings et al. 1973), good candidates for 'black holes'. Note that none of the three sources exhibits X-ray eclipses.

3.    OPTICAL CHARACTERISTICS OF LOW-MASS X-RAY BINARIES (LMXB)

3.1   General Properties

Here the companion star is a low-mass low-luminosity object, transferring matter by Roche-lobe overflow. The group comprises the bright bulge sources, globular cluster sources, soft X-ray transients, the X-ray bursters and the long-thought-to-be archetype of this class, Sco X-1. The X-ray emission is

generally soft  and  very few of these systems pulse in X-rays.
The  optical counterparts are intrinsically faint objects whose
spectra show  flat  continua with a few characteristic emission
lines implying an origin  for  this  continuum in the accretion
disk, where X-rays are reprocessed into optical radiation.  The
X-ray to optical luminosity ratio $L_x/L_{opt}$  may  be as high as
$10^3$ .

In this class of systems, which according to their spatial
distribution  belong  to  Population  II,  the  accretion  disk
plays  a  dominant  role.   The  hallmarks  of  LMXB  are  the
following:

Average  de-reddened colours indicate  a  flat  spectrum
(U-B =  -0.97 $\pm$  0.20  and  B-V = 0.01 $\pm$ 0.29), in other words
constant spectral intensity throughout the optical band  or  an
equivalent  black-body  temperature  $\gtrsim$ 15000 K (Fig. 6).  The
average absolute magnitude  is  about  $M_v \sim$ 1.0 (van Paradijs
1981).  Typical  spectra  show  continua  with  very few or
no  absorption lines and several characteristic emission lines:
$\lambda$ 4686 He II, $\lambda\lambda$ 4630-4650  C  III - N III, and sometimes the
Balmer  lines  (Fig.  7).  They are characterized by  a  rather
narrow range of $L_x/L_{opt}$ .  With six exceptions the difference
between the B magnitude  and an "X-ray magnitude" introduced by
van Paradijs (1981) as -2.5 log $F_x$  (in $\mu$Jy) is rather narrowly
clustered around 21.5 $\pm$ 1.1 mag.

With some exceptions (1626-67, HZ  Her and GX 1+4) none of
these  sources pulse in X-rays.  No X-ray  eclipses  are  seen
except some  partial  eclipses  in  three sources which will be
discussed below.  No strong evidence of  any  heating effect is
found on any of these sources (except HZ  Her)  in spite of the
rather large X-ray to optical luminosity ratios typical of LMXB
$(L_x/L_{opt} \sim 10^3$ ).

When  there  is  X-ray variability, the optical flux often
shows correlated variability.  In fact optical  variability  is
a  hallmark  of  LMXBs.  Orbital periods range from 19h to 42m.
In  the  cases  where  the optical companion is not a dwarf but
rather a subgiant, as in Her  X-1,  Cyg  X-2 and 0921-630, the
orbital  periods  are  correspondingly  longer:  1.7d, 9.0d and
9.8d.  Transient X-ray sources belonging to the LMXB class show
accompanying "nova-like" intensity increases in the optical (by
5-6  magnitudes)  with  subsequent  parallel  slow  decay.  At
quiescence  the optical spectra of several of these sources are
typical of  mid-K  dwarfs:  0620-00 (Oke  1977), Cen X-4 (van
Paradijs  et  al.  1980),  Aql  X-1 (Thorstensen  et  al.
1978) and 1743-28 (Murdin et al. 1980).

The above properties can be understood in the following framework: this class of source consists of close binary systems where the compact object (neutron star or 'black hole') is accompanied by a low-mass, low luminosity, low-temperature companion (Joss & Rappaport 1979). A large-sized accretion disk surrounds the compact object and is illuminated by the central X-ray source (Milgrom 1978). A small fraction of this flux is absorbed in the disk and is re-emitted as optical radiation (about 2%). The late-type companion tends to be shielded most of the time from the X-rays by the flared-up edges of the disk (Milgrom). In transient sources and in a few others we see at certain times (when either the mass transfer stops or a favourable occultation takes place) stellar absorption features from the companion. The predominant absence of X-ray pulsation might be taken at face value to indicate lower magnetic fields in these Pop II neutron stars. Table 2 presents an up-to-date (March 1984) list of low-mass binaries with optical identification. Excluded from this list are the globular cluster sources.

## 3.2 Types of Optical Variability in LMXB

Among low-mass X-ray binaries we can distinguish several types of optical variability.

### Periodic Phenomena Due to Orbital Motion.

Photometric. There is modulation of the light due to varying visibility conditions of the disk and there may be in addition eclipses and/or dips. In some cases the late-type companion might become visible during the minima as in the case of 0921-63 (Chevalier & Ilovaisky 1981, 1982). The optical modulations seen in the transient 0620-00 during the quiescent state (7.8h, McClintock et al. 1983), and in the bursters 1636-53 (3.8h, Pedersen et al. 1981) -Fig. 9b- and 1735-44 (4.3h, McClintock & Petro 1981) have not yet been clearly demonstrated to be due exclusively to orbital motion.

Spectroscopic. There may be a periodic modulation of the radial velocities of the emission lines coming from the accretion disk as in the case of Sco X-1 (Cowley & Crampton 1975). When the late-type companion is sufficiently luminous and becomes detectable, a modulation of the radial velocities of the absorption line is seen, as in 0921-630 (Cowley et al. 1982) and Cyg X-2 (Cowley et al. 1977). The reported 4.3h photometric modulation in 1735-44 (Mclintock & Petro) is in conflict with the report of a spectroscopic modulation at 2.86h (Hutchings et al. 1983). Note that Smale et al. (1984) see spectroscopic evidence for binary motion but do not confirm either period.

## Periodic Phenomena Due to Rotation of Neutron Star

Pulsations. Low-amplitude optical pulses have been seen in Her X-1 (Middleditch 1983) and 1626-67 (Ilovaisky et al. 1981). They are due to reprocessing fo X-ray pulses in the accretion disk and/or in the facing hemisphere of the companion. These reprocessed pulses have been used to derive in the case of HZ Her the mass of the neutron star (Middleditch & Nelson 1976) and in the case of 1626-67 its 41min orbital period (Middleditch et al. 1981) as well as to establish the prograde spin of this pulsar.

## Non-Periodic and Quasi-Periodic Phenomena Taking Place in the Disk

Bursts. In X-ray bursters there are very often optical bursts coincident with X-ray bursts. This is the case for 1636-53 (Pedersen et al. 1983), 1735-44 (McClintock et al. 1979) and Ser X-1 (Hackwell et al. 1979). This is due to reprocessing of X-ray bursts in the accretion disk. In all cases the optical bursts follow the X-ray bursts by 2 to 3 seconds and appear to be smeared. The time delays can be used to study the structure of the accretion disk. It appears that the radius of the disk is > 1.5 light sec, the temperature of the optical reprocessor varies from 25000 K at quiescence to 50000 K at burst maximum and that at most 3% of the total X-ray energy is converted into optical radiation at all wavelengths (Lawrence et al. 1983). A fairly complete review on the optical emission from X-ray bursters can be found in Pedersen (1984).

Quasi-oscillations. In 1626-67 (Li et al. 1980, Middleditch et al. 1981) -Fig. 9a- and sometimes in GX 339-4 (Motch et al. 1981) -Fig. 11a- there are large amplitude oscillations of the optical and X-ray fluxes which can amount to 50% of the total flux. These quasi-periods (1000s and 20s respectively) are not well understood but are thought to be due to inhomogeneities (a bulge) or instabilities in the accretion disk.

Flickering. The most common example of a source exhibiting this behaviour is Sco X-1. In this source and in others (for example 1626-67, see Middleditch et al. 1981 and Ilovaisky et al. 1981) there is optical flickering associated with similar X-ray behaviour (Fig. 8). In Sco X-1 the two can be strikingly correlated when the X-ray source is bright, but the optical flickering may still persist when the X-ray source is in a low and quiet state (Ilovaisky et al.

1980).   Optical flickering activity does not appear to exhibit time structure  faster  than 20 sec, indicative of reprocessing in an extended region  (Petro et al.  1981).   The very rapid flaring  of GX 339-4 when very bright (Motch et  al.  1981)  is unique and will be described later.

## Transient Behaviour Due to Instabilities in the Mass Transfer

There  are  large-amplitude  (5-6  magnitude) increases in the optical flux correlated with  the X-ray intensity increases in  sources  exhibiting  transient  behaviour,  with subsequent parallel decay. The light is produced  in  the  accretion disk through reprocessing of X-ray  photons  as  in  other  low-mass binaries.

During the  decline phase of the transient source 0620-00, X-ray  and optical  modulations  at  7.8d  have  been  observed (Matilsky et  al.  1976,  Chevalier et al.  1980).  This may be  a  behaviour  analogous to that observed in Her X-1 and LMC X-4, where an 'apparently'  precessing  disk  modulates the X-ray and optical fluxes (see Section on LMC  X-4).  Long X-ray periods have been reported for Aql X-1 and 1915-05 (Pridehorsky & Terrell 1984), 0614+09 (Marshall & Millit 1981), GX17+2 and GX349+2 (Ponman 1982), but  their  exact  significance  is uncertain.

## 3.3  Extended X-ray Sources

There are three low mass binary source for which the X-ray light curves observed show  partial  eclipses. All three have orbital periods of about 5h  (Fig. 10).

**1822-37.** This source has a 5.57h  period  and  a  1  mag optical modulation,  independent  of  wavelength  (Mason et al. 1980).  There is a sinusoidal X-ray  intensity  variation and a simultaneous  X-ray/optical  shallow  minimum.  The  optical spectrum contains He  II  $\lambda$ 4686,  $\lambda$ 3815, C III/N III $\lambda\lambda$ 4640-50  and  O  III  $\lambda$ 3765 in emission  and  Ca  II  H+K  in absorption.

**2129+47.** The  orbital period is  5.2h  and  the  optical modulation is 1.5  mag  deep,  with  associated  colour  changes (0.25m in B-V and 0.5m in U-B). The  light curve is remarkably similar  to  that  of  Her  X-1 (Thorstensen et  al.  1979, McClintock et  al.  1981).  The  X-ray  light  curve  shows a sinusoidal  variation plus  a  shallow minimum.  The  optical spectrum shows  He  II  $\lambda$  4686,  C  III/N  III  $\lambda\lambda$ 4640-50 in emission; H $\beta$,  Ca II H+K often show up in  absorption.  We note that this  source  has  been recently observed to be in an X-ray  OFF  state  (Pietsch et  al.  1983)  and  that  the large-amplitude optical modulation has virtually disappeared.

**Cyg X-3.** The orbital period is 4.8h. This outstanding X-ray and radio object remains undetected at optical wavelengths but is seen in the infrared. Both the IR and the X-ray fluxes show a sine-like modulation (Becklin et al. 1973, Mason et al. 1976).

The observed X-ray light curves of 1822-37 and 2129+47 and possibly Cyg X-3 are the result of an eclipse by the companion star of an extended X-ray scattering cloud, presumably a highly ionized corona above the central part of the accretion disk produced by evaporation of material from the disk surface (White & Holt 1982). The X-ray source itself is not directly seen, as it is shielded by the accretion disk. Inclinations lie between 70 deg. and 78 deg. The disks have angular thicknesses (as seen from the neutron star) of $>$ 10 deg. The companion stars are not completely shielded from X-rays but a polar cap is visible from the neutron star above the accretion disk. A small heating effect is expected. A problem remains: why then does 2129+47 show such a large X-ray 'heating' effect?

### 3.4  A Special Case: GX339-4 (4U1658-48)

**Introduction.** A reduced error box for this source obtained by Doxsey et al. (1979) by combining Uhuru, Ariel 5 RMC and HEAO-1 A3 MC observations led to an optical identification with a blue star which had already been selected on the basis of variability as a likely candidate by Penston et al. (1975) in their search of the Uhuru error box. Grindlay (1979) reported on optical observations which showed this variable blue star to exhibit a reddened continuous spectrum with emission lines reminiscent of Circinus X-1 (Whelan et al. 1977).

**X-ray properties.** The observation by Samimi et al. (1979) of very rapid X-ray variability in GX339-4, together with the existence of 'hard' and 'soft' states (Markert et al. 1973), led Doxsey et al. and Grindlay to speculate, by analogy with Cygnus X-1 which also exhibits these two characteristics (see Oda 1977), that GX 339-4 might contain a massive collapsed object (a 'black hole'). This similarity was strengthened further by reports of rapid X-ray variations in Circinus X-1 (Toor 1977 & Sadeh et al. 1979). Furthermore, it was speculated that GX 339-4 and Circinus X-1 were, like Cygnus X-1, massive binary systems with OB supergiant companions.

**Optical properties.** The early observations of Penston et al. showed the star to vary by a least 0.5 mag, while

Grindlay found the star to increase in brightness by 0.2-0.3 mag in a 3-day interval. Its colours (V=16.6, B-V=+0.8, U-B=-0.2) were consistent with a reddened early-type continuum. An investigation of the Harvard plate archives by Grindlay showed the object to vary between photographic magnitude 16 and 18.5, with no apparent periodicity. The spectrum showed essentially a red continuum with strong H-alpha emission, weak He II $\lambda$4686 and C III/N III$\lambda\lambda$4640-50 emission, plus some absorption lines, presumably interstellar.

In late February and early March 1981 the optical object associated with GX 339-4 went into a previously unknown low state (Hutchings et al. 1981, Ilovaisky & Chevalier 1981), dropping down to about 21st magnitude. X-ray observations with Hakucho in April an 'off' X-ray state ($\lesssim$ 15 $\mu$Jy), similar to that reported for 1972 by Markert et al. (1973) from OSO-7 data. Such behaviour suggested right away that the optical counterpart could not be a massive OB supergiant as imagined earlier, but was probably a low-mass, low-luminosity companion, with most of the light coming from an accretion disk (Motch et al. 1981a). Note in this respect the mounting evidence (Nicolson et al. 1980, Argue & Sullivan 1982) that the optical/IR emission from the Cir X-1 candidate does not originate in a heavily reddened OB supergiant but may instead be also due to an accretion disk.

**Correlated X-ray/optical observations.** Observations carried out at ESO in May (Motch et al. 1981b) showed the optical object to have increased in brightness by 5-6 magnitudes in less than 2 months, a behaviour reminiscent of X-ray 'soft' transient sources. High-speed photometry of the star revealed an unprecedented degree of optical activity in an X-ray binary: very numerous, strong and fast 'flares', some amounting to a full magnitude and lasting only 20 milliseconds (Fig. 11b), superposed on large-amplitude (up to 50%), quasi-periodic 20 sec oscillations (Fig. 11a). The power spectrum of the optical activity of GX 339-4 turned out to be remarkably like that found by Nolan et al. (1981) for the X-ray activity of Cygnus X-1, with the exception of the broad peak at 20s. (Fig. 12b).

X-ray observations made in late May 1981 with the Ariel 6 satellite, some of them fortuitously simultaneous with the optical observations (Motch et al. 1983), show the X-ray source to have been at that time in one of its 'hard' states with X-ray flaring activity (Fig. 12a). Analysis of the simultaneous data reveals 20 sec quasi-periodic X-ray (1-13 KeV) oscillations anticorrelated to the optical oscillations. Substantial power is also found in 10 sec quasi-periodic oscillations. There is marginal evidence that the harder

X-rays (13-20 keV) were correlated. No case of a simultaneous X-ray and optical fast 'flare' was found.

Further Hakucho observations (Maejima et al. 1984) show the X-ray source to have remained in the 'hard' state until about the end of June, a transition to the 'soft' state taking place between June 26 and July 4. Long-term optical photometry (unpublished) shows the star remained bright until at least June 23 and then dropped in brightness by July 2. Again this is clear evidence of an anti-correlation between the soft X-rays (3-6 keV) and the optical, or, in other words, of a correlation between the optical brightness and the hardness ratio I(6-10 keV)/I(3-6 keV). Ariel 6 spectra of the source taken during both the 'hard' and 'soft' states (Ricketts 1983) shows the 'pivot' energy (that at which the intensity stays constant during a transition) to be around 6 keV, as in Cygnus X-1 (Sanford et al. 1976).

**A possible scenario, many questions and a surprise.** the rapid activity observed in May could not have originated in a thermal source, as the brightness temperature of the 'flares' reached $5 \times 10^9$ K. As suggested by Fabian et al. (1982), these flares might have been due to optically thick electron cyclotron emission from extremely hot gas at $10^9$ degrees in the inner regions of the accretion disk around a massive collapsed object. Such hot gas is also needed to explain the X-ray emission from Cyg X-1 (Guilbert & Fabian 1982).

The March 1981 low state, probably due to a drop in the accretion rate, was followed by renewed accretion leading to re-formation of the accretion disk, which during a certain time might have been small and ring-like. The quasi-periodic oscillations could have been due to a density enhancement in this ring.

The similarity in X-ray behaviour between GX 339-4 and Cygnus X-1 shows that the same 'X-ray machine' must be at work in both systems. The different long-term optical behaviour and appearance is explained by the differences in the nature of the optical companions: a massive OB supergiant furnishing a steady stream of matter to the compact object in Cyg X-1 and a low-mass, low-luminosity dwarf, which experiences periods of no mass loss in GX 339-4.

The very recent (November-December 1984) discovery with EXOSAT of hybrid X-ray behaviour in the transient source V0332+53 (Davelaar et al. 1983, Stellar & White 1984, White et al. 1984 and Parmar et al. 1984), where very rapid (10 ms) time variability, of the Cygnus X-1 and GX 339-4 type, coexists with regular 4.4 second pulsations - a hallmark of magnetized rotating neutron stars - shows that the existence of very rapid X-ray variability can no longer be used as an indication of the presence of a 'black hole'. Apart from direct optical radial velocity studies, the only remaining 'black hole' characteristic might perhaps be the two-state ('hard'-'soft') spectral behaviour (White 1984) also seen in active galactic nuclei (White et al. 1984).


4.    CONCLUSIONS

In this review I have tried to highlight the important role played by accretion disks in massive X-ray binaries and to emphasize the increasing amount of circumstantial evidence for the existence of massive collapsed objects ('black holes') in Cyg X-1, LMC X-3 and LMC X-1. The existence of 'tilted', apparently precessing disks may be more widespread than previously believed (LMC X-4, Cyg X-1, 1907+09) and still poses basic theoretical problems. In the more luminous systems the presence of accretion disks may be inferred only from spectroscopic observations.

The X-ray heated accretion disk is the dominant source of light in low-mass X-ray binaries and explains the optical activity of these systems (pulses, bursts, flickering) in terms of reprocessed X-rays. The absence of X-ray eclipses in these systems has been interpreted in terms of thick disks which shield the companion star. The existence of 'partial' X-ray eclipses in three sources confirms this view, but poses the problem of the origin of the large optical modulations observed. In some transient sources long-period modulations may be due to 'precessing' disks. The study of the companion star, in particular its mass, is of great interest but is handicapped by the extreme faintness of these objects. For some transient sources in quiescence, spectra indicate these companions are mid-K dwarfs. The nature of the compact source is of utmost importance. For some systems (1626-67, Her X-1) a neutron star appears well established while for others (GX 339-4) a 'black hole' may still be a possibility, but recall that recent EXOSAT observations of VO332+53 show the very rapid X-ray activity to no longer be an indicator of 'black hole' candidacy.

TABLE 1

Massive X-Ray Binaries

| Source | Name | V mag | Spect. | $P_O(d)$ | $P_S(s)$ | $P_l(d)$ | Comment |
|--------|------|-------|--------|------|------|------|---------|
| | | | **Class I companions** | | | | |
| 0115-737 | SMC X-1 | 13.3 | B0I | 3.9 | 0.7 | | On-Off states |
| 0532-664 | LMC X-4 | 14 | O7III-V | 1.4 | 13 | 30.4 | X-ray flares |
| 0538-641 | LMC X-3 | 16.9 | B3V | 1.7 | --- | | Soft x-ray spect. |
| 0540-697 | LMC X-1 | 14.5 | O7 | 3.9 | --- | | Soft x-ray spect. |
| 0900-403 | Vela X-1 | 6.9 | B0.5Ib | 9.0 | 283 | | |
| 1119-603 | Cen X-3 | 13.4 | O6-8(f)p | 2.1 | 4.8 | | On-Off states |
| 1145-616 | --- | 13.1 | B2Iae | 5.6 | 297 | | |
| 1223-624 | Wray 977 | 10.8 | B1.5Ia | 41.4 | 699 | | |
| 1516-569 | Cir X-1 | 22 | cont,em | 16.6 | --- | | Rapid var. |
| 1538-522 | QV Nor | 14.5 | B0I | 3.7 | 529 | | |
| 1700-377 | HD153919 | 6.6 | O6.5f | 3.4 | --- | | |
| 1907+097 | --- | 16.4 | OB? | 8.4 | 438 | 41.6 | |
| 1956+350 | Cyg X-1 | 8.9 | O9.7Iab | 5.6 | --- | 294 | Very rapid var. |
| | | | **Class II (Be star) companions** | | | | |
| 0050-727 | SMC X-3 | 15 | O9III-Ve | --- | --- | | Transient |
| 0053-739 | SMC X-2 | 16 | B1.5Ve | --- | --- | | Transient |
| 0053+604 | γ Cass | 1.6 | B0.5II-Ve | --- | --- | | High. variable |
| 0114+65 | LSI+65°010 | 11.0 | B0.5IIIe | --- | --- | | Variable |
| 0115+634 | V635Cass | 14-16 | Be | 24.3 | 3.6 | | Rec.Transient |
| 0332+53 | --- | 16 | Be? | 34.2 | 4.4 | | Rec.Transient |
| 0352+309 | X Per | 6-6.7 | O9.5III-Ve | --- | 835 | | |
| 0535-668 | LMC Tran | 13-15 | B2IVe | 16.7 | 0.069 | | Rec.Transient |
| 0535+262 | HDE245770 | 9.1 | O9.7IIIe | 111 | 104 | | Rec.Transient |
| 0726-260 | --- | 11.6 | B0Ve | --- | --- | | |
| 1118-616 | He3-640 | 12.1 | O9.5IVe | --- | 405 | | Transient |
| 1145-619 | Hen 715 | 9.0 | B1Vne | 188 | 292 | | |
| 1258-613 | GX304-1 | B2Vne | 132 | 272 | | | |
| 1417-624 | --- | 16 | BI/Be? | --- | 17.6 | | |
| 1553-542 | --- | ? | ? | 30.6 | 9.3 | | |
| 1735-28 | Hen3-1450? | 11.2 | Be | --- | --- | | Transient |

Note: Most data have been taken from Bradt and McClintock (1983) but new results are from Tanaka et al. (1983), White et al. (1984), Priedhorsky and Terrell (1983a,b, 1984) and Corbet and Mason (1984).

## TABLE 2. Low-Mass X-Ray Binaries

| Source | Name | V mag | Spect. | $P_o$ | $P_s$ | $P_l$ | Comment |
|--------|------|-------|--------|-------|-------|-------|---------|
| **Companion seen at times or dominates spectrum** | | | | | | | |
| 0323+022 | --- | 16 | cont/G? | --- | | | |
| 0921-630 | --- | 16 | F-GIII,em | 9.0d | | | Opt.eclipses/dips |
| 1656+354 | HZ Her | 13v | A9 | 1.7d | 1.4s | 35d | |
| 1728-247 | GX1+4 | 19 | M6III+? | --- | 120s | | |
| 1813-140 | GX17+2 | 17 | G ? | --- | | | |
| 1822-371 | V691CrA | 15 | K5 ? | 5.6h | | | Partial X-Ray ecl. |
| 2129+470 | V1727Cyg | 16 | late K? | 5.2h | | | Partial X-Ray ecl. |
| 2142+380 | Cyg X-2 | 15 | F2III+cont | 9.8d | | | X-ray dips |
| **Disk dominates spectrum** | | | | | | | |
| 0521-720 | LMC X-2 | 18 | cont,em | --- | | | at LMC ? |
| 0543-682 | LHG 83 | 17 | cont,em | --- | | | burst? |
| 0614+091 | V1055Ori | 18 | cont,em | --- | | 4.9d | burst? |
| 1254-690 | --- | 19 | cont | --- | | | Rapid variable |
| 1617-155 | Sco X-1 | 12v | cont,em | 0.8d | | | |
| 1627-673 | KZ Tra | 18 | cont | 41m | 7.7s | 1000s | |
| 1636-536 | V801 Ara | 17 | cont,em | 4h | | | Burster |
| 1658-298 | --- | 18-22 | cont,em | 7.1h | | | Burster |
| 1659-487 | GX 339-4 | 15-21 | cont,em | --- | | | very rapid var. |
| 1728-169 | GX9+9 | 17 | cont,em | 4h?? | | | Burster |
| 1735-444 | V926 Sco | 17 | cont | 4.4h | | | Soft spect/dips |
| 1755-338 | --- | 19 | cont | --- | | | Burster |
| 1837+049 | Ser X-1 | 19 | cont,em | 50m | | 199d? | Burster |
| 1916-053 | --- | 22 | cont | --- | | | Burster |
| 1957+115 | --- | 19 | cont | --- | | | Burster |
| 2030+407 | Cyg X-3 | --- | IR | 4.8h | | | Radio/gamma source |
| **Soft Transients (mostly off)** | | | | | | | |
| 0620-003 | V616 Mon | 11-18 | K5-K7V | 7.8h | | 7.8d | Rec.Trans. |
| 1455-314 | Cen X-4 | 13-19 | K3-7 | 8.2h? | | | Rec.Trans. |
| 1524-617 | KY TrA | 19 | --- | --- | | | |
| 1543-475 | --- | 15-16 | --- | --- | | | Rec.Trans. |
| 1608-522 | QX Nor | 18-20 | --- | --- | | | |
| 1705-250 | V2107 Oph | 16-21 | --- | --- | | | |
| 1743-288 | --- | 18? | K3V | --- | | | |
| 1908+005 | Aql X-1 | 15-19 | K0V | 1.3d? | | 124d? | Bursts/Rec.Tran. |

Note: Most data have been taken from Bradt and McClintock (1983). Recent results are from Pakull et al. (1984), Priedhorsky and Terrell (1984), Cominsky and Wood (1984).

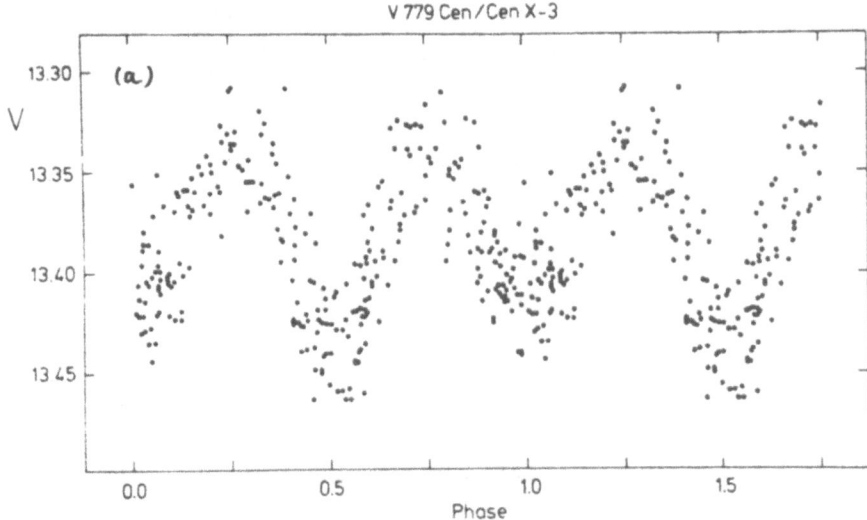

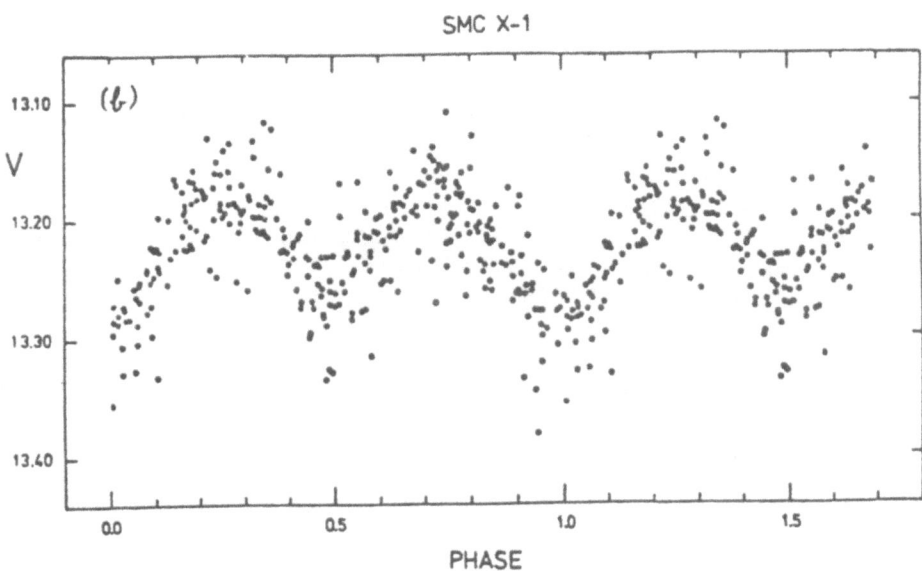

Figure 1. Typical curves for two massive binaries. (a) Cen X-3, taken from van Paradijs et al. (1983) showing a normal deep mid-phase minimum. (b) SMC X-1, taken from van Paradijs and Kuiper (1984), which exhibits a filled-in mid-orbital phase minimum due to X-ray heating.

Figure 2.   Radial velocity curves for the HeII 4686 emission line
in four massive X-ray binaries:  LMC X-4 (Hutchings et al. 1978),
4U1538-52 (Crampton et al. 1978), SMC X-1 (Hutchings et al. 1977)
and Cen X-3 (Mouchet et al. 1980).  In three cases the radial
velocity curve for the absorption lines of the primary are shown.
For Cen X-3 the solid line is the X-ray source Doppler curve and
the dashed line the curve for the Lagrangian point $L_1$ .

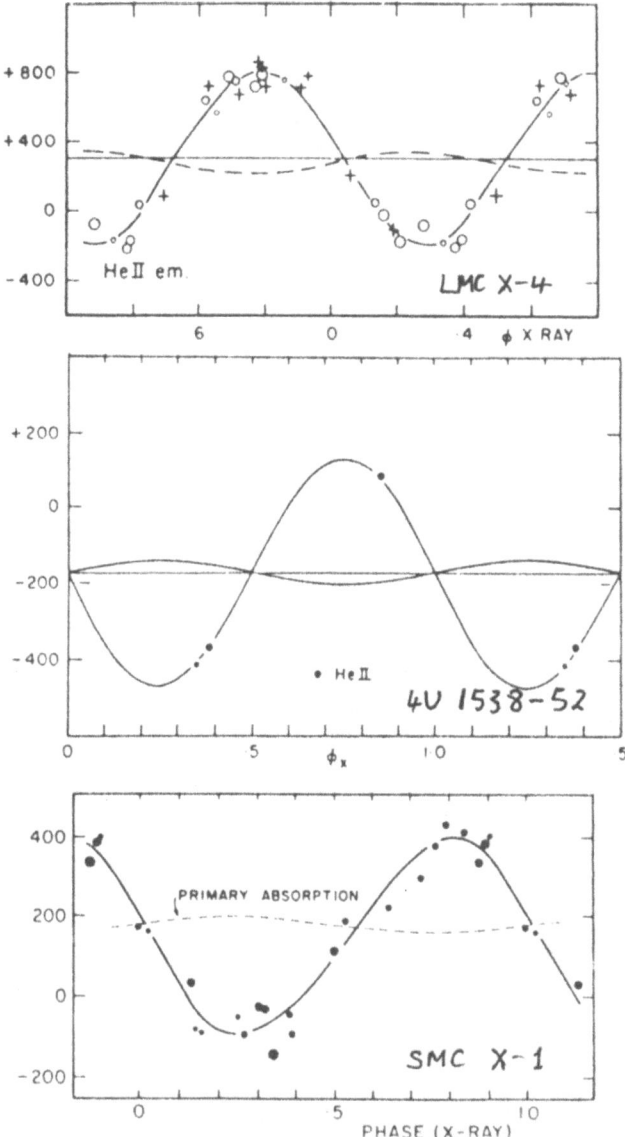

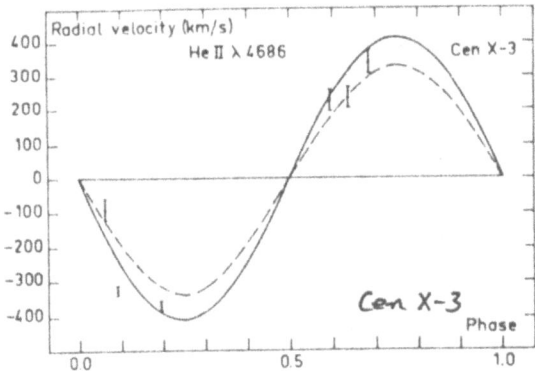

Figure 3a. The ultraviolet continua from massive binaries LMC X-4 and SMC X-1 taken from Bonnet-Bidaud et al. (1981) and derived from IUE short and long wavelength spectra. The continuous lines are Kurucz model atmosphere fits (30000K, E$_{B-V}$ = 0.05 and log g = 4.0 for LMC X-4 and 25000K, E$_{B-V}$ = 0.09 and log g = 3.0 for SMC X-1).

(a)

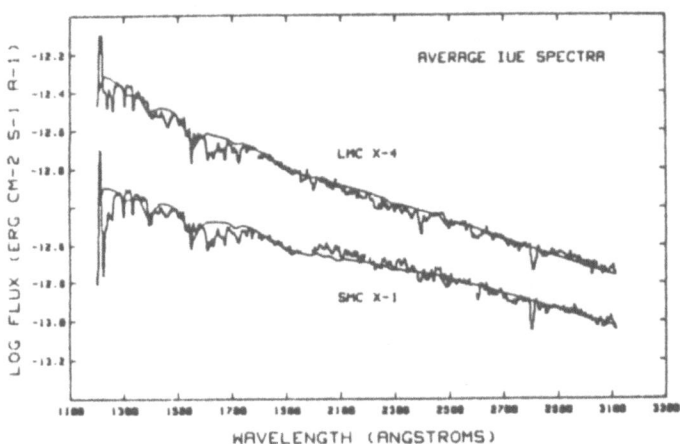

(b)

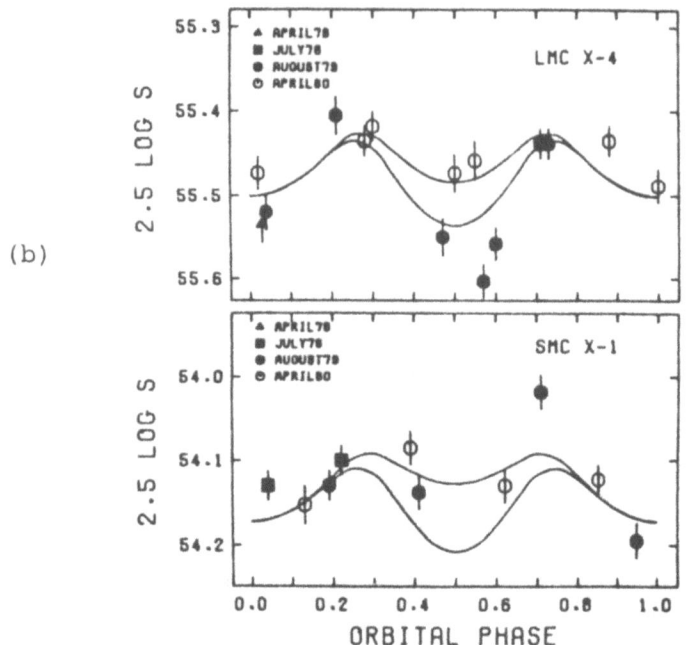

Figure 3b. Ultraviolet light curves for LMC X-4 and SMC X-1 taken from van der Klis et al. (1982). Points shown are the normalization constants of the best fit model atmospheres. Continuous lines are calculated ellipsoidal light curves with (upper) and without (lower) X-ray heating.

(c)

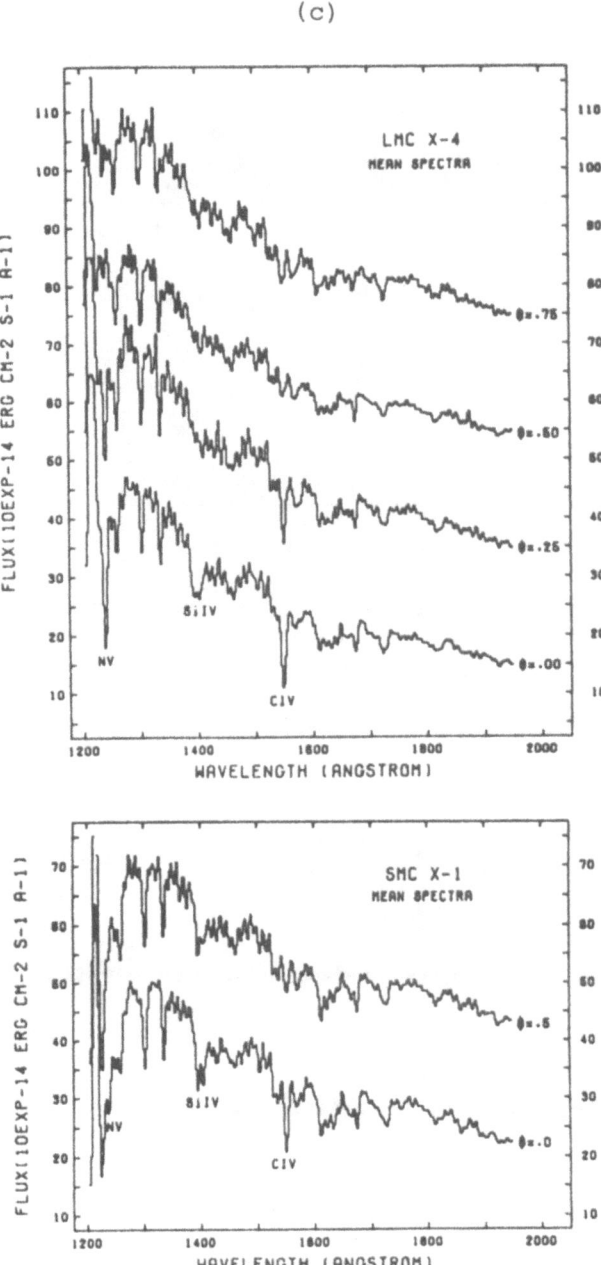

Figure 3c. Mean low-dispersion spectra of LMC X-4 and SMC X-1 taken from van der Klis et al. (1982). Binary phases are indicated at right. Note the dramatic changes in the C IV resonance absorption line.

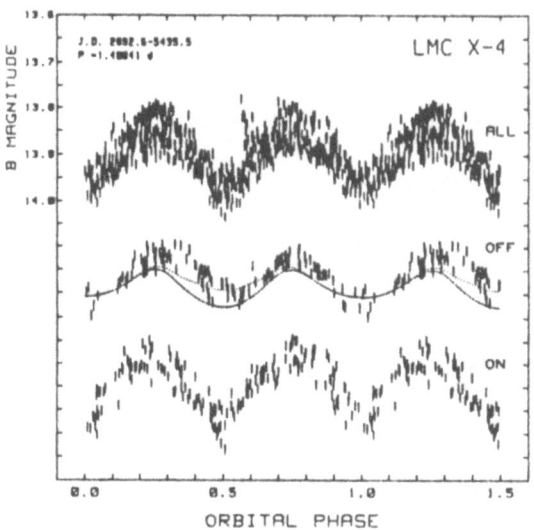

Figure 4a.   B-filter light curves for the LMC X-4 massive binary,
taken from Ilovaisky et al. (1984). TOP:  All data.  MIDDLE:
Data corresponding to the X-ray OFF phase (the lower dashed curve
is a pure ellipsoidal model while the upper dotted curve shows the
effect of X-ray heating). BOTTOM:  Data corresponding to the
X-ray ON phase.

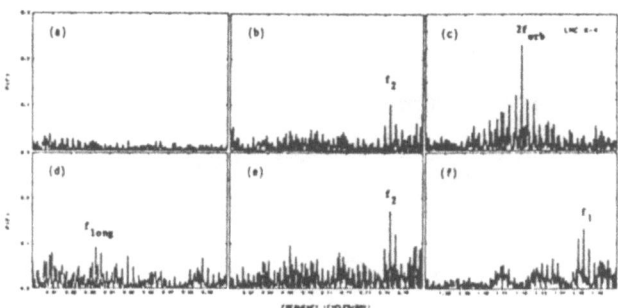

Figure 4b.   Periodograms for the LMC X-4 optical data taken from
Ilovaisky et al. TOP panel:  for the raw data around the 30-day
frequency, $f_{long}$ , the orbital frequency $f_{orb}$  and half-orbital
frequency $2f_{orb}$ .  BOTTOM panel:  the same frequency intervals
but for the data after demodulation at the dominant half-orbital
frequency.  The peaks at $f_1$ and $f_2$ are the frequency sums
$f_1 = 2f_{orb} + f_{long}$ and $f_2 = f_{orb} + f_{long}$ .

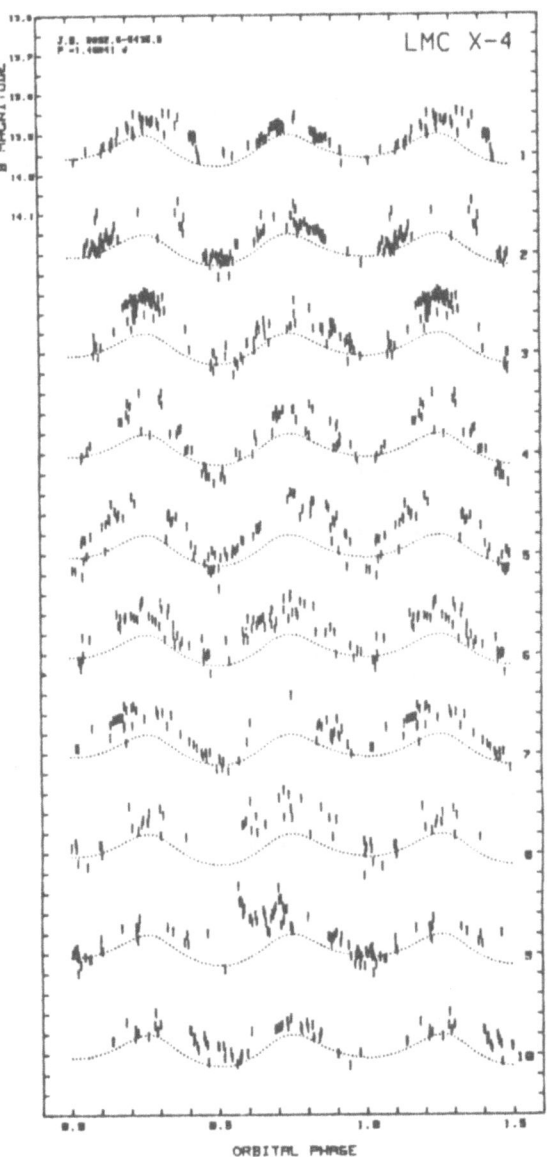

Figure 4c.  The B filter data of LMC X-4 folded with the 1.408-day orbital period and split into 10 elementary light curves each corresponding to a consecutive 30-day cycle pahse bin.  Taken from Ilovaisky et al. (1984).  Curves numbered 1 through 10 correspond to bins 0.0-0.1, ..., 0.9-1.0.  The dotted curve superposed onto all light curves as a reference is the pure ellipsoidal model (no heating) shown in (a) as the lower dashed line.  X-ray OFF phases are from 0.9-0.1 and X-ray ON phases from 0.35-0.60.

LMC X-1

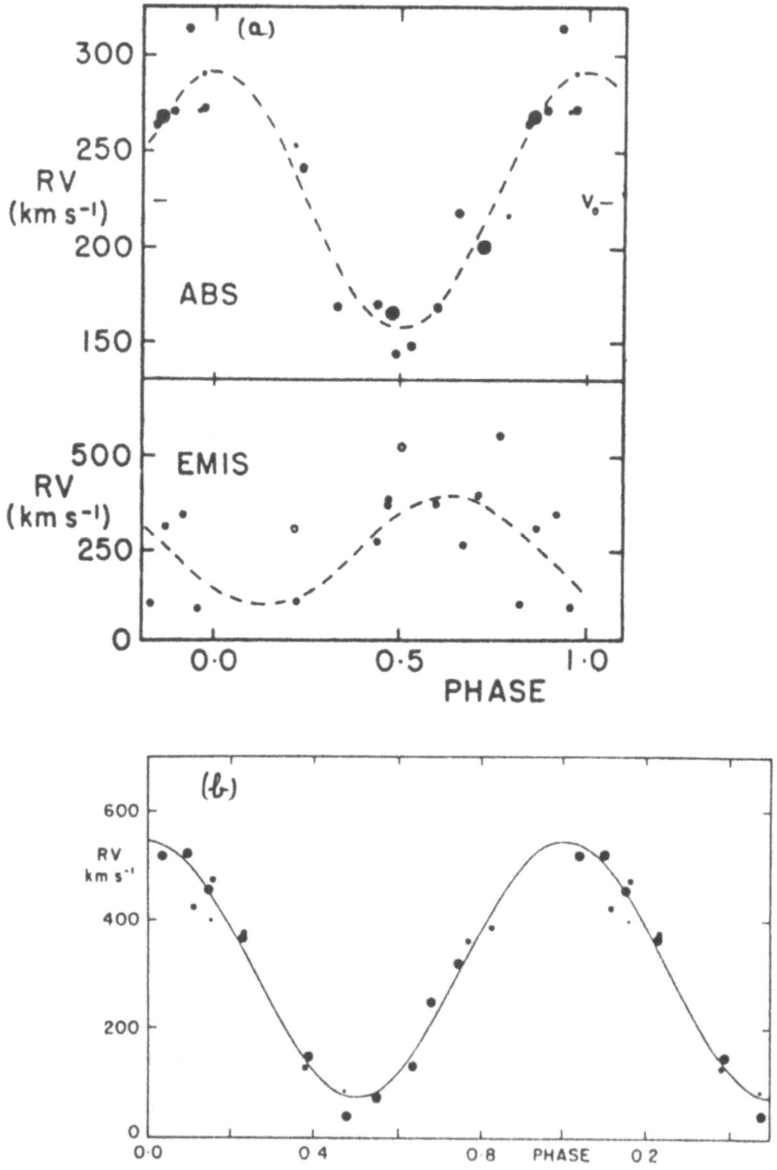

Figure 5.

LMC X-3

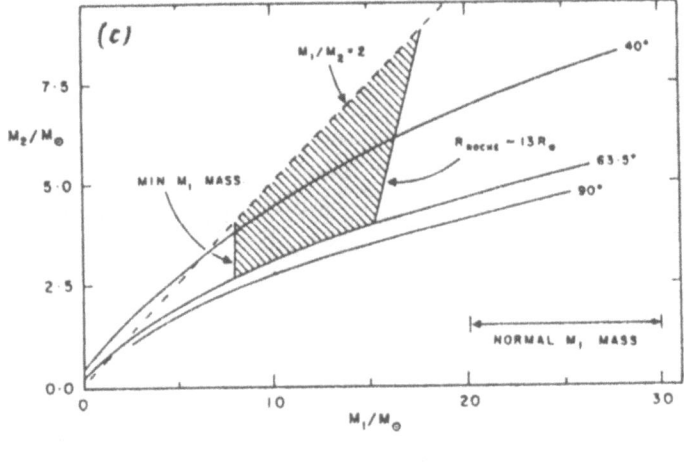

LMC X-1

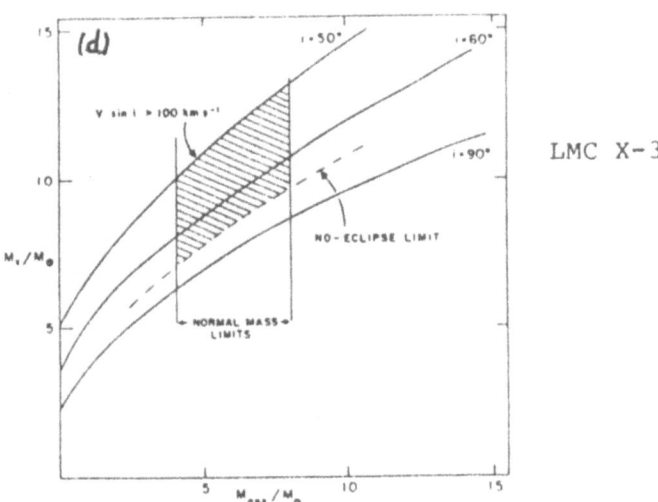

LMC X-3

Figure 5.  Radial velocity curves for (a) the absorption lines and
4640 emission line from the LMC X-1 optical candidate, taken from
Hutchings et al. (1983) and (b) the absorption line velocities for
the LMC X-3 optical counterpart taken from Cowley et al. (1983).

Mass diagrams for the (c) LMC X-1 binary candidate, taken from
Hutchings et al. (1983) and for (d) the LMC X-3 optical counterpart taken from Cowley et al. (1983).  Allowed region is hatched.

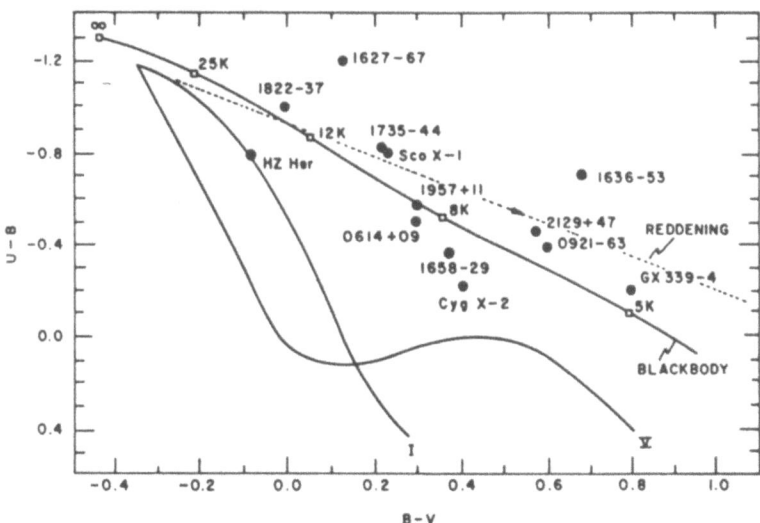

Figure 6. Colour-colour plot on the UBV photometric systems showing the location of a number of low-mass X-ray binaries, taken from Bradt & McClintock (1983). Shown also are the main sequence (V) and supergiant (I) branches. Several of the sources lie near the predicted locus for black-body emitters, which is represented by a solid line with temperature in units of 1000K. The dashed line is the reddening trajectory of an early O star.

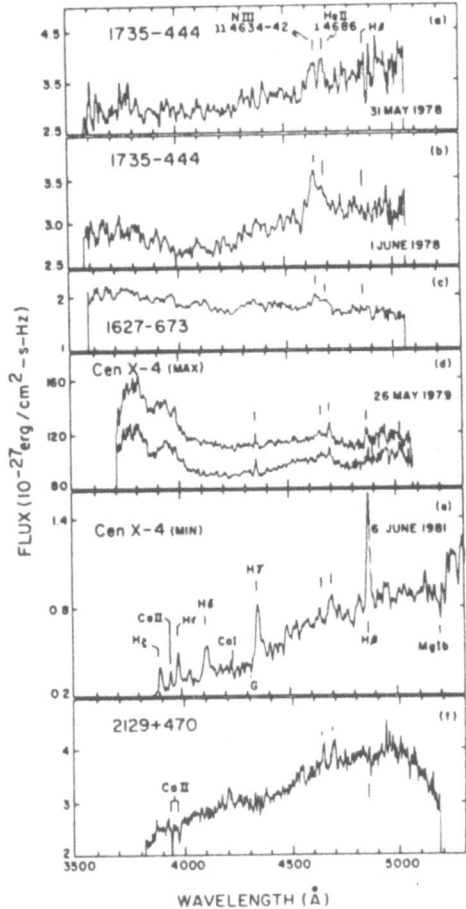

Figure 7. Spectrophotometric data for four low-mass binaries, taken from Bradt & McClintock (1983). 1735-44 on two consecutive nights, 1626-67, Cen X-4 at maximum and minimum and 2129+47. Prominent emission lines are indicated. For other details see Bradt & McClintock.

(a)

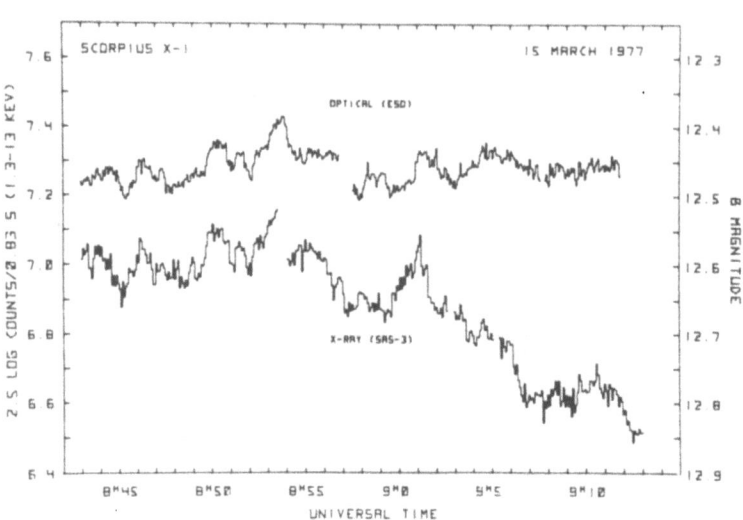

(b)

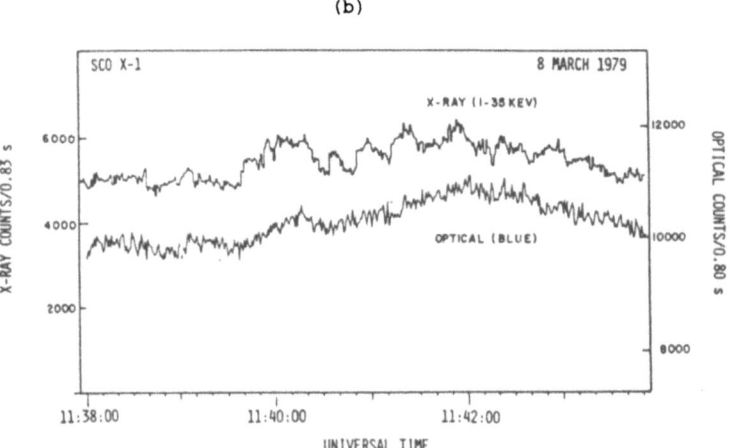

Figure 8.  Sample simultaneous X-ray/optical observations of
Sco X-1 taken from (a) Ilovaisky et al. (1980) and (b) Petro et al.
(1981).  X-ray data are from the SAS-3 satellite and the optical
data from the ESO 1m telescope for (a) and the Mount Wilson 2.5m
telescope for (b).

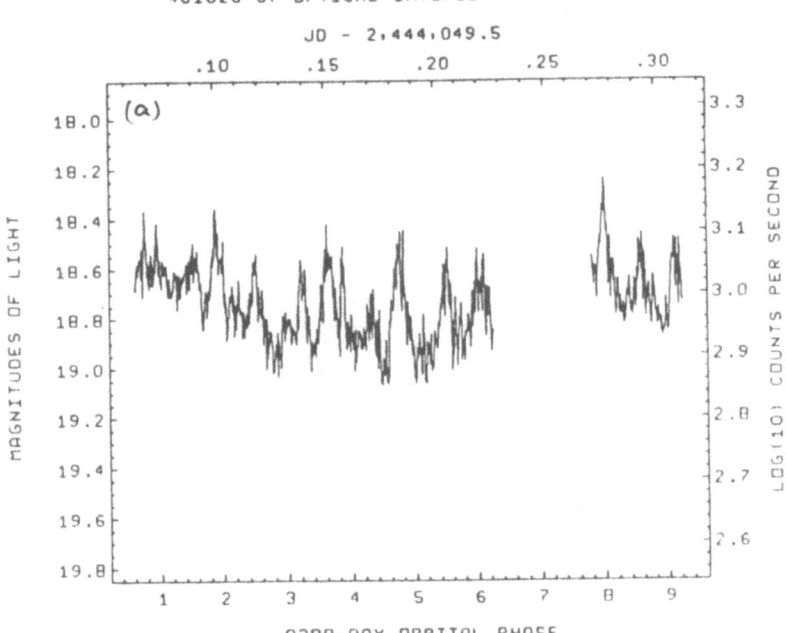

Figure 9. Examples of optical variability in low-mass binaries:
(a) the 1000 sec oscillations of 1626-67, taken from Middleditch
et al. (1981) and (b) the 4h modulation in 1636-53 from Pedersen
et al. (1982). In the latter the lower trace is for the comparison
star.

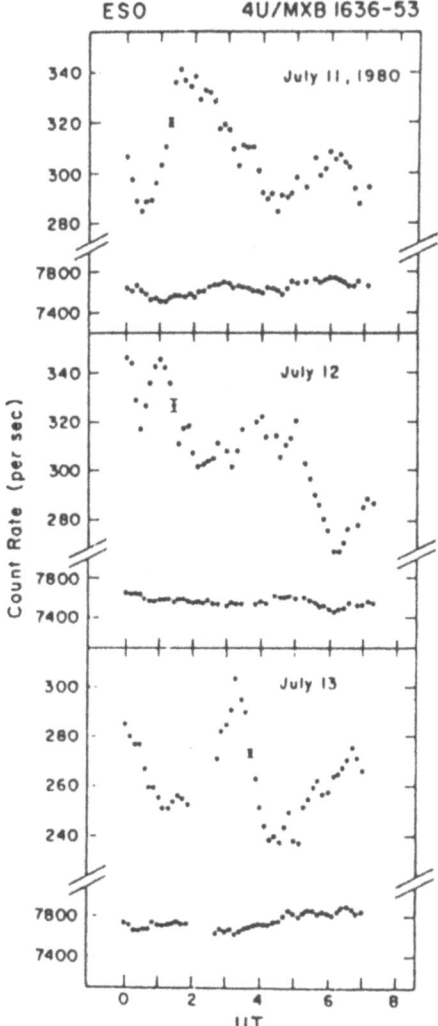

Figure 9(b).

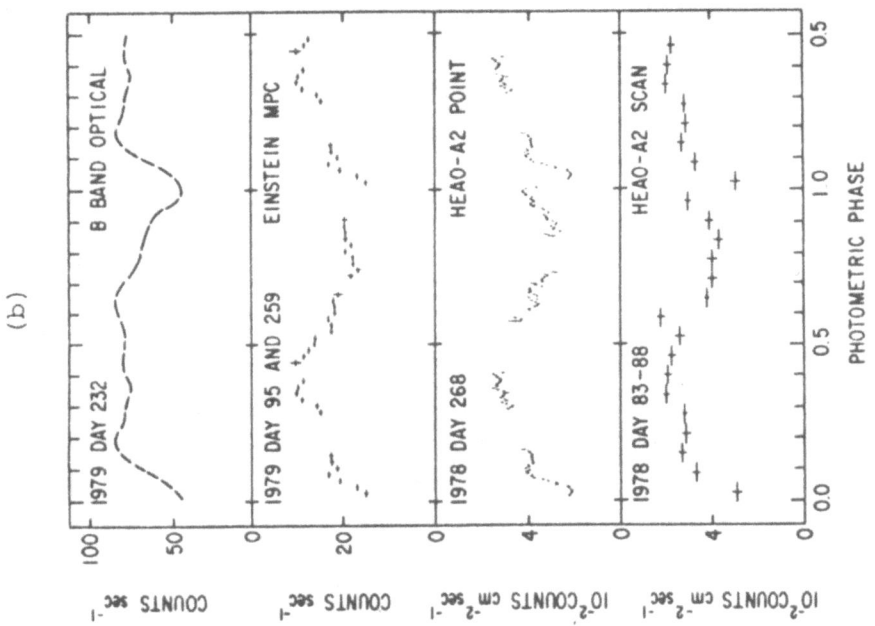

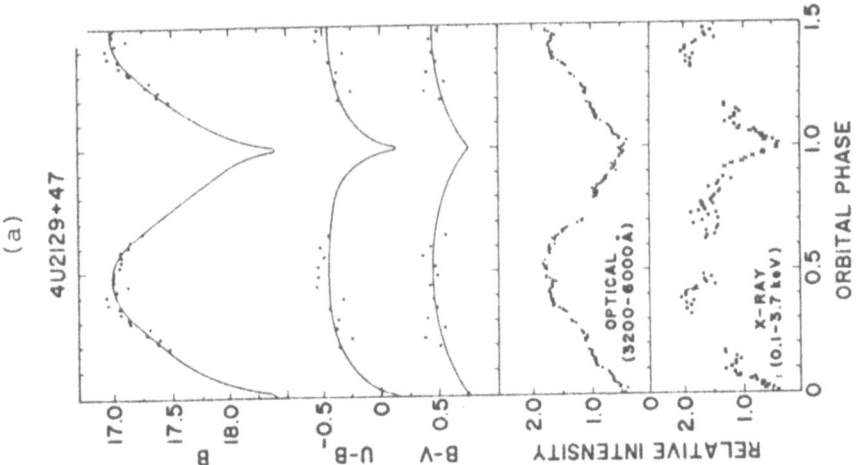

Figure 10

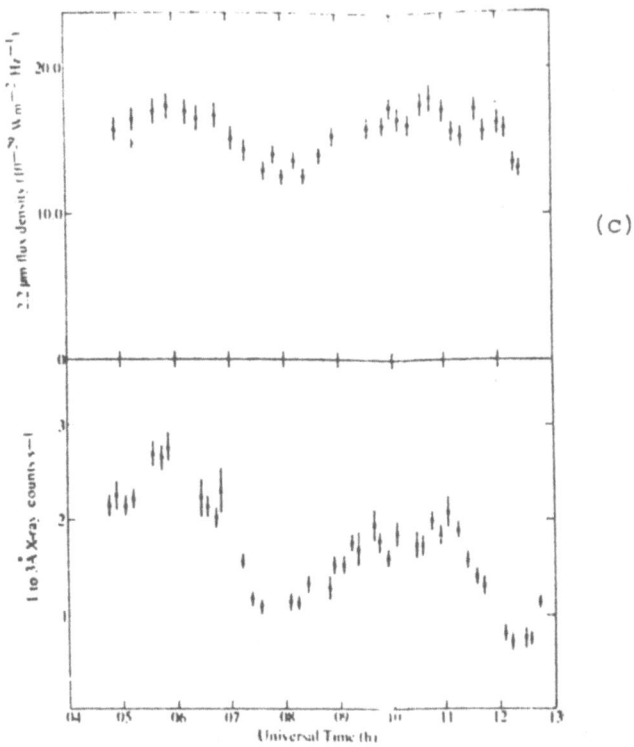

(c)

Figure 10.   Optical and X-ray light curves for the three
'extended' X-ray sources discussed in the text: (a) 2129+47,
taken from McClintock et al. (1982); (b) 1822-37, taken from
White et al. (1981); and Cyg X-3, taken from Becklin et al. (1973).

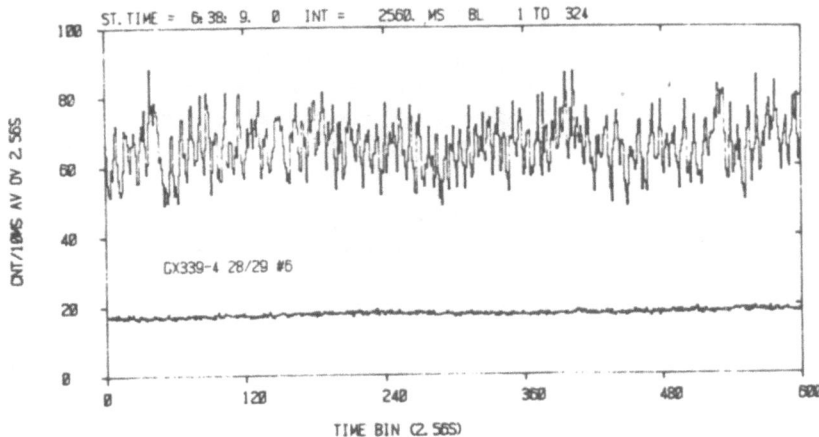

Figure 11a. Optical light curve for GX 339-4 on 28/29 May, 1981 showing the 20 sec quasi-periodic oscillations. The lower trace is that for the sky. Data were taken in white light with 2-channel photometer at the ESO 1.54m Danish telescope at ESO (see Motch et al. 1981).

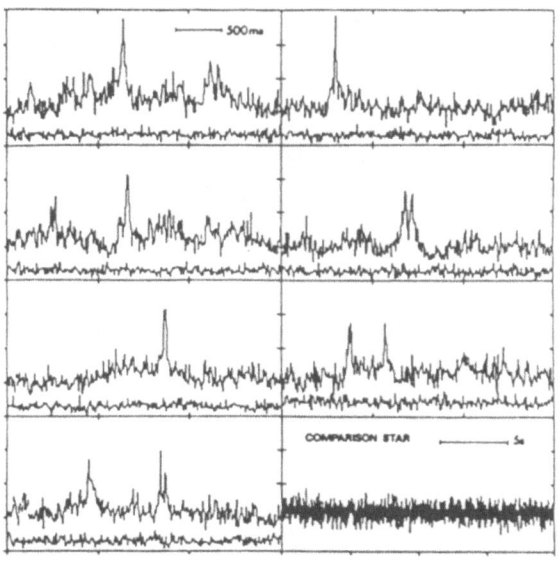

Figure 11b.  A sample of GX 339-4 flares picked up in data from
data with 10 ms time resolution taken from Motch et al. (1981).
The lower trace in each is the sky.  The last (lower-right) trace
is for the comparison star, shown with a compressed time scale.
Counting rates range from 0 to 240 counts/10 ms.  Time scales are
shown.

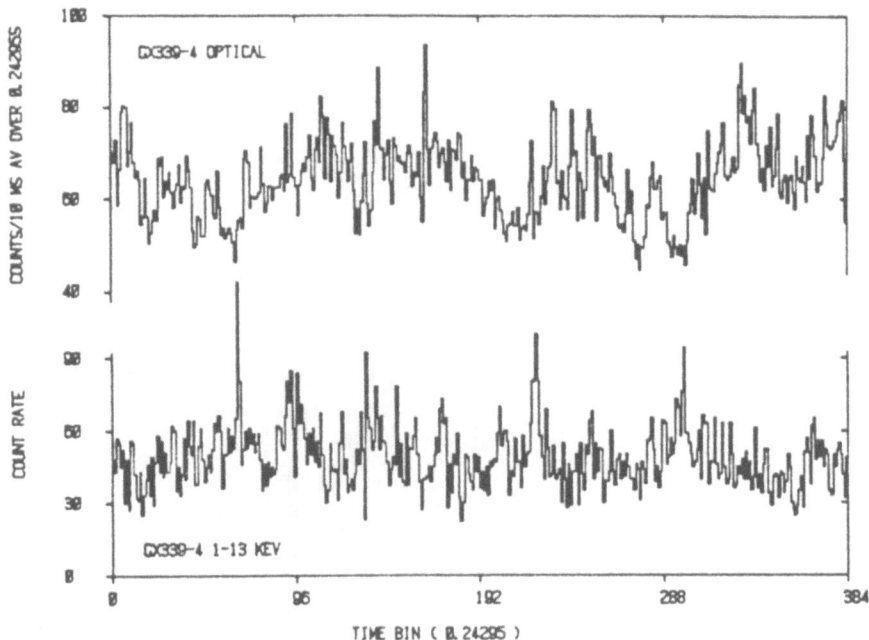

Figure 12a.   Simultaneous X-ray/Optical observations of GX 339-4
taken from Motch <u>et al</u>. (1983).   The optical are from the 1.54m
telescope at ESO and the X-ray data from the Ariel 6 satellite.
Data are shown in 0.243 sec bins.

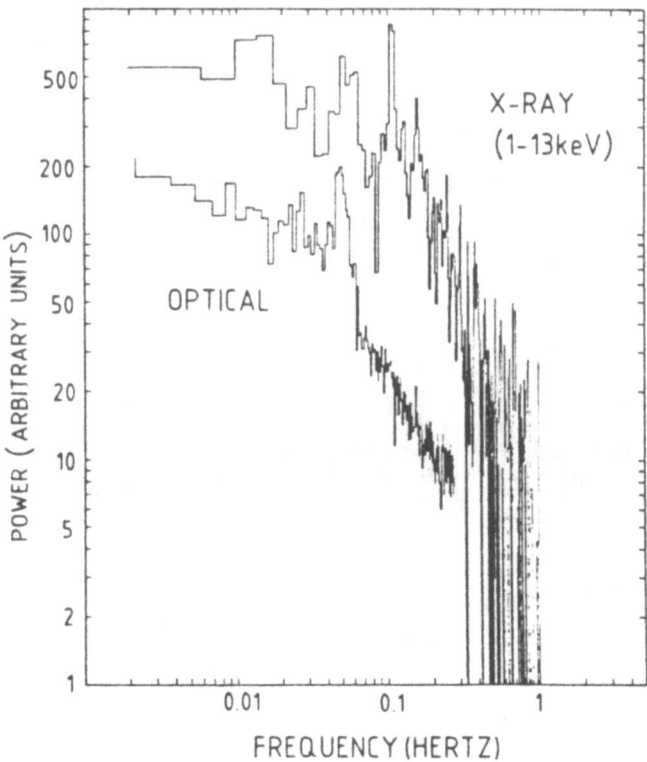

Figure 12b.  Power spectra for the entire X-ray and optical data sets for GX 339-4 during the 'hard' flaring state taken from Motch et al. (1983). Note the prominent peak at 20 sec in the optical data and the 10 sec peak in the X-ray data, the flat region below those peaks and the 1/f slope above.

## ACKNOWLEDGEMENTS

I am happy to thank C. Motch, C. Chevalier, J.M. Bonnet-Bidaud, M. Mouchet, M. Pakull, J. van Paradijs, M. van der Klis, A. Fabian and N. White for useful discussions and colleagues elsewhere for sending preprints of their work. Research at Besançon is supported by the CNRS (ERA 904), MEN, INAG and CNES.

## REFERENCES

Argue, A.N. & Sullivan, C., 1982. Observatory, **102**, 4.
Becklin, E.E., Neugebauer, G., Hawkins, F.J., Mason, K.O., Sanford, P.W., Mathhews, K., Wyn-Williams, C.G., 1973. Nature, **245**, 302.
Bolton, C.T., 1972. Nature, **240**, 124.
Bolton, C.T., 1975. Ap.J., **200**, 269.
Bonnet-Bidaud, J.M., Ilovaisky, S., Mouchet, M., Hammerschlag-Hensberge, G., van der Klis, M., Glencross, W.M., & Willis, A.J., 1981. Astr.Ap., **101**, 184.
Boynton, P.E., Crosa, L.M. & Deeter, J.E., 1980. Ap.J., **237**, 169.
Bradt, H.V.D. & McClintock, J.E., 1983. Ann.Revs.Astr.Ap., **21**, 13.
Charles, P., 1982. In "Accreting Neutron Stars" (W. Brinkman and J Trumper, Eds.) Proc. M.P.E. Workshop, July 19-23, 1982, Garching, p.1.
Chevalier, C. & Ilovaisky, S.A., 1977. Astr.Ap., **59**, L9.
Chevalier, C., Ilovaisky, S.A., Motch, C., Pakull, M., Lub, J. & van Paradijs, J.A., 1981. Sp.Sci.Revs., **30**, 405.
Chevailier, C., Janot-Pacheco, E., Mauder, H. & Ilovaisky, S.A., 1980. Astr.ap., **81**, 368.
Cominsky, L. & Wodd, K., 1984. Preprint.
Corbert, R.H.D. & Mason, K.O., 1984. Preprint.
Cordova, F. & Mason, K.O., 1983. In "Accretion-Driven Stellar X-ray Sources", (W.H.G. Lewin and E.P.J. van den Heuvel, Eds), Cambridge Univ. Press.
Cowley, A.P. & Crampton, D., 1975. Ap.J. (Lett.), **201**, L65.
Cowley, A.P., Crampton, D., & Hutchings, J.B., 1978. Astr.J., **83**, 1619.
Cowley, A.P. Crampton, D., & Hutchings, J.B., 1979. Ap.J., **231**, 539.
Cowley, A.P., Crampton, D., Hutchings, J.B., Remillard, R. & Penfold, J.E., 1983. Ap.J., **272**, 118.
Cowley, A.P., Crampton, D. & Hutchings, J.B., 1982. Ap.J., **256**, 605.
Crampton, D., Hutchings, J.B. & Cowley, A.P., 1978. Ap.J. (Lett.), **225**, L63.
Crosa, L. & Boyton, P.E., 1980. Ap.J., **235**, 999.

Davelaar, J., Blissett, R., Stella, L., McKay, M., White, N.E.
    & Bleeker, J., 1983.  IAUC No. 3893.
Deeter, J., Crosa, L., Gerend, D. & Boynton, P.E., 1976.
    Ap.J., **206**, 861.
Doxsey, R., Grindlay, J., Griffiths, R., Bradt, H., Johnston, M.,
    Leach, R., Schwartz, D. & Schwartz, J., 1979.  Ap.J., (Lett.),
    **228**, L67.
Dupree, A.K., Gursky, H., Black, J.H., Davis, R.J., Hatmann, L.,
    Matilsky, T., Raymond, J.C., Hammerschlag-Hensberge, G.,
    van den heuvel, E.P.J., Burger, M., Lamers, H.J.G.,L.M.,
    vanden Bout, P.A., Morton, D.C., de Loore, C., van Dessel, E.L.,
    Menzies, J.W., Whitelock, P.A., Watson, M., Sanford, P.W.
    & Pollard, G.S.C., 1980.  Ap.J., **238, 969.**
Epstein, A., Delvaille, J., Helmken, H., Murray, S., Schnopper, H.,
    Doxsey, R. & Primini, F., 1977.  Ap.J., **216**, 103.
Fabian, A.C., Guilbert, P.W., Motch, C., Ricketts, M.,
    Ilovaisky, S.A. & Chevalier, C., 1982.  Astr.Ap., **111**, L9.
Gerend, D. & Boynton, P.E., 1976.  Ap.J., **209**, 562.
Griffiths, R.E. & Seward, F.D., 1977.  Mon.Not.R.astr.Soc.,
    **180**, 75P.
Grindlay, J.E., 1979.  Ap.J. (Lett.), **232**, L33.
Guilbert, P.W. & Fabian, A.C., 1982.  Nature, **296**, 226.
Hackwell, J.A., Grasdalen, G.L., Gehrz, R.D., van Paradijs, J.,
    Cominsky, L. & Lewin, W.H.G., 1979.  Ap.J. (Lett.), **233**, L115.
Hatchett, S. & McCray, R., 1977.  Ap.J., **211**, 552.
Hutchings, J.B., 1982.  In "Galactic X-Ray Sources", Sanford, P.,
    Laskarides, P., Salton, J. (Eds.), John Wiley and Sons, N.Y.
Hutchings, J.B., Crampton, D., Glaspey, J. & Walker, G.A.H.,
    1977.  Ap.J., **182**, 549.
Hutchings, J.B., Crampton, D., Cowley, A.P. & Osmer, P., 1977.
    Ap.J., **217**, 186.
Hutchings, J.B., Crampton, D. & Cowley, A.P., 1978.  Ap.J.,
    **225**, 548.
Hutchings, J.B.,Cowley, A.P. &  Crampton, D., 1981.  IAUC
    No. 3585.
Hutchings, J.B.,Cowley, A.P. &  Crampton, D., 1983.  Pub.astr.
    Soc.Pacific., **95**, 23.
Hutchings, J.B. Crampton, D. & Cowley, A.P., 1983.  Ap.J.
    (Lett.), **275**, L43.
Ilovaisky, S.A., 1982.  In "Galactic X-Ray Sources", Sanford, P.,
    Laskarides, P., Salton, J. (Eds.) John Wiley and Sons, N.Y.
Ilovaisky, S.A., 1983.  Physica Scripta (Nice Workshop
    Proceedings).
Ilovaisky, S.A., Motch, C. & Chevalier, C., 1978.  Astr.Ap.,
    **70**, L19.
Ilovaisky, S.A., Chevalier, C., White, N.E., Mason, K.O.,
    Sanford, P.W., Delvaille, J.P. & Schnopper, H.W., 1980.
    mon.Not.R.astr.Soc., **191**, 81.
Ilovaisky, S.A. & Chevalier, C., 1981.  IAUC No. 3586.

Ilovaisky, S.A., Chevalier, C. & Motch, C., 1981.  Space Sci.
     Revs., **30**, 415.
Ilovaisky, S.A., Chevalier, C., Motch, Ch., Pakull, M.,
     van Paradijs, J. & Lub, J., 1984. Astr.Ap. (submitted).
Joss, P.C. & Rappaport, S.A., 1979.  Astr.Ap., **71**, 217.
Joss, P.C. & Rappaport, S.A., 1984.  Ann.Rev.Astr.Ap., **22** (in
     press).
Kemp, J.C., Barbour, M.S., Henson, G.D., Kraus, D.J., Nolt, I.G.,
     Radostitz, J.V., Priedhorsky, W.C., Terrell, J. & Walker,
     E.N., 1983. Ap.J. (Letters), **271**, L65.
van der Klis, M., Hammerschlag-Hensberge, G., Bonnet-Bidaud, J.M.,
     Ilovaisky, S.A., Mouchet, M., Glencross, W.M., Willis, A.J.,
     van Paradijs, J.A., Zuiderwijk, E.J. & Chevalier, C., 1981.
     Astr.Ap., **106**, 339.
van der Klis, M., Tjemken, S. & van Paradijs, J., 1983. Astr.Ap.,
     **126**, 265.
Lang, F.L., Levine, A.M., Butz, M., Hauskins, S., Howe, S.,
     Primini, F.A., Lewin, W.H.G., Baity, W.A., Knight, F.K.,
     Rothschild, R.E. & Patterson, J.A., 1981.  Ap.J. (Letters),
     **246**, L21.
Lawrence, A., cominsky, L., Engelke, C., Jernigan, G., Lewin,
     W.H.G., Matsuoka, M., Mitsuda, K., Oda, M., Ohashi, T.,
     Pederson, H. & van Paradijs, J., 1983.  Ap.J., **271**, 793.
Li, F.K., Joss, P.C., McClintock, J.E., Rappaport, S.A. &
     Wright, E.L., 1980.  Ap.J., **240**, 628.
Li, F., Rappaport, S. & Epstein, A., 1978. Nature, **271**, 37.
Maejima, Y., Makishima, K., Matsuoka, M., Ogawara, Y., Oda, M.,
     Tawara, Y. & Doi, K., 1984.  Preprint.
Marshall, N. & Millit, J.M., 1981.  Nature, **293**, 379.
Mason, K.O., Becklin, E.E., Blankenship, L., Brown, R.L.,
     Elias, J., Hjellming, R.M., Matthews, K., Monse, M.,
     Murdin, P.G., Neugebauer, G., Sanford, P.W. &
     Willner, S.P., 1976. Ap.J., **207**, 78.
Mason, K.O., Middleditch, J., Nelson, J.E., White, N.E.,
     Seitzer, P., Tuohy, I.R. & Hunt, L.K., 1980. Ap.J.
     (Lett.), **242**, L109.
Mason, K.O. & Corbett, B., 1984.  Preprint.
Matilsky, T., Bradt, H.V., Buff, J., Clark., G.W., Jernigan, J.,
     Joss, P., Laufer, B., McClintock, J.E. & Zubrod, D., 1976.
     Ap.J., (Lett.), **210**, L127.
McClintock, J.E., Canizares, C.R., van Paradijs, J., Cominsky, L.,
     Li, F., Lewin, W.H.G. & Grindlay, J.E., 1979. Nature, **279**, 47.
McClintock, J.E., London, R.A., Bond, H.E. & Grauer, A.D., 1982.
     Ap.J., **258**, 245.
McLintock, J.E. & Rappaport, S.A., 1984.  In "7th North
     American Workshop on Cataclysmic Variables and Low-Mass
     Binaries", in press.
McClintock, J.E., Remillard, R.A. & Margon, B., 1981.  Ap.J.,
     **243**, 900.
McClintock, J.E. & Petro, L.D., 1981.  IAUC No. 3615.

McClintock, J.E., Petro, L.D., Remillard, R.A. & Ricker, G.R.,
     1983. Ap.J. (Lett.), **266**, L27.
Merritt, D. & Petterson, J.A., 1980. Ap.J., **236**, 255.
Middleditch, J., 1983. Ap.J., **275**, 140.
Middleditch, J. & Nelson, J.E., 1976. Ap.J., **208**, 567.
Middleditch, J., Mason, K.O., Nelson, J.E. & White, N.E.,
     1981. Ap.J., **244**, 1001.
Milgrom, 1978. Astr.Ap. **67**, L25.
Motch, C., Ilovaisky, S.A. & Chevalier, C., 1981. Astr.Ap.,
     **109**, L1.
Motch, C., Ricketts, M., Page, C., Ilovaisky, S.A. &
     Chevalier, C., 1983. Astr.Ap., **119**, 171.
Mouchet, M., Ilovaisky, S.A. & Chevalier, C., 1980. Ap.J.,
     **90**, 113.
Murdin, P.G., Allen, D.A., Morton, D.C., Whelan, J. & Thomas,
     R.M., 1980. Mon.Not.R.astr.Soc., **192**, 709.
Nicolson, G.D., Feast, M.W. & Glass, I.S., 1980. Mon.Not.
     R.astr., Soc., **191**, 293.
Nolan, P.L., Gruber, D.E., Matteson, J.L., Peterson, L.E.,
     Rothschild, R.E., Doty, J.P., Levine, A.M., Lewin, W.H.G.
     & Primini, F.A., 1981. Ap.J., **246**, 494.
Oda, M., 1977. Space Sci.Rev., **20**, 757.
Oke, J.B., 1977. Ap.J., **217**, 181.
Pakull, M., Ilovaisky, S.A. & Chevalier, C., 1984.
     In preparation.
Papaloizou, J.C.B. & Pringle, J.E., 1983. Mon.Not.R.astr.
     Soc., **202**, 1181.
van Paradijs, J., 1981. Astr.Ap., **103**, 140.
van Paradijs, J., 1983. In "Accretion-Driven Stellar X-Ray
     Sources", (W.H.G. Lewin and E.P.J. van den Heuvel, Eds.),
     Cambridge University Press.
van Paradijs, J. & Zuiderwijk, E., 1977. Astr.Ap., **61**, L19.
van Paradijs, J., Verbunt, F., van der Linden, T., Pedersen, H.,
     Wamsteker, W., 1980. Ap.J. (Lett.), **241**, L161.
Parmar, A.N., Blissett, T., Courvoisier, T. & Chiappetti, L.,
     1984. IAUC No. 3906.
Pedersen, H., 1984. Invited talk 7th European Regional
     Meeting, Florence, December 1983.
Pedersen, H., van Paradijs, J. & Lewin, W.H.G., 1981. Nature,
     **294**, 725.
Penston, M.V., Penston, M.J., Murdin, P. & Martin, W., 1975.
     Mon.Not.R.astr.Soc., **172**, 313.
Petro, L.D., Bradt, H.V., Kelley, R.L., Horne, K. & Gomer, R.,
     1981. Ap.J. (Lett.), **251**, L7.
Petterson, J.A., 1975. Ap.J. (Letters), **201**, L61.
Petterson, J.A., 1977. Ap.J., **218**, 783.
Petterson, J.A., 1978. Ap.J., **224**, 625.
Pietsch, W., Steinle, H. & Gottwald, M., 1983. IAUC No. 3887.
Ponman, T., 1982. Mon.Not.R.astr.Soc., **200**, 351.
Priedhorsky, W.C. & Terrell, J., 1983a. Nature, **303**, 681.

Priedhorsky, W.C. & Terrell, J., 1983b.  Ap.J., **273**, 709.
Priedhorsky, W.C. & Terrell, J., 1984.  Ap.J., (in press).
Priedhorsky, W.C., Terrell, J., & Holt, S.S., 1983.  Ap.J.,
    **270**, 233.
Rapley, C.G. & Tuohy, I.R., 1974.  Ap.J. (Letters), **191**, L113.
Rappaport, S.A. & Joss, P.C., 1983.  In "Accretion-Driven
    Stellar X-Ray Sources" (W.H.G. Lewin and E.P.J. van den
    Heuvel, Eds), Cambridge Univ. Press.
Rappaport, S.A. & van den Heuvel, E.P.J., 1983.  In IAU
    Symposium No. 98 (Be Stars), D. Reidel Publ. Co.
Ricketts, M., 1983.  Astr.Ap., **118**, L3.
Sadeh, D., Meidav, M., Wood, K., Yentis, D., Smathers, H.,
    Meekins, J., Evans, W., Byram, E.T., Chubb, T.A. &
    Friedman, H., 1979.  Nature, **278**, 436.
Samimi, J., Share, G.H., Wood, K., Yentis, D., Meekins, J.,
    Evans, W.D., Shulman, S., Byram, E.T., Chubb, T.A. &
    Friedman, H., 1979.  Nature, **278**, 434.
Sanford, P.W., Ives, J.C., Bell-Burnell, S.J., Mason, K.O.
    & Murdin, P.G., 1975.  Nature, **256**, 109.
Skinner, et al. 1980.  Ap.J., **240**, 619.
Smale, A.P., Charles, P.A., Tuohy, I.R., & Thorstensen, J.R.,
    1984.  Mon.Not.R.astr.Soc., **207**, 29P.
Stella, L. & White, N.E., 1983.  IAUC No. 3902.
Tanaka, Y. & the Tenma Team 1983.  IAUC No. 3882.
Thorstensen, J., Charles, P. & Bowyer, S., 1978.  Ap.J. (Lett.),
    **220**, L131.
Thorstensen, J., Charles, P., Bowyer, S., Briel, U.G., Doxsey,
    R.E., Griffiths, R.E., &  Schwartz, D.A., 1979.  Ap.J.
    (Lett.), **233**, L57.
Treves, A., Chiapetti, L., Tanzi, E.G., Tarenghi, M., Gursky, H.,
    Dupree, A.K., Hartman, L.W., Raymond, J., Davis, R.J.,
    Black, J., Matilsky, T., vanden Bout, P., Sanner, F.,
    Pollard, G., Sanford, P.W., Joseph, R.D., Meikle, &  W.P.S.,
    1980.  Ap.J., **242**, 1114.
Tuohy, I.R. & Rapley, C.G., 1975.  Ap.J. (Letters), **198**, L69.
Warren, P.R. & Penfold, J.E., 1975.  Mon.Not.R.astr.Soc.,
    **172**, 41P.
Watson, M., 1976.  Mon.Not.R.astr.Soc., **176**, 19P.
Weisskopf, M.C, Kahn, S.M., Darbo, W.A., Elsner, R.F., Grindlay,
    J.E., Naranan, S., Sutherland, P.G. & Williams, A.C., 1983.
    Ap.J. (Lett.), **274**, L65.
Whelan, J.A.J., Mayo, S.K., Wickramasinghe, D.T., Murdin, P.G.,
    Peterson, B.A., Hawarden, T.G., Longmore, A., Haynes, R.F.,
    Goss, W.M., Simons, L., Caswell, J., Little, A.G. &
    McAdam, W.B., 1977.  Mon.Not.R.astr.Soc., **181**, 259.
White, N.E., 1978.  Nature, **271**, 38.
White, N.E., 1984.  Preprint.
White, N.E., Fabian, A.C. & Mushotzky, R.F., 1984.  Preprint.
White, N.E., Becker, R.H., Boldt, E.A., Holt, S.S., Serlemitsos,
    P.J. & Swank, J.H., 1981.  Ap.J., **247**, 994.

White, N.E. & Holt, S.S., 1982. Ap.J., **257**, 318.
White, N.E. & Marshall, F.E., 1984. Ap.J.,(Lett.), submitted.
White, N.E. Swank, J.H. & Holt, S.S., 1983. Ap.J., **270**, 711.
White, N.E., Davelaar, J., Parmar, A.N., Stella, L. & van der
     Klis, M., 1984. IAUC No. 3912.

# MASSIVE X-RAY BINARIES

N.E. White

Space Science Department of The European Space Agency
European Space Research and Technology Centre
Noordwijk, The Netherlands

## 1.  INTRODUCTION

An OB star loses mass via a stellar wind and any collapsed binary companion to such a star will capture a fraction of that material, releasing gravitational potential energy predominantly as X-ray emission. If in addition the OB star fills its critical potential lobe, then material will also spill onto the compact star via an accretion disk. About one quarter of the 100 or so known galactic X-ray sources with luminosities greater than $10^{32}$ erg s$^{-1}$ have been identified with such systems. Fourteeen contain X-ray pulsars and a further three have, on the basis of mass estimates for the X-ray source, been suggested as black hole candidates. For many, X-ray pulse timing and radial velocity measurements in conjunction with X-ray eclipse durations have allowed the orbital parameters to be determined.

The systems containing O or B supergiants, where mass loss rates of $10^{-6}$ M$_\odot$ yr$^{-1}$ are typical, have binary separations of order 1.5 to a few stellar radii such that the compact object is well embedded in the stellar wind of the primary (cf. Conti 1978). In contrast the systems containing B stars still close to the main sequence tend to have longer orbital periods and wider binary separations. This paper concentrates on the seven supergiant systems listed in Table 1, although many of the properties of these systems also occur in the B-star main sequence X-ray binaries. Detailed reviews of the main sequence systems can be found in Rappaport & van den Heuvel (1982) and White et al. (1982). The orbital periods of the massive systems range from 2.1 days to 41.5 days with

*P. P. Eggleton and J. E. Pringle (eds.), Interacting Binaries, 249–287.*
© *1985 by D. Reidel Publishing Company.*

all but GX 301-2 and Cyg X-1 showing X-ray eclipses by the
companion star. The X-ray pulsars Cen X-3, 4U0900-40, SMC X-1,
GX 301-2 and 4U1538-52 have periods ranging between 0.7 s and
700 s and it is generally accepted that these represent the
rotation periods of accreting neutron stars (cf. Rappaport &
Joss 1977). X-ray pulsations have yet to be detected from
4U1700-37 although this too is most likely an accreting neutron
star (White, Kallman & Swank 1983). Radial velocity
measurements of the optical counterpart to Cyg X-1 suggest
that the X-ray source has a mass of 10 $M_\odot$ and that is an
accreting black hole (for example Bolton 1982).

While the X-ray properties show great diversity from
source to source, there are common elements and underlying
trends in the group as a whole. Three representative orbital
light curves are given in Figure 1. The degree of variability
seen on timescales of a few minutes to hours from a particular
object is related to the time average X-ray luminosity $L_x$ ,
with the lowest luminosity systems showing the most activity.
Outside of eclipse the X-ray flux from 4U1700-37 where $L_x \sim$
$10^{36}$ erg s$^{-1}$ continuously varies on a timescale of 5 to 10
minutes by up to a factor of 10 (Figure 1a, taken from Mason,
Branduardi & Sanford 1976, see also Jones et al. 1973).
Similar activity dominates the light curves of 4U900-40 (Vela
X-1, $L_x \sim 10^{36}$ erg s$^{-1}$ ; Charles et al. 1978, Watson &
Griffiths 1977) and GX 301-2 (4U1223-62, $L_x \sim 10^{36}$ to $10^{37}$
erg s$^{-1}$, Watson, Warwick & Corbett 1982, White, Mason & Sanford
1976). This flaring is not accompanied by any significant
change in the spectrum of the underlying continuum. The
non-eclipsing systems Cyg X-1 where $L_x \sim 10^{37}$ erg s$^{-1}$
undergoes $\sim 20\%$ variability on a timescale of a few hours
(Figure 1b taken from Mason et al. 1974). SMC X-1 where
$L_x$ on occasions reaches $10^{39}$ erg s$^{-1}$ (Figure 1c taken from
Marshall, White & Becker 1983, see also Bonnet-Bidaud & van
Klis 1981) and Cen X-3 with $L_x$ up to $10^{38}$ erg s$^{-1}$ (Figure 2a,
taken from Schreier et al. 1976; Bonnet-Bidaud & van der
Klis 1979) both display little, if any, flaring activity.

Table 1:   The Supergiant OB X-Ray Binaries

| Source | Primary | $P_{orb}$ | A | $R_*$ | Log $L_x$ | Log $\dot{M}$ | $V_t$ |
|--------|---------|-----------|------|------|-----------|---------------|-------|
| Cen X-3 | 06.5III | 2.1 | 1.52 | 12.5 | 37-38 | -6.0 | 2200 |
| 4U1700-33 | 06 f | 3.4 | 1.45 | 20 | 35.6 | -5.0 | 2600 |
| 4U1538-53 | B0 I | 3.7 | 1.8 | 20 | 36.6 | -6.0 | 1500 |
| SMC X-1 | B0 I | 3.9 | 1.73 | 15 | 38.7 | -6.3 | 1600 |
| Cyg X-1 | 09.7 Iab | 5.6 | 2.39 | 18 | 36.8 | -5.6 | 2000 |
| 4U0900-40 | B0.5 Iab | 9.0 | 1.51 | 35 | 36.0 | -5.7 | 1700 |
| GX301-2 | B2 I | 41.5 | 2.6 | 45 | 37.0-36.3 | -5.7 | 1500 |
|  |  | day | $R_*$ | $R_\odot$ | erg s$^{-1}$ | $M_\odot$yr$^{-1}$ | km s$^{-1}$ |

The X-rays as  they  propagate through the stellar wind of the   primary   may   undergo  photo-electric  absorption. Phase-related orbital variations in absorption are commonly seen, with additional variations on shorter and longer timescales. 4U0900-40 and 4U1700-37 undergo a general increase in absorption after $\phi$ = 0.5, where $\phi$ = 0 defines the mid-eclipse of the X-ray source (Figure 2b taken from Mason, Branduardi & Sanford 1976; see also Charles et al. 1978). This asymmetry in absorption about $\phi$ = 0.5 is superimposed on top of that caused by viewing close to the limb of the OB star either side of X-ray eclipse. Absorption dips lasting from a few minutes to an hour are also a common feature from the non-eclipsing system Cyg X-1 around $\phi$ = 0, when the X-ray source is viewed over the limb of the primary (Figure 1b; Mason et al. 1974; Li & Clark 1974). Absorption dips also occur in 4U0900-40 preferentially around $\phi$ = 0.2 to 0.3 (Becker et al. 1978) and from SMC X-1 close to X-ray eclipse (Marshall, White & Becker 1983). Variations in absorption up to a few $10^{23}$ H cm$^{-2}$ on a timescale of a day have been reported from GX301-2 (Swank et al. 1976).

The more luminous systems Cen X-3 and SMC X-1 vary by a factor of ten (or more) on a timescale of order 50 to 100 days between "low" and "high" states. Two distinct types of low state have been identified from Cen X-3 by Schreier et al. (1976). In one the reduction in flux is caused by a decrease in the flow of material to the compact object. The other is the result of a large increase in absorption ($N_H \gtrsim 10^{24}$ H cm$^{-2}$) "snuffing out" the X-ray source. Transitions out of the latter are very distinctive with first a strong spike appearing at $\phi$ = 0.5 that on subsequent orbit cycles broadens out to the normal eclipsed light curve (Figure 2a from Schreier et al. 1976; see also Pounds et al. 1975). When Cen X-3 is in a high state and/or between states an asymmetric increase in absorption is also sometimes seen from this source after $\phi$ =

0.5 (Pounds et al. 1976; Tuohy & Cruise 1975). Only low
states  associated with decreases in the accretion rate have to
date been identified from SMC X-1.

Pulse timing measurements have allowed the eccentricity of
the  orbits to be determined to high  precision.   This  ranges
from effectively  zero  in   the   shortest orbital period system
Cen X-3 (e = 0.0008; Fabbiano  &  Schreier 1977) to e = 0.47 in
the longest (GX 301-2; White & Swank 1984;  Kelley, Rappaport &
Petre  1980).  A regular outburst is seen from GX  301-2  every
41.5 days,  around  the  time  of  periastron  passage (Watson,
Warwick  & Corbett 1982 and refs. therein).  Secular changes in
the pulse periods of all these systems indicate the net angular
momentum vector of the accreted material to be non-zero.

In the   next   section   the observed X-ray luminosities are
compared with those expected from  stellar  wind capture by the
compact object.  This analysis is  extended in Section 3 to use
the observed X-ray absorptions to investigate the properties of
the stellar wind in the line of sight.  Section 4 shows how the
compact objects in these  systems can be used to probe the fine
structure  of  the stellar wind.   Section   5  considers  the
implications of observed  pulse  period changes for the stellar
wind accretion model.  The final section (Section 6) summarises
the current view of these systems and the outstanding problems.


2.    STELLAR WIND ACCRETION

Davidson & Ostriker (1973) showed that the stellar wind of
an OB supergiant could provide sufficient material to  power an
accretion driven X-ray source.  For  a  cylindrically symmetric
flow past a compact star the gravitational  deflection  of  the
flow  will  cause  all  material within the "capture radius" $R_c$
given by

$$R_c \sim \frac{2GM_x}{V_x^2 + V_w^2}$$

2.1

to  collide  and subsequently be accreted (Bondi & Hoyle 1944).
$M_x$ and $V_x$  are  the  mass  and  orbital velocity of the X-ray
source, and $V_w$ is the wind velocity in  the  vicinity  of  the
X-ray  source.   Accretion  onto  a  neutron star or black hole
releases  $\sim 10\%$ of the rest  mass energy and this combined with
the accretion cross section of $\sim \pi R_c^2$   yields $L_x \sim \dot{M}.V_w^{-4}$ (for
$V_w \gg$    $V_x$ ) such that  the  X-ray  luminosity  is  extremely
sensitive to the assumed wind velocity.

Lamers, van den Heuvel & Petterson (1976), Conti (1978), and Petterson (1978) have all attempted to use the known properties of stellar winds of OB stars to determine whether stellar wind capture can account for the observed X-ray luminosities. In such an investigation two assumptions are usually made: (1) that the stellar wind is radiatively driven with $\dot{M} \propto L_{BOL}^{-1.8}$, where $L_{BOL}$ is the bolometric luminosity of the OB star and (2) the terminal velocity, $V_t$ is three times the escape velocity from the surface of the supergiant. These two assumptions are based on the observed optical and ultraviolet properties of OB stars (see for example Abbott 1982 and refs. therein). A radiatively driven wind is expected to have a velocity law of the form (Castor, Abbott & Klien 1975)

$$V(R) = V_t \left(1 - \frac{R_*}{R}\right)^{\beta},$$

2.2

where $R/R_*$ is the distance from the primary in stellar radii and $\beta$ is a constant usually set to 0.5, although this has not yet been observationally confirmed. Figure 3a shows that at the orbit of the X-ray source in these systems the stellar wind has, according to this law, reached between 50 and 70% of its terminal value. For $V_x \ll V_w$ we can write

$$\frac{\dot{M}}{V_t^4} = \left(\frac{L_x}{\eta c^2}\right) \cdot \left(\frac{A}{GM_x}\right)^2 \left(1 - \frac{R_*}{A}\right)^2,$$

2.3

where A is the binary separation and $\eta$ the efficiency of energy production ($\sim$ 0.1). The binary parameters and X-ray luminosities for the X-ray binaries given in Table 1 can be input into equation 2.3 to compare $(\dot{M}/V_t^4)_{req}$, the ratio of the stellar wind values required to power the X-ray emission with $(\dot{M}/V_t^4)_{exp}$ the ratio of those expected (or measured) for the OB stars in question. As concluded by Lamers et al. 81976), and later by Conti (1978) and Petterson (1978) 4U0900-40, 4U1700-37 and Cyg X-1 can all be powered by pure stellar wind capture, but Cen X-3 and SMC X-1 (and possibly the less well studied 4U1538-52) all require larger accretion rates than can be plausibly provided by an undisturbed stellar wind.

The above papers discuss various ways of providing extra material that centre on the fact that the primary may be at or close to its critical potential lobe. If the primary fills its critical lobe then it must, in order to avoid a catastrophically high mass transfer rate, be expanding on a nuclear rather than thermal timescale such that the supergiant is still undergoing core nuclear burning. This is probably the case for all but 4U0900-40 (Savonije 1978). Another possiblity is that the wind velocity at the X-ray source is much lower than expected, perhaps because the radiatively driven wind is disrupted as a result of the primary coming close to its critical potential lobe (Friend & Castor 1982) or because the X-ray source ionizes away the line transitions responsible for radiatively accelerating the wind (Hatchett & McCray 1977).

The determination of the 41.5 day orbital period of GX301-2 from periodic outbursts (Watson, Warwick & Corbett) allows further insight into this problem since the highly eccentric (e = 0.47) orbit samples the stellar wind over a range of distances from the primary. At no point in this orbit does WRA977 come close to filling its critical potential lobe (Kelley, Rappaport & Petre 1980). The stellar wind capture model can account for the 41.5 day outburst cycle as due to enhanced accretion at periastron (Watson, Warwick & Corbett 1982; Kelley, Rappaport & Petre 1980; White & Swank 1984). GX301-2 lies in the same area of Figure 3b as Cen X-3 and SMC X-1, i.e. the nominal stellar wind parameters for WRA977 do not provide sufficient material to power the overall level of X-ray emission.

Figure 4 shows the expected 41.5 day X-ray light curve for accretion by a neutron star of material in the stellar wind from the primary WRA977 (taken from White & Swank 1984). This modelling includes the effects of the X-ray source velocity and photoelectric absorption by the stellar wind (the latter will be discussed in more detail in paragraph 3). The nominal values for $\dot{M}$ of $2 \times 10^{-6}$ $M_\odot$ yr$^{-1}$ and $V_t$ of 1200 km s fail to reproduce not only the overall luminosity, but also the ratio of the minimum to maximum flux. The latter can only be accommodated by a factor of three reduction in the velocity of the wind within the X-ray orbit. Parkes et al. (1976) in measuring the P Cygni profiles of WRA977 also find evidence for low wind velocities within a few stellar radii. This still does not bring the range of predicted luminosity into agreement with that observed which requires $\dot{M}$ to be increased by a factor of 5 to 10, close to the maximum possible when all the momentum of the radiation field is transferred to the wind. Thus the stellar wind of WRA977 is extreme when compared to that of a single star of the same type. Unlike Cen X-3 and SMC X-1 this cannot be due to the primary coming close to its critical potential lobe.

## 3.    PHOTOABSORPTION BY THE STELLAR WIND

The X-ray spectrum below a few keV may be severely attenuated by photoelectric absorption by the K and L shells of the more abundant low Z elements (Buff & McCray 1974). The atomic number density $n_x$ of the stellar wind at the orbit of the X-ray source should be of order $10^9 - 10^{11}$ cm$^{-3}$. Integrating along the line of sight through the wind when $\phi = 0.5$ gives, for solar abundances, a range of equivalent hydrogen column densities $N_H$ between $10^{21}$ and $10^{23}$ H cm$^{-2}$ (the optical depth at 1 keV is unity for a column density of $10^{22}$ H cm$^{-2}$). The column density scales as $\dot{M}.v^{-1}$, making it three powers less sensitive to the velocity of the wind than the X-ray luminosity (paragraph 2). While it is simple to estimate the absorption as a function of orbital phase, there is the additional complication that the X-ray source will photoionise the wind and reduce its opacity.

For an optical thin gas in local ionisation and thermal balance irradiated by a point source of X-rays with a given spectral shape and luminosity $L_x$, a single parameter $\xi$ can be used to determine the ionisation state at any distance $r_x$ from the X-ray source (Tarter, Tucker & Salpeter 1969) such that

$$\xi = \frac{L_x}{n \, r_x^2} .$$

3.1

When $\xi > 10^3$ most of the low Z elements will be completely ionised (Hatchett, Buff & McCray 1976). "Stromgren spheres" of constant $\xi$ in the stellar wind can be defined for a constant velocity wind by introducing a second parameter q (Hatchett & McCray 1977) such that $(\text{for } \beta = 0)$

$$q = \frac{n_x A^2}{L_x} . \xi = \frac{r_*^2}{r_x^2}$$

3.2

where $r_*$ is the distance from the primary. Figure 5, taken from Hatchett & McCray (1977) illustrates that when q is >1.0 the surfaces of constant $\xi$ are closed spheres, but that as q decreases the ionised region expands until when q < 1.0 the surface is open. An order of magnitude estimate of q when $\xi = 10^3$ gives a useful estimate of the fraction of the wind that is completely photoionised. When a wind accelerating according to the velocity law 2.2 is included the surfaces of constant $\xi$ move towards the X-ray source such that they no longer intersect the primary (see Hatchett & McCray 1977); the order of magnitude estimate given for the constant velocity case, none-the-less, provides a useful indicator of the ionisation state of the wind.

The line of sight absorption $N_H$ will be a minimum at $\phi = 0.5$ and increase symmetrically either side of this. The abrupt increase in $N_H$ after $\phi = 0.5$ seen from many of these systems is not expected and, as will be discussed later, is the result of the passage of the X-ray source disrupting the wind. The value of $N_H$ measured by the Einstein SSS, HEAO-1 A2 and OSO-8 for several X-ray binaries between $\phi = 0.25$ and 0.5 are compared in Table 2 with the values of $N_H$ and q predicted by the nominal stellar wind parameters. PHA spectra of 4U0900-40 and 4U1700-37 (before and after $\phi = 0.5$) are shown in Figure 6 (taken from Kallman & White 1982; White, Kallman & Swank 1983). The increased low energy absorption after $\phi = 0.5$ is evident in both, with the absorption measured prior to $\phi = 0.5$

Table 2:   Absorption Values

| Source | q | log $(N_H)$ pred. | log $(N_H)$obs. |
|---|---|---|---|
| 4U0900-40 | 60 | 22.3 | 22.5 to 23.0 |
| 4U1700-37 | 530 | 23.0 | 23.0 to 23.2 |
| Cyg X-1 | 8 | 21.9 | 21.5 |
| SMC X-1 | 0.03 | 21.9 | 21.6 |
| GX301-2 | 30 to 6 | 22.0 | 23.0 [a] |
| Cen X-3 | 3 to 0.3 | 22.2 | ? |

a) See Figure 5

in good agreement with that expected from an undisturbed stellar wind. SMC X-1 and Cyg X-1 only give upper limits of $4 \times 10^{21}$ H cm$^{-2}$ (Pravdo et al. 1980; Marshall, White & Becker 1983), although in both the stellar wind of the OB star is relatively weak and the predicted $N_H$ only a factor of two above the upper limit; the low values of q for both suggests that this discrepancy can be accounted for by X-ray photoionisation of the wind.

The previous section discussed how the 41.5 day outburst cycle of the eccentric system GX301-2 can be reproduced using a stellar wind accretion model. Figure 4 also gives the computed values of $N_H$ as a function of orbital phase along with, in the bottom right panel, the values of $N_H$ measured at a variety of orbital phases by HEAO-1 A2 and OSO-8 (from White & Swank 1984). The nominal stellar wind parameters (bottom left panel)

not only fail to reproduce the light curve (Section 3) but also
by an order of magnitude the measured range of absorption.
While decreasing the velocity of the wind by a factor of 3
gives reasonable agreement with the minimum to maximum
intensities of the light curve, the absorption is still not in
agreement, which is a consequence of $N_H \propto v^{-1}$ whereas $L_X \propto$
$v^{-4}$. Only by increasing $\dot{M}$ by an order of magnitude does the
predicted absorption approach the measured values. There is a
dip in the observed light curve at the time when the X-ray
source is viewed over the limb of the OB star and absorption
by the wind is a maximum. The general decrease in absorption
is, as predicted, observed as the X-ray source moves out from
behind the primary. One observation made close to periastron
passage shows much higher absorption than expected by this
model. This observation was made after the X-ray source passed
in front of the primary and the increased absorption may be
similar to that seen from 4U0900-40, 4U1700-37 and Cen X-3
after $\phi = 0.5$.

Variations in the rate of mass loss and velocity of the
wind of the primary have been invoked to explain the low and
high states of Cen X-3. The distinctive expanding spike that
marks the transition from a low to high state (Figure 2a) can
be interpreted in terms of a change in the wind density causing
the surfaces of constant $\xi$ to expand and open up to expose the
X-ray source (Schreier et al. 1976; Hatchett & McCray
1977). The low energy turnover of Cen X-3 during its high
state has not been accurately measured, although it is
certainly less than a few $10^{22}$ H cm$^{-2}$. The observations made
by Einstein SSS found the source to be in a low state. The
orbital light curve seen by the Einstein SSS (in the 0.5 to
4.5 keV band) and MPC (2-10 keV) plus a hardness ratio from MPC
are shown in Figure 7a folded over the 2.1 day orbital period
(repeated for half a cycle for clarity). The eclipse centered
on $\phi = 0.0$ seen in the MPC is not total with the hardness
ratio indicating a softening of the spectrum. The SSS count
rate shows little evidence for any eclipse at $\phi = 0$. The PHA
spectra taken during eclipse reveal a hard spectrum with no
substantial low energy turnover (Figure 7b). Outside of the
eclipse the MPC spectrum (not shown in Figure 7b) shows an
additional hard component that is heavily cut off below 6 keV.

If the low state is caused by a dramatic increase in
absorption by the stellar wind, then the nature of the
unabsorbed eclipsed flux is puzzling. The hard X-ray spectrum
of the eclipsed emission is reminiscent of the uneclipsed high
state spectrum of the pulsar (also shown in Figure 7b) and
suggests that the residual eclipsed flux may be emission from
the pulsar being scattered around an absorbing medium. This
could either be an accretion disk, or a non-spherically

symmetric wind from the primary. The former would mean abandoning the variable stellar wind model for the low and high states, with no obvious alternative. A non-spherically-symmetric wind has been discussed by Friend & Castor (1982) who include the modifications to a radiatively driven stellar wind caused by an OB star in a binary system approaching its critical potential lobe. They find that as the primary approaches its tidal lobe the wind is enhanced in the direction of the compact object, with an associated drop in the terminal velocity. Quoting from Friend & Castor "as the star expands to fill its critical potential lobe, the wind gets stronger and slower and sharply peaked towards the X-ray source, until it strongly resembles Roche lobe overflow". The density profile is very sharply peaked in the orbital plane such that the X-ray source may be able to ionise a "Stromgren Tunnel" perpendicular to the orbital plane through which unabsorbed X-rays scatter to the observer. The calculations of Friend & Castor did not include the ionising effects of the X-ray source.

A radiatively driven wind depends on the absorption and scattering of photospheric continuum radiation in the ultraviolet resonance lines of abundant ions in the wind (Castor, Abbott & Klein 1975). The magnitude of this force is sensitive to the ionisation and thermal state of the wind, both of which will be distorted by the ionising radiation from the X-ray source. Hatchett & McCray (1977) predicted that the terminal velocity of the wind will be lowered because the X-rays will photoionise the wind in its vicinity, removing all UV resonance transitions and causing the wind to coast. This effect was observationally confirmed by Dupree et al. (1980) who used IUE to show that the terminal velocity given by the P Cygni profiles of the companion to 4U0900-40 is lower when the X-ray source is in front of the OB star. The degree to which this effect modifies the stellar wind has been considered in more detail by MacGregor & Vitello (1982). They find that for modest X-ray luminosities of $\sim 10^{34}$ erg s$^{-1}$ the number of UV resonance lines are increased and the stellar wind is slightly enhanced, but when $L_x \gtrsim 10^{35}$ erg s$^{-1}$ radiative acceleration is no longer possible because most of the wind illuminated by the X-ray source is too highly ionised. MacGregor and Vitello do not include in their calculations the enhanced outflow predicted by Friend & Castor which, as noted by the latter authors, may increase the X-ray luminosity at which radiative acceleration is curtailed. Nontheless it is rather surprising that for 4U0900-40 and 4U1700-37 both the observed luminosities and absorptions (prior to $\phi = 0.5$) are in reasonable agreement with that expected from a simple unmodified stellar wind model. This is even more surprising for Cyg X-1 where the entire wind exposed to the X-ray source should be so highly ionised that radiative acceleration cannot be at all relevant and yet the

observed luminosity is still close to that expected from pure
stellar wind accretion. An additional contribution to the mass
flow in this case may be the evaporation of material from the
surface of the primary by the X-ray source itself (McCray &
Hatchett 1974), although the good agreement with the stellar
wind capture model would then be coincidental.

The adverse effect of the X-ray emitting compact object on
the stellar wind is most clearly demonstrated by the increase
in absorption seen after the X-ray source has passed in front
of the OB star. Fransson & Fabian (1980) have suggested that
this is because the ionised part of the wind that is coasting
will, after the X-ray source has passed, then interact with the
unaffected faster moving material coming up behind it, causing
shocks, turbulence and other complications. This model is,
however, hard to follow in detail and seems to be in conflict
with the calculations of MacGregor & Vitello who find a much
larger region of the wind to be affected. The supersonic bow
shock model of Jackson (1975) where an accretion wake is formed
following the compact object cannot produce an extended region
of absorption between $\phi = 0.5$ and X-ray eclipse, although the
passage of the bow shock through the line of sight might
explain localised dips sometimes seen from Cen X-3 and
4U0900-40 (Livio, Shara & Shaviv 1979; Eadie et al. 1975).
If the primary is not in synchronous rotation with the orbit,
then any substantial gas stream flowing through the inner
Lagrangian point will not have sufficient angular momentum to
catch up with the compact object and the stream will fan out
trailing the compact object on its way out of the system
(Petterson 1978). There is some evidence for such a stream
from enhanced optical line emission from 4U0900-40 and
4U1700-37 at $\phi = 0.7$ (Conti & Cowley 1975; Bessell, Vidal &
Wickramasinghe 1975). However if the primary does fill its
critical potential lobe then the substantial distortion of the
radiatively driven wind discussed by Friend & Castor (1982) is
not reflected in the X-ray properties. At present there seems
to be no satisfactory quantitative explanation for the
increased $N_H$ after $\phi = 0.5$.

## 4.    PROBING THE STELLAR WINDS OF OB SUPERGIANTS

The preceeding two sections have assumed the winds of OB
stars to be be homogeneous and spherically symmetric; however
there is increasing evidence to the contrary. The most
compelling is the discovery of coronal emission from many OB
stars which indicates the presence of a $10^6$ -$10^7$ K plasma that
must somehow co-exist with the cooler outflowing wind. The
location of the hot plasma is as yet not known, but the fact
that the coronal X-ray luminosity is proportional to the

bolometric luminosity of the OB star (Pallavicini et al. 1981) suggests a link between it and the radiatively driven stellar wind. The X-ray spectra of the coronal emission do not show the substantially low energy cut-off that might be expected if it were located at the base of the wind (Long & White 1980 , Cassinelli & Swank 1983). Two models have been proposed that move the coronal emission out into the wind. Lucy & White (1980) find that a radiatively driven wind is unstable and breaks up into radiatively driven blobs that shock heat against an ambient medium. Alternatively, Underhill (1983) has suggested that magnetic fields may play an important role in the acceleration of the wind, which brings to mind analogies with the solar corona. In either event the compact object in the OB supergiant binary systems should be well embedded in the coronal region and can be used as a probe to test the above models.

If 4U0900-40, 4U1700-37 and GX301-2 are driven by pure stellar wind accretion, then the distinctive X-ray variability from these objects on timescales of minutes to a few hours must be caused by inhomogeneities in the stellar wind. Figure 8 shows two Einstein MPC observations of 4U1700-37 made at $\phi \sim 0.29$ and $\phi \sim 0.75$ by White, Kallman & Swank (1983) which demonstate that the flaring activity on timescales of 10-30 minutes is accompanied by increases in spectral hardness. The range of variation in hardness ratio corresponds to changes in absorption of up to $6 \times 10^{22}$ H $cm^{-2}$. Closer examination of Figure 8 indicates that this correlation is rather loose, and a formal cross-correlation analysis shows only 50% of the events to be correlated. Assuming a wind velocity of $10^3$ km $s^{-1}$ and a timescale for the inhomogeneities to pass the X-ray source of $\sim 10^3$ s gives a dimension of $10^{11}$ cm. The density of the wind in the vicinity of the X-ray source can be no more than $10^{11}$ $cm^{-3}$ (from the luminosity of the X-ray source), which gives a maximum change in absorption of $10^{22}$ H $cm^{-2}$. This is a factor of 6 below that observed and suggests that the inhomogeneities or blobs are elongated in the direction of the line of sight. Since both observations were made at quadrature, the blobs must be are elongated perpendicular to the velocity vector i.e. they compressed in the direction of motion. Another factor that suggests elongation of the blobs is that the X-ray source will completely ionise around it a $10^{11}$ cm region. The loose correlation between $N_H$ and intensity can be understood in terms of a random distribution of blobs passing either side of the X-ray source. The quasi-simultaneous variations in $N_H$ and intensity rule out magnetospheric instabilities as the cause of the flaring activity (Brinkman 1981), since the material within the magnetosphere will be too highly ionised to give any appreciable absorption.

Inhomogeneities in the wind also provide a plausible explanation for the absorption events observed on many occasions from Cyg X-1 close to $\phi$ = 0. Around this orbital phase we view tangentially over the limb of the primary such that if the blobs are compressed in their direction of motion the variation in column density caused by a blob passing through the line of sight is maximised. In addition the X-ray source ionises a substantial fraction of the wind (paragraph 3) and only when we view close to the limb of the primary where the densities are high enough to prevent complete ionisation will be possible to see absorption events. SSS spectra taken during the absorption events indicate that the ionisation state of the absorbing medium is anomalous (Pravdo et al. 1980). Absorption dips have also been detected from SMC X-1, another system with a low value of q, but again only when we view close to the limb of the primary (Marshall, White & Becker 1983).

## 5.  ANGULAR MOMENTUM CONSIDERATIONS

Secular variations in the pulse periods of the X-ray pulsars indicate that angular momentum is being exchanged between the neutron star and the rest of the binary system. The pulse period histories of four of the the pulsars in OB giant systems are given in Figure 9. The neutron stars in the two most luminous systems Cen X-3 and SMC X-1 are both spinning up with a characteristic timescale ($P/\dot{P}$) of $\sim$ $10^3$ and $10^4$yr, although Cen X-3 undergoes substantial deviations from a strictly linear trend. Initial measurements of the pulse period of Vela X-1 showed that it too followed the same systematic tendency to decrease, but between 1980 and 1982 this trend reversed (Figure 9; Nagase et al. 1983). GX301-2 shows local changes in its pulse period that if continued would cause an overall period change by many tens of seconds over several years. These variations reverse sense on a timescale of the order of 100 to 200 days such that over seven years of observations no systematic trend to spin up or spin down has yet occurred. Two points are raised by these results: (1) Can sufficient angular momentum be extracted from the stellar wind of the primary to account for the magnitude of the period changes and (2) What causes the direction of the period variations to change sense?

Shapiro & Lightman (1976) estimate for a constant velocity stellar wind form a non-rotating primary that assymmetries in density across an accretion face when $V_w \gg V_x$ will cause a net angular momentum flux $h_x$ to be captured of

$$h_x \sim \tfrac{1}{2} R_c^2 \Omega$$

4.1

where $\Omega$ is the orbital angular velocity of the binary system. The inclusion of a rotating primary and accelerating wind does not alter this estimate by more than a factor of two (Savonije 1980). The additional angular momentum will cause the period of rotation of a neutron star P to change at a rate $\dot{P}$ given by (Lamb, Pethick & Pines 1973).

$$\frac{\dot{P}}{P} = - \dot{M}_x h_x \cdot \frac{P}{2\pi} \cdot I^{-1} \quad s^{-1},$$

4.2

where I is the moment of inertia of a neutron star ($10^{45}$ gm cm$^2$) and $\dot{M}_x$ is the mass flow to the neutron star. Combining 4.1 and 4.2 with 2.1 gives

$$-\frac{\dot{P}}{P} \sim 10^{-14} V_3^{-8} P A_{30}^{-2} P_d^{-1} \dot{M}_{-6} \quad s^{-1},$$

4.3

where $V_3$ is the wind velocity in units of 1000 km s$^{-1}$, $A_{30}$ is the binary separation in units of 30 $R_\odot$, $P_d$ is the orbital period in days and $\dot{M}_{-6}$ is the mass loss rate of the primary in units of $10^{-6}$ $M_\odot$ yr$^{-1}$. Davies & Pringle (1980) have pointed out that the simple Bondi-Hoyle accretion capture radius (formula 2.1) is based on the collision of stellar wind particles after they have been deflected by the compact object and by definition if accretion is to proceed the net angular momentum must be zero. This is clearly in conflict with the observed changes in pulse period, but does mean that the stellar wind capture model has an internal inconsistency that remains to be be resolved.

As with the X-ray luminosities and absorptions, the observed values of $\dot{P}/P$ can be compared with those expected from the simple stellar wind accretion model (Savonije 1980). Clearly the predicted values of $\dot{P}/P$ are going to be extremely sensitive to the assumed wind velocity (equation 4.3). In Table 3 the predicted values of $\dot{P}/P$ are given for $V_3 = 1$ with, in addition, the values of $V_3$ computed from 4.3 required to give the observed $\dot{P}/P$. For the time being the discussion is confined to magnitude of P, not its sign. For 4U0900-40 the predicted value of $V_2$ is 0.75 which, given all the various uncertainties, is in reasonable agreement with the model. For Cen X-3 and SMC X-1 the required velocities of 500 and 300 km s$^{-1}$ respectively are similar to the orbital velocities of the X-ray sources, the point where 4.1 breaks down. Similarly the maximum period changes seen from GX301-2 also require very low wind velocities. The requirement of low velocities is similar to that found from the earlier comparison of the X-ray luminosities and absorptions with the stellar wind capture model.

Table 3: Observed and Predicted Period Changes

| Source | Log $(\dot{P}/P)_{exp}$ [a] | Log $(\dot{P}/P)_{obs}$ | Required $V_x$ |
|--------|-------------------|------------------|----------------|
| SMC X-1 | - 15.05 | - 10.70 | 280 |
| Cen X-3 | - 13.30 | - 11.05 | 520 |
| GX301-2 | - 12.70 | - 9.0 | 330 |
| 4U0900-40 | - 12.52 | - 11.52 | 750 |
|  | (s$^{-1}$) | (s$^{-1}$) | (km s$^{-1}$) |

a) assumes $V_x = 10^3$ km s$^{-1}$

Wang (1981) has extended the Shapiro and Lightman work to the case where $V_w \sim V_x$ and also to include the effects of gradients and inhomogeneities in the wind. Wang attempts to account for the reversals in the sign of $\dot{P}$ in terms of changes in the velocity and density grdients across the accretion face. He finds that when $V_w \sim V_x$ the net trend is still to spin up with a magnitude similar to that predicted by 4.3, but that any negative velocity and/or density gradients in the wind that might be caused by inhomogeneities in the wind, could lead to localised spin down episodes. It is however difficult to see how this mechanism can sustain spin down episodes for up to 2 years.

Ghosh & Lamb (1978) find that if the Alfvén radius, $R_a$ (the point at which matter first co-rotates with the neutron star) is comparable to the co-rotation radius $R_\Omega = G.M_x.P^2/4\pi^2$, then a braking torque may be applied to the neutron star either by the centrifugal ejection of disk material ($R_\Omega < R_a$) or by magnetic coupling between the disk and the magnetosphere of the neutron star ($R_\Omega \sim R_a$). For 4U0900-40 and GX301-2 the observed luminosities and rotation periods combined with typical surface fields of $10^{12}$ G gives $R_\Omega$ to be at least one order of magnitude larger than $R_a$. Since $R_a \propto L_x^{-2/7}$, a major reduction in flux should accompany any spin down episode; no such reduction has

been observed (White & Swank 1984; Nagase et al. 1984).
Another possibility is that the internal structure of the
neutron star is responding to the additional angular momentum
and material received from the accretion flow (Lamb, Pines &
Shaham 1978a,b). But in this case it is difficult to
understand how such an effect can give more than a short-lived
glitch, or oscillation in the pulse period.

Sustained intervals of "spin reversed accretion" i.e
reversals in the net angular momentum vector of the accreted
material seem to provide the only plausible explantion for the
spin down episodes (Wang 1981). To sustain such episodes for
months to years requires a modification to the stellar wind
accretion model. One possibility discussed in White & Swank
(1984) based on Wang's calculations is that the stellar wind
does not conserve angular momentum as it flows out. This would
occur if the wind were forced to co-rotate with the primary by
fairly modest magnetic fields of ~1 G at the X-ray source orbit
or 10-100 G on the stellar surface. The changes in the sense
of $\dot{P}$ could then arise from cyclic variations in the field
strength caused by an underlying magnetic cycle, analogous
perhaps to a sun spot cycle.

A fundamental concern when considering the accretion
process onto the compact object is whether the flow is mediated
by Keplerian accretion disks. This will only occur when the net
specific angular momentum of captured material, $h_x$ is equal to
or larger than that of the last Keplerian orbit before the
magnetosphere in the case of a neutron star, or the event
horizon for a black hole (Shapiro & Lightman 1976). The
observed values of P of the X-ray pulsars allow an estimate of
$h_x$ by turning around equation 4.2 so that

$$h_x = 6 \times 10^{17} \dot{P}_{-11} L_{37}^{-1} P^{-2} \quad cm^2 \ s^{-1}$$

4.4

$\dot{P}_{-11}$ is the rate of period change in units of $10^{-11}$ s s$^{-1}$ and
$L_{37}$ is the observed luminosity in units of $10^{37}$ erg s$^{-1}$. This
can be compared with $h_m$, the specific angular momentum of the
last Keplerian orbit before the disk material is forced to
co-rotate with the magnetosphere where

$$h_m = 1 \times 10^{17} \, L_{37}^{-1/7} \quad cm^2 \, s^{-1}.$$

4.5

For disk accretion $h_x$ will not be greater than $h_m$ since the disk must dissipate angular momentum if material is to flow through it. Table 4 compares $h_x$ with $h_m$ and also with the ratio of $R_\Omega/R_a$ which gives an indication of the importance of braking torques. The values of $\dot{P}$ for both Cen X-3 and SMC X-1 show h to be only a factor of two to four below that expected. Given the uncertainties in the neutron star parameters and that for both sources $R_\Omega/R_a$ is close to unity such that magnetic braking torques may be important,

Table 4: Angular Momentum Considerations

| Source | $h_x(cm^2 \, s^{-1})$ | $h_m(cm^2 \, s^{-1})$ | $R_\Omega/R_a$ |
|---|---|---|---|
| SMC X-1 | $3 \times 10^{16}$ | $6 \times 10^{16}$ | 2 |
| Cen X-3 | $2 \times 10^{16}$ | $8 \times 10^{16}$ | 4 |
| 4U0900-40 | $6 \times 10^{15}$ | $1 \times 10^{17}$ | 20 |
| GX301-2 | $2 \times 10^{17}$ | $1 \times 10^{17}$ | 60 |

it seems likely that accretion disks mediate the flow in these systems. In contrast 4U0900-40 has a value of $h_x$ that is at least 15 times less than that required for a disk to be present. The values of $h_x$ in Table 4 assume the maximum value of $\dot{P}$ observed so that this inadequacy becomes much worse between these extremes. For GX301-2 a disk may just be possible at the maximum value of $\dot{P}$, but again not between these values. The lack of any lag between the 41.5d outburst cycle and the time of periastron passage provides further evidence against an accretion disk in the GX301-2 system (Figure 4). The processing of material in an accertion disk may smooth out any inhomogeneities in the wind and accretion disks mediating in the flow in Cen X-3 and SMC X-1 would explain why there is no notable flaring activity from these sources.

The interaction of the inflowing material with the magnetosphere of the neutron star determines the size and shape of the X-ray emission region on the surface of the neutron star. The angular momentum content of the wind and the rotation period of the neutron star play crucial roles in determining the way the material interacts with and penetrates the magnetosphere. When the flow is essentially spherical onto a slowly rotating neutron star then the material will be shocked at the point where the gas pressure equals that of the magnetosphere. Further flow inwards occurs via the Rayleigh-Taylor instability inverting the post shock material with the magnetic field. The penetrating material is not threaded onto the field lines by this process and since the free-fall times are short, it is possible that most of the material falls directly to the surface. When an accretion disk is present the interaction between the inner edge of the disk and the "pinched" magnetosphere is determined by the Kelvin-Helmholtz instability. This rapidly disrupts the inner disk and threads the flow directly into field lines and hence to the magnetic poles. In the latter case strong pulsations are to be expected, whereas in the case of spherical accretion, little if any modulation may be present. These processes were first applied to accretion onto neutron stars by Arons & Lea (1980), Elsner & Lamb (1977) and Ghosh & Lamb (1979). The dichotomy in accretion modes appears to be reflected in the properties of the X-ray pulsars. The pulsars in the widely separated binaries where pseudo-spherical accretion occurs show weak modulations when compared to the more luminous disk accretors (White, Swank & Holt 1983).

In systems where an accretion disk is not possible the flow still may have considerable residual specific angular momentum left when it reaches the magnetosphere and it is not clear which of the aforementioned instabilities is important. Variations in the pulse profile of GX301-2 shows evidence that the accretion process is very finely balanced between the two instabilities (White & Swank 1984). Figure 10 shows the intensity profile of a 4 day observation of GX301-2 taken from White & Swank along with the pulse profile averged over half day intervals. The light curve is punctuated by active periods that last for approximately one day. During the first two quiescent intervals A and C, the amplitude of the double-peaked pulse is close to zero. During the third quiescent interval F the pulsations remain unchanged. The X-ray spectrum shows the X-ray absorption to have increased by $\sim 10^{23}$ H $cm^{-2}$ during all three quiescent intervals (White & Swank 1983). Within each active period (B,D, E and G) similar flaring activity to that seen from 4U1700-37 and 4U0900-40 is present.

The one day duration of the quiescent and active intervals suggest that in addition to the inhomogeneities discussed earlier, there exists larger structure in the wind with a typical size of a stellar radius. The changes in the pulse profile indicate that this larger scale structure is associated with changes in the specific angular momentum captured from the wind. Another complicating factor is that the interaction between the accretion flow and the magnetosphere is also sensitive to the magnitude of any magnetic fields contained in the accretion flow (Wang & Welter 1982). The variations in the captured angular momentum from the wind implied by the variable pulse profile may be intimately connected with the reversals in sign and variations in P.

This discussion has concentrated on the properties of the systems containing neutron stars since the effect of the accreted material on the spin of the neutron star readily conveys information on the captured specific angular momentum from the wind. For Cyg X-1 where the compact object seems to be a good candidate for an accreting black hole Shapiro & Lightman (1976) have shown that for an undisrupted stellar wind there should be sufficient captured angular momentum for an accretion disk to form. It is interesting to note that Shapiro & Lightman suggested spin reversed accretion as the mechanism that produces high and low spectral states from Cyg X-1. These high and low states are characterised by a major change in the shape of the continuum spectrum, not by absorption variations. While spin reversed accretion is clearly now viable, in the intervening time other models that involve instabilities in the accretion process onto a black hole have been proposed that seem to provide a more likely mechanism for the high and low states (e.g. Guilbert & Fabian 1982).


6.   SUMMARY AND CONCLUSIONS

This paper has emphasized how the compact X-ray sources in massive X-ray binaries can be used to investigate the properties of the stellar winds of the OB supergiant companions. The X-ray luminosity, absorption and, for the pulsars, variations in pulse period provide three independent probes. The large scale structure of the winds of the primaries in these binaries range from surprisingly normal to extreme. The X-ray luminosities and photo-electric absorptions measured for both 4U0900-40 and 4U1700-37 prior to superior conjunctions are in reasonable agreement with the X-ray emission being driven and subsequently absorbed by the stellar wind of the primary. This agreement is despite the fact that the radiatively driven wind should have been modified by X-ray illumination ionising away the UV line transitions responsible for the acceleration process (Hatchett & McCray 1977; MacGregor & Vitello 1982). Orbital variations in the terminal velocity given by C IV and Si IV P Cygni lines in the spectrum of the companion to 4U0900-40 confirm that such effects are important (Dupree et al. 1980). Similarly in the case of Cyg X-1 a large fraction of the wind exposed to the X-ray source is highly ionised, and yet the X-ray luminosity is still close to that expected from unmodified stellar wind capture. It is important to note that the luminosity and absorption provide two independent means of probing the wind, with the latter three powers less dependent on the velocity of the wind. For 4U1700-37 and 4U0900-40 where X-ray illumination of the wind has little effect on the photo-electric absorption of X-rays, the ratio of $\dot{M}$ to $V_w$ can not be varied without loosing the good agreement for one of the two observables.

The high X-ray luminosities of Cen X-3 and SMC X-1 require at least one order of magnitude more accreted material than can be provided by an unperturbed stellar wind. Increased absorption $\gtrsim 10^{24}$ H cm$^{-2}$ seen during extended low states of Cen X-3 indicates that the X-ray source on occasions becomes immersed in an optically thick wind from the primary. A second type of low state seen from Cen X-3 and also from SMC X-1 is characterised by a reduction in the accretion flow. It has not as yet been possible to establish any particular sequence or pattern to the two different types of low state. The eccentric (e = 0.47) orbit of GX301-2 about its companion WRA977 results in an X-ray outburst every 41.5 days close to the time of periastron passage. The whole 41.5 d light curve can only be modelled by the stellar wind capture model if the stellar wind velocity within the X-ray orbit is at least a factor of three lower than expected, and is comparable to the velocity of the source. In addition the overall X-ray luminosity and absorption indicate a mass loss rate of $10^{-5}$ M$_\odot$ yr$^{-1}$ , close to the maximum permissible if all the momentum of the radiation field is given up to the stellar wind.

The GX301-2/WRA977 system is the only one of the above where we can be reasonably confident that the primary at no point in the orbit comes close to filling its critical potential lobe (Kelley, Rappaport & Petre 1980). There has been much debate (see for example Petterson 1978) as to whether the primaries in the other systems fill their critical potential lobes. The conclusion of previous workers has been that there is no need to invoke a second flow through the inner Lagrangian point for 4U0900-40, 4U1700-37 and Cyg X-1, although it is questionable that a radiatively driven wind can be maintained in the presence of X-ray illumination (Macgregor & Vitello 1982). If any of the primaries in these systems do fill their critical potential lobes then they must be undergoing core hydrogen burning and expanding on a nuclear timescale, or be in the very short-lived phase just prior to the OB supergiant becoming a red giant. As discussed by Conti (1978) and Savonije (1978) the former is probably the case for all except 4U0900-40 (and GX301-2?). Friend & Castor (1982) show that a radiatively driven wind will become focussed

towards the compact object when the primary approaches within a
few percent or less of its critical potential lobe, although
the effects of the X-ray illumination of the wind suppressing
the acceleration process was not included in the calculations.
Instabilities in such behaviour (perhaps caused by the X-ray
source) might account for the high and low states of Cen X-3
and SMC X-1. This behaviour cannot account for the extreme
stellar wind of WRA977.

Ubiquitous flaring activity on a timescale of tens of
minutes from 4U0900-40, 4U1700-37 and GX301-2 most probably
arise from inhomogeneities in the stellar wind. The most
compelling evidence for this comes from 4U1700-37 where
quasi-simultaneous absorption events are associated with the
flaring (White, Kallman & Swank 1983). The observed variations
in column density and timescale of the events indicate that the
inhomogeneities are pancake shaped with a thickness in the
direction of motion of $\sim 10^{10}$ cm. These inhomogeneities may
be the same as those proposed by Lucy & White (1980) to explain
the coronal X-ray emission seen from OB stars in general. In
the more luminous systems Cyg X-1 and SMC X-1 the stellar wind
exposed to the X-ray source is almost completely ionised except
in the high density regions close to the primary. In this case
the absorption events are confined to orbital phase intervals
when the X-ray source is viewed close to the limb of the
primary. Flaring activity is much reduced in these systems,
probably because the flow is mediated by an accretion disk. In
addition to this structure, quiescent intervals lasting of
order one day from GX301-2 indicate large "holes" the wind with
a typical size of the order of a stellar radius where the
density is reduced and more homogenous. It has not been
possible as yet to determine if these "holes" are azimuthally
distributed around, or radially moving out from the primary.
Dramatic reductions in the amplitude of the X-ray pulsations
during the quiescent intervals suggests that they are
associated with changes in the specific angular momentum
captured from the wind.

The enhanced absorption seen from 4U0900-40, 4U1700-37 and
Cen X-3 after superior conjuction of the X-ray source must be
caused by the X-ray source in some way disrupting the stellar
wind. However despite many interesting suggestions the
mechanism responsible for this disruption remains elusive.
The original and often quoted accretion wake model of Jackson
(1975) is incapable of producing more than localised increases
in absorption after superior conjuction, not the extended broad
event observed for more than one quarter of a binary cycle.
Another model involving disruption of the radiation driven wind
(Fransson & Fabian 1980) only gives a qualitative description
of the phenomenon and is hard to apply in detail.

The rate of change of pulse period of the pulsars provides a measure of the specific angular momentum content of the captured material. Sustained spin-downs in the rotation period of 4U0900-40 and GX301-2 that can last for up to a year or more are not understood (Nagase et al. 1984, White & Swank 1984). Stellar wind capture, even including the effects of inhomogenities in the wind should still cause spin-ups, punctuated by short-lived spin-down episodes (Wang 1981). Set against this is the point made by Davies & Pringle (1980) that the Bondi-Hoyle accretion radius is based on the deflection and collision of particles and as such does not allow the capture of angular momentum. This indicates the need for a fundamental re-appraisal of the assumptions made about the nature of the stellar wind. One possibility is that the wind does not conserve angular momentum and it is being forced to co-rotate out to several stellar radii by a 10-100G magnetic field on the surface of the OB star, thereby causing the net angular momentum vector of the flow around the neutron star to switch sign (Wang 1981). It is difficult to reconcile the presence of such magnetic fields with the current view of radiatively driven stellar winds.

## ACKNOWLEDGEMENTS

The author thanks Tim Kallman for many stimulating discussions that have helped crystalise the important issues in the study of OB supergiants. The high Energy Astrophysics Group at the GSFC has played an important role in providing many of the recent observations. The receipt of an ESA fellowship is gratefully acknowledged.

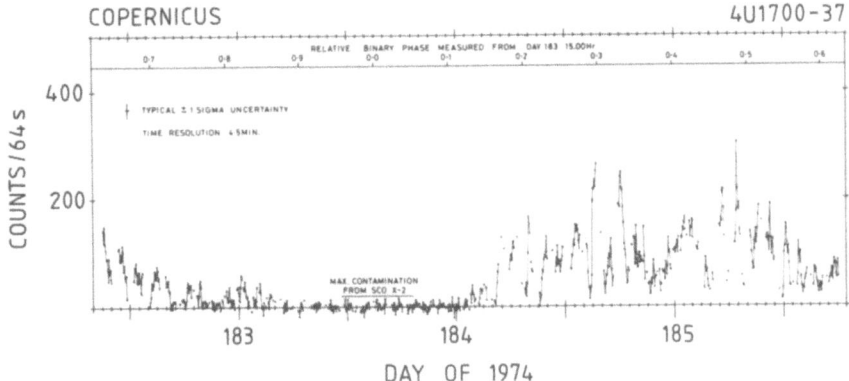

Figure 1a. A Copernicus observation of one complete orbital cycle of 4U1700-37 with a time resolution of 4.5 min (taken from Mason, Branduardi & Sanford 1976).

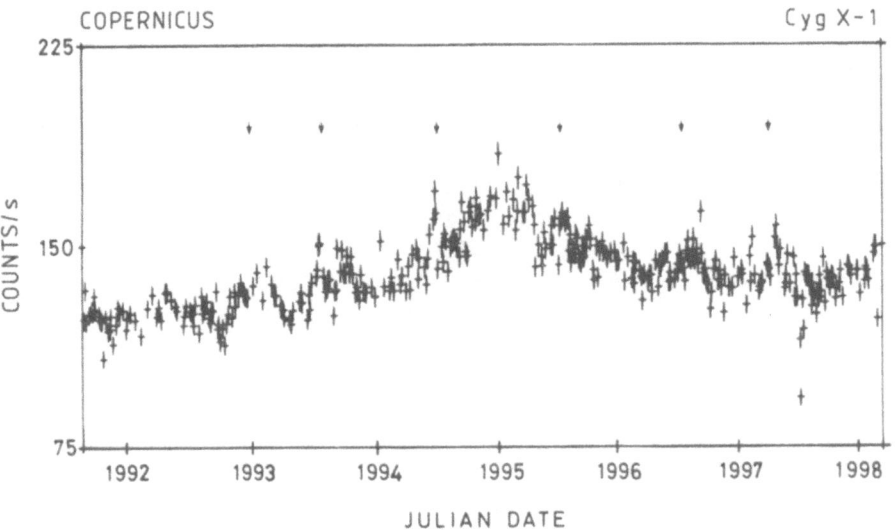

Figure 1b. A Copernicus observation of one complete orbital cycle of Cyg X-1 with a time resolution of 14 min (taken from Mason et al. 1974). An absorption dip at $\phi$ = 0.0 can be seen on day 1997.

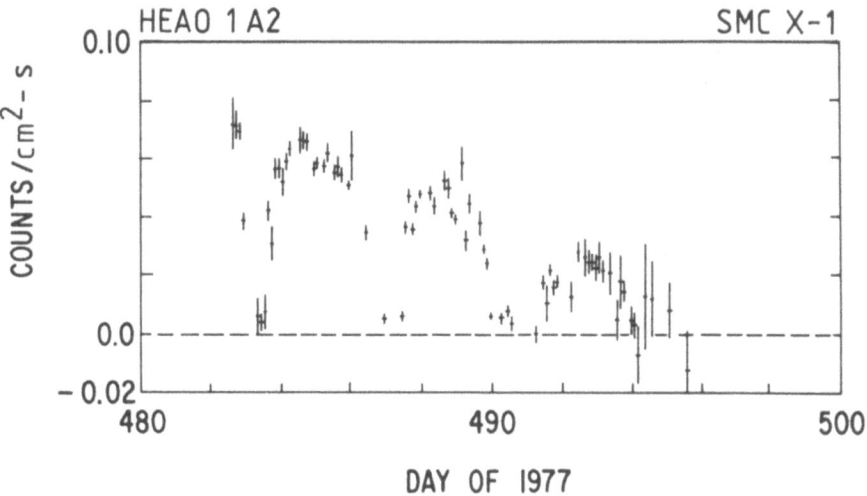

Figure 1c.  The light curve of SMC X-1 over three orbital cycles
as seen by HEAO-1 A2 from scanning observations (taken from
Marshall, White & Becker 1983).

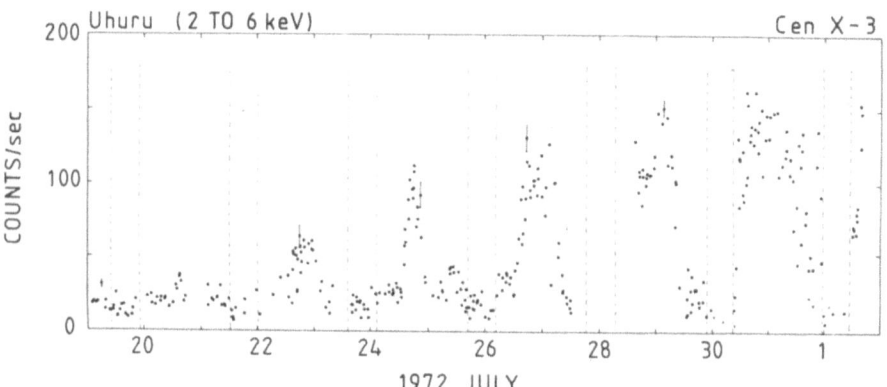

Figure 2a.  UHURU observations of a turn-on of Cen X-3 (taken
from Schreier et al. 1976).  Each point represents a 20 s accumulation.  The times of eclipse entry and exit are indicated by
dashed lines.

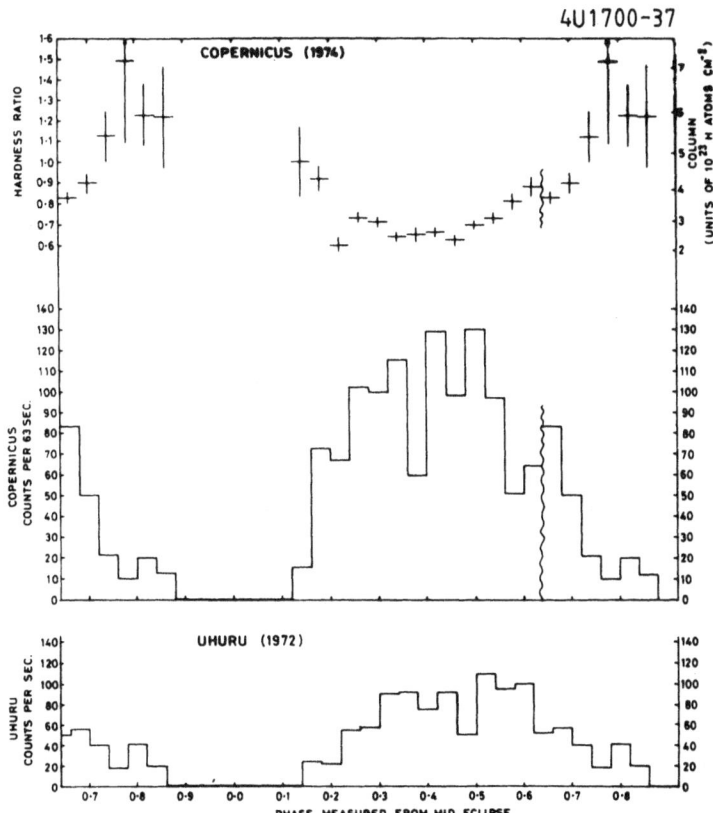

Figure 2b. The average <u>UHURU</u> and <u>Copernicus</u> orbital light curves of 4U1700-37 (taken from Mason, Branduardi & Sanford 1976). The spectral hardness ratio and equivalent absorbing column are given for the <u>Copernicus</u> data.

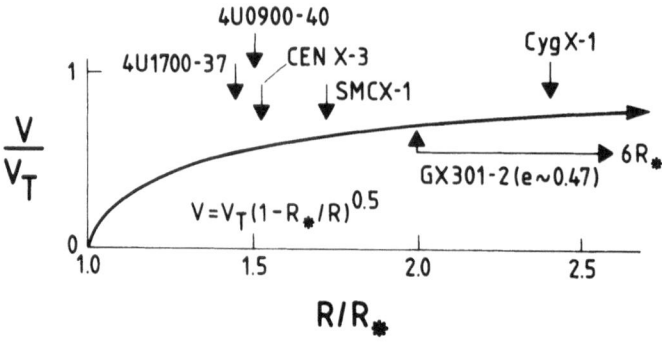

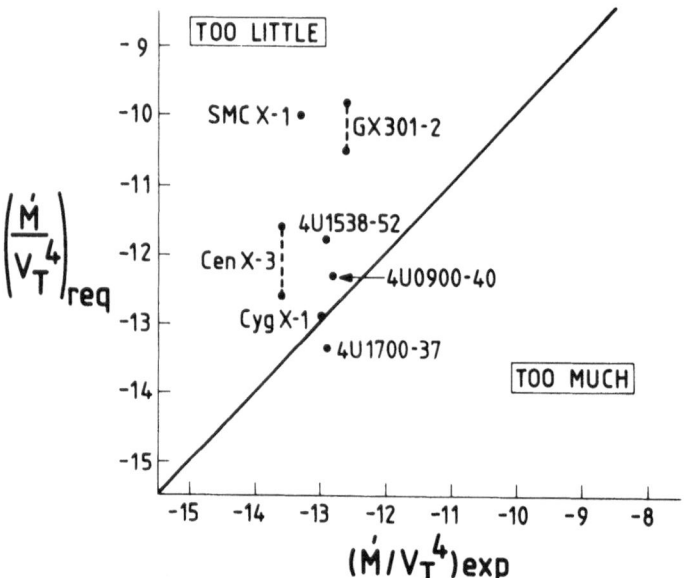

Figure 3a.  This shows the expected velocity of a stellar wind if it follows a velocity law of the form $V = V_T (1-R_*/R)^{0.5}$ .  The binary separations of the OB supergiant X-ray are indicated.

Figure 3b.  The expected and required values of the ratio $\dot{M}/V_T^4$ . The former is calculated using the parameters given in Table 1, and the latter using Table 1 and equation 2.3.  Sources that lie to the upper left of the solid line require more material than can be captured from an unperturbed stellar wind.

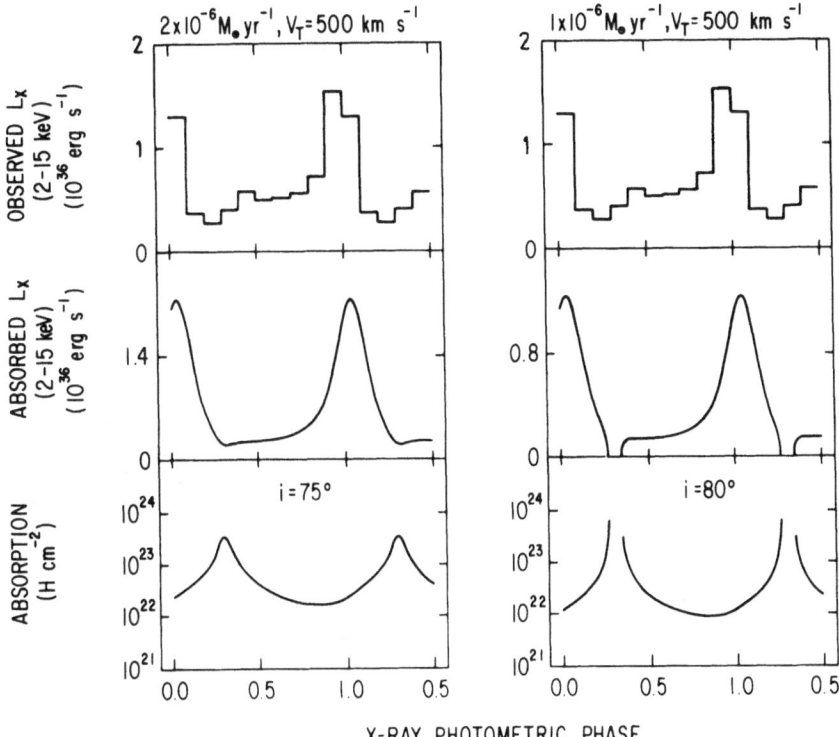

Figure 4.

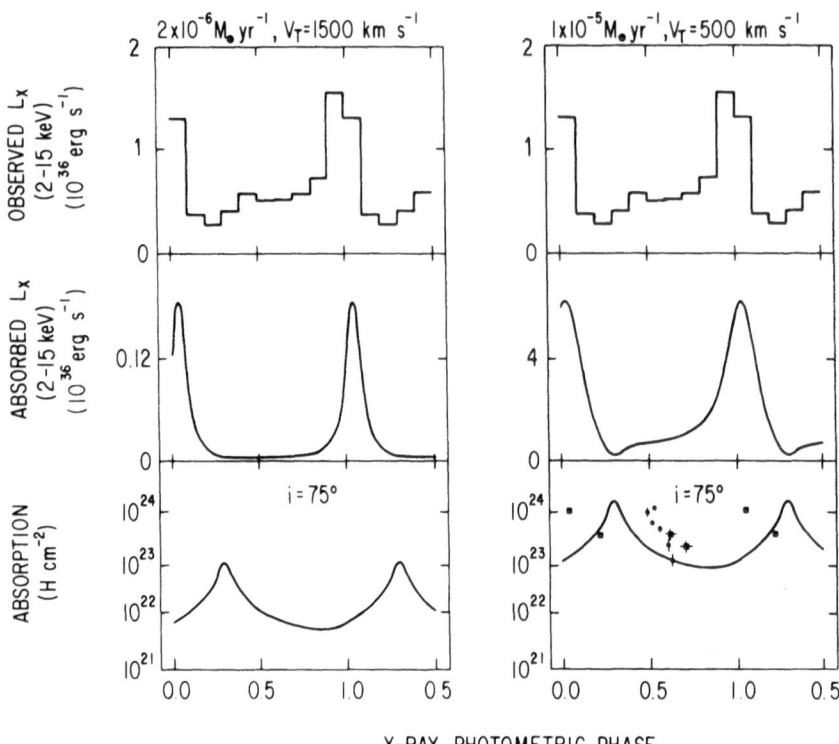

Figure 4. The predicted X-ray light curve from an eccentric orbit
of a neutron star through the stellar wind of WRA977 for four
different sets of stellar wind parameters (from White & Swank
1984). The inclination i, the mass loss rate in $M_\odot$ $yr^{-1}$ and the
terminal velocity $V_T$ used for each model are indicated. The Ariel
V SSI light curve of GX301-2 is given for comparison. The X-ray
phase is in this case only defined as photometric maximum, occurring at $\phi$ = 0.0.

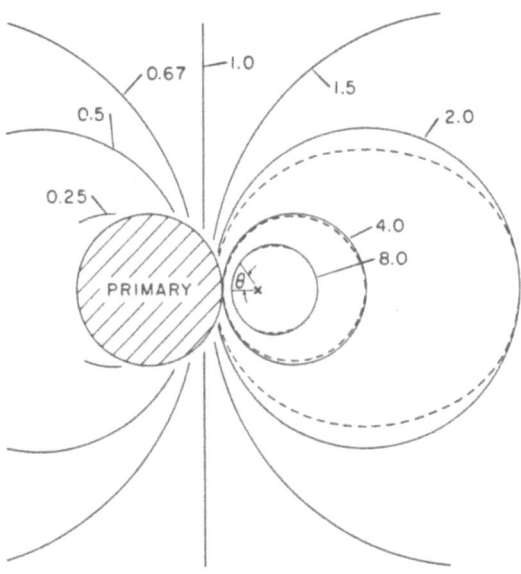

Figure 5. Lines of constant ξ (solid lines) and column density
(dashed lines) for an X-ray illuminated stellar wind with a
constant velocity (taken from Hatchett & McCray 1977). The former
are labelled in terms of the parameter  q  (equation 3.2).

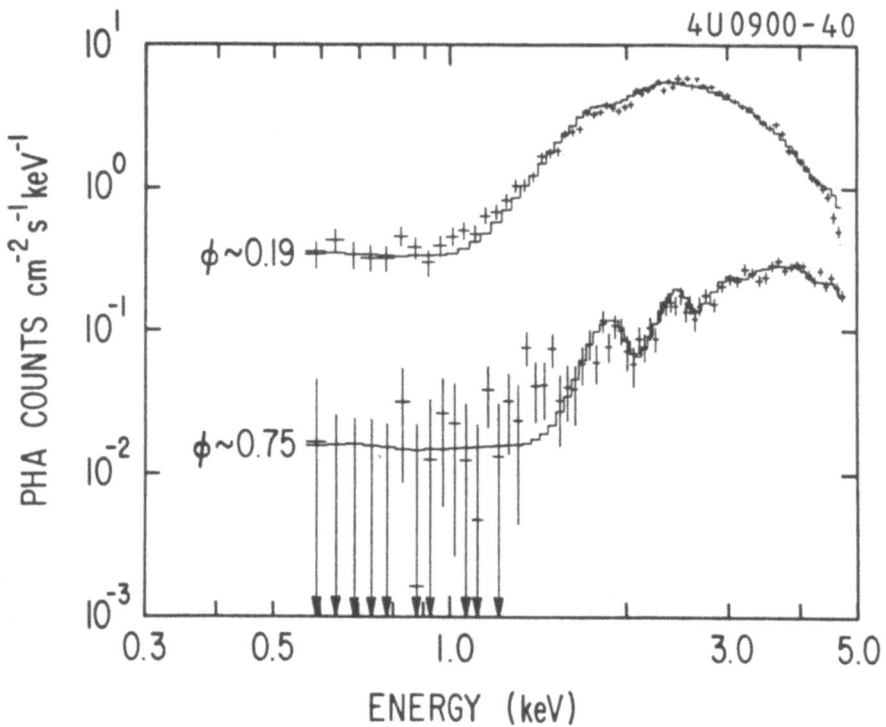

Figure 6.

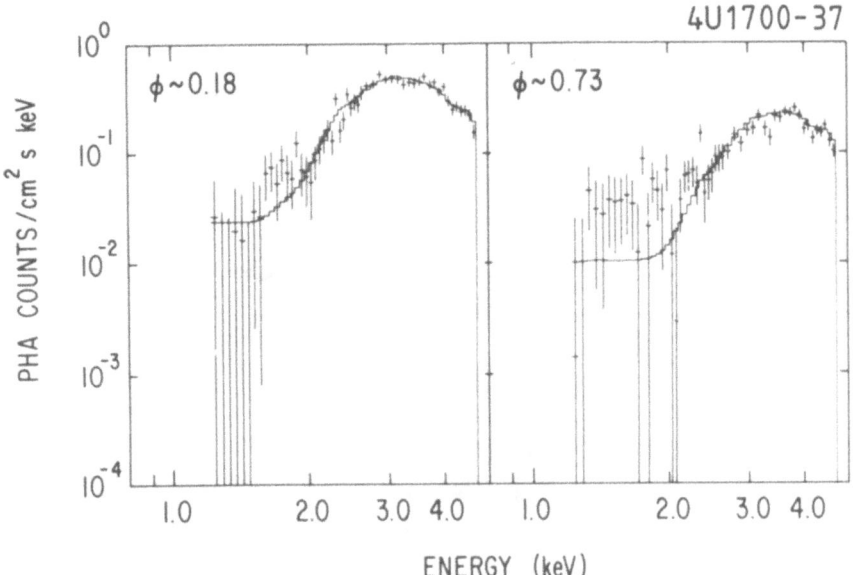

Figure 6.  The PHA spectra of 4U1700-37 and 4U0900-40 obtained by
the _Einstein_ SSS each at two different orbital phases (taken from
Kallman & White 1982 and White, Kallman & Swank 1983).  The histograms represent the best fitting model convolved through the
detector response.

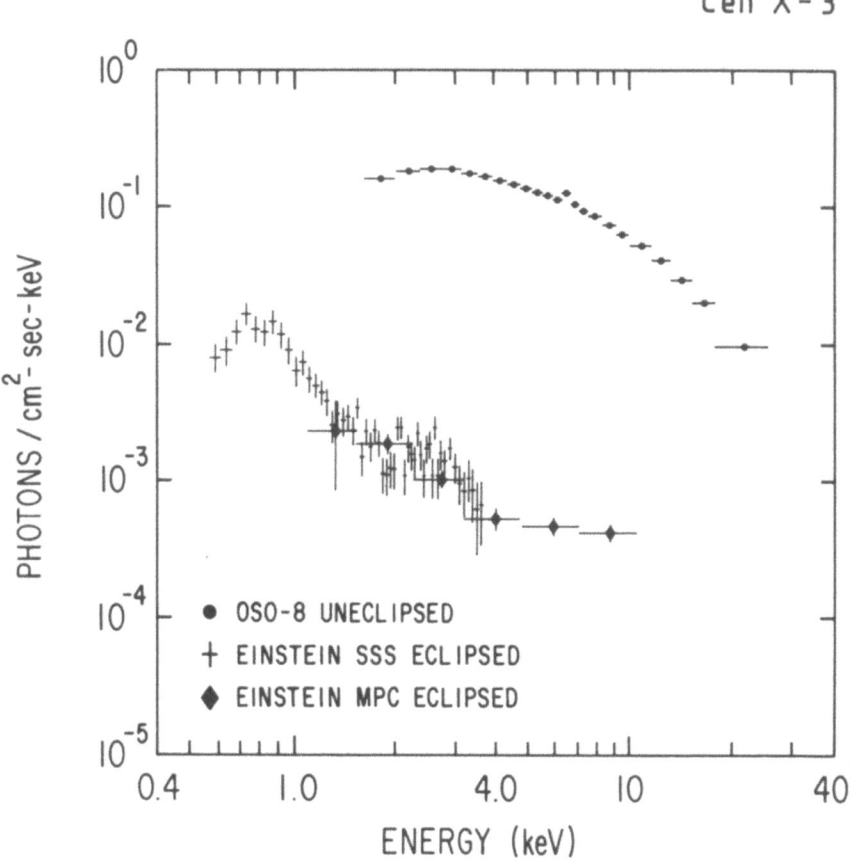

Figure 7a.    The orbital light curve of Cen X-3 recorded by the
SSS and MPC detectors on <u>Einstein</u>.    The data are folded over the
orbital period of 2.1 days and repeated for half a cycle.    A
spectral hardness ratio from the MPC is also given.

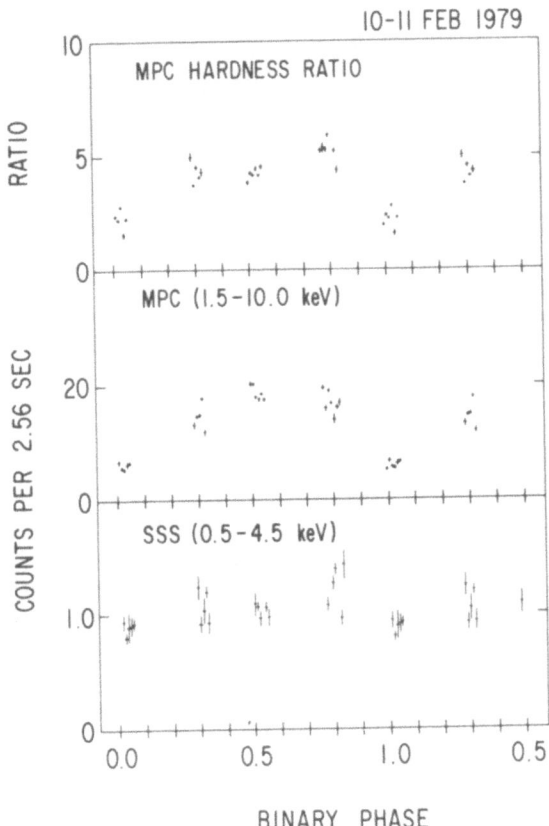

Figure 7b.  The SSS and MPC spectra of the residual flux seen from Cen X-3 during eclipse.  Also given for comparison is an uneclipsed high state spectrum taken by OSO-8.

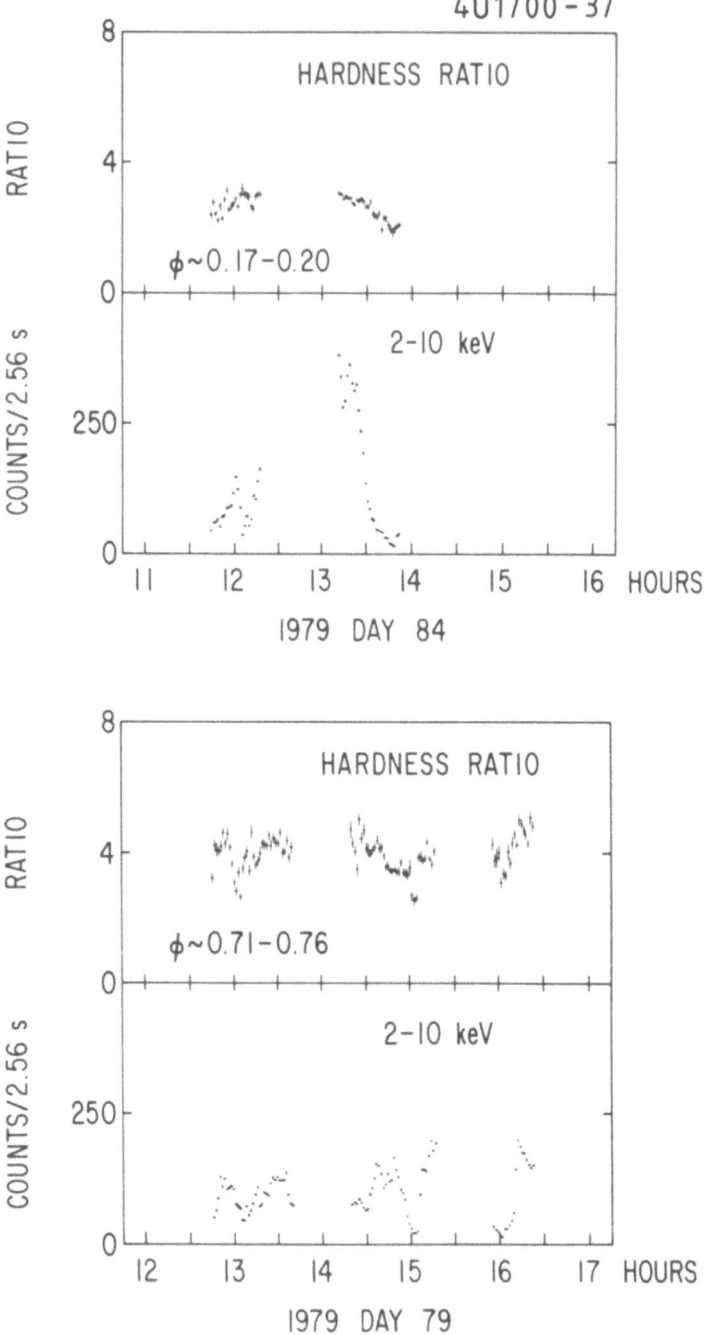

Figure 8. The 2-10 keV intensity of 4U1700-37 from the MPC in 50 s accumulation intervals along with the corresponding spectral hardness ratio (from White, Kallman & Swank 1983).

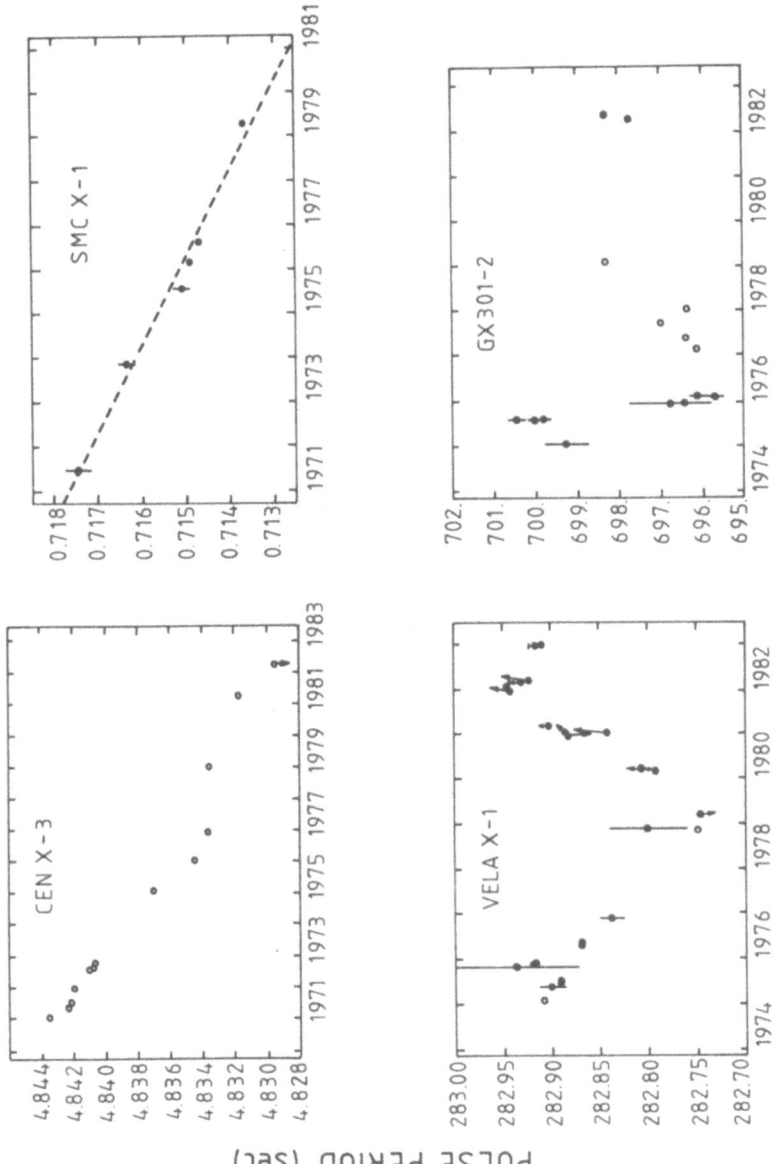

Figure 9. The pulse period histories of the X-ray pulsars in OB supergiant systems SMC X-1, Cen X-3, 4U0900-40 and GX301-2 (from e.g. Nagase et al. 1983).

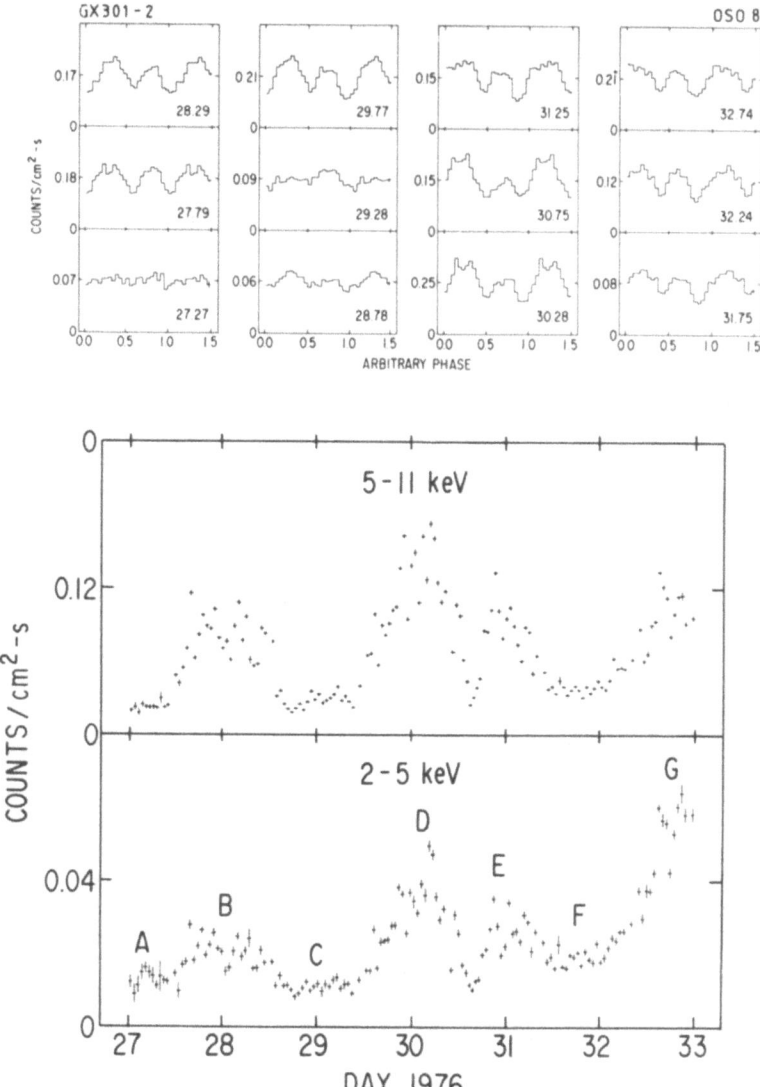

Figure 10.  Top - The pulse profile of GX301-2 (4U1223-62) aver-aged over 0.5 day intervals.

Bottom - The intensity profile of the OSO-8 observation used to obtain the pulse profiles.  The accumulation time is 2 pulse periods.

REFERENCES

Abbott, D., 1982. Ap.J., **259**, 282.
Arons, J. & Lea, S.M., 1980. Ap.J., **235**, 1016.
Burnard, D.J., Lea, S.M., Arons, J., 1983. Ap.J., **266**, 175.
Becker, R.H., Rothchild, R.E., Holt, S.S., Pravdo, S.H.,
    Serlemitsos, P.J. & Swank, M.H., 1978. Ap.J., **221**, 912.
Bessell, M.S., Vidal, N.W. & Wickramasinghe, A., 1975. Ap.J.,
    **195**, L117.
Bolton, C.T., 1982. Ap.J., **260**, 240.
Bondi, H. & Hoyle, F., 1944. M.N.R.A.S., **104**, 273.
Bonnet-Bidaud, J.M. & van der Klis, M., 1979. Astr.Ap.,
    **73**, 90.
Bonnet-Bidaud, J.M. & van der Klis, M., 1981. Astr.Ap.,
    **97**, 134.
Brinkman, W., 1981. Astr.Ap., **94**, 323.
Buff, J. & McCray, R., 1974. Ap.J., **188**, L37.
Cassinelli, J.P. & Swank, J.H., 1983. Ap.J., **271**, 681.
Castor, J.I., Abbott, D.C. & Klein, R.I., 1975. Ap.J.,
    **195**, 157.
Charles, P.A., Mason, K.O., White, N.E., Culhane, J.L.,
    Sanford, P.W. & Moffatt, A.J.F., 1978. M.N.R.A.S.,
    **183**, 813.
Conti, P.S., 1978. Astr.Ap., **63**, 225.
Conti, P.S. & Cowley, A.P., 1975. Ap.J., **200**, 133.
Davidson, K. & Ostriker, J.P., 1973. Ap.J., **179**, 585.
Davies, R.E. & Pringle, J.E., 1980. M.N.R.A.S., **191**, 599.
Dupree, A.K., et al. 1980. Ap.J., **238**, 969.
Eadie, G., Peacock, A., Pounds, K.A., Watson, M., Jackson,
    J.C. & Hunt, R., 1975. M.N.R.A.S., **172**, 39P.
Elsner, R.F. & Lamb, K.F., 1977. Ap.J., **215**, 897.
Fabbiano, G. & Schreier, E.J., 1977. Ap.J., **214**, 235.
Fransson, C. & Fabian, A.C., 1980. Astr. & Ap., **87**, 102.
Friend, D.B. & Castor, J.I., 1982. Ap.J., **261**, 293.
Ghosh, P. & Lamb, F.K., 1978. Ap.J., **223**, L83.
Ghosh, P. & Lamb, F.K., 1979. Ap.J., **232**, 259.
Guilbert, P.W., Fabian, A.C., 1982. Nature, **296**, 226.
Hatchett, S., Buff, J., McCray, R., 1976. Ap.J., **206**, 847.
Hatchett, S., McCray, R., 1977. Ap.J., **211**, 683.
Jackson, J.C., 1975. M.N.R.A.S., **172**, 483.
Jones, C., Forman, W., Tananbaum, H., Schreier, E., Gursky, H.,
    Kellogg, E. & Giacconi, R., 1973. Ap.J., **181**, L43.
Kallman, T.R. & White, N.E., 1982. Ap.J., **261**, L35.
Kelley, R., Rappaport, S. & Petre, R., 1980. Ap.J., **238**, 699.
Lamb, F.K., Pethick, C.J. & Pines, D., 1973. Ap.J., **184**, 271.
Lamb, F.K., Pines & Shaham, J., 1978a. Ap.J., **224**, 969.
Lamb, F.K., Pines & Shaham, J., 1978b. Ap.J., **225**, 582.
Lamers, H., van den Heuvel, E.P.J. & Petterson, J.A., 1976.
    Astr.Ap., **49**, 372.
Li, F.K. & Clark, G.W., 1974. Ap.J., **191**, L27.

Livio, M., Shara, M.M. & Shaviv, G., 1979. Ap.J., **233**, 704.

Long, K.S. & White, R.L., 1980. Ap.J., **239**, L65.

Lucy, L.B. & White, R.L., 1980. Ap.J., **241**, 300.

MacGregor, K.B. & Vitello, P.A.J., 1982. Ap.J., **259**, 267.

Marshall, F.E., White, N.E. & Becker, R.H., 1983. Ap.J.,
    **266**, 814.

Mason, K.O., Branduardi, G. & Sanford, P.W., 1976. Ap.J.,
    **203**, L29.

Mason, K.O., Hawkins, F.J., Sanford, P.W., Murdin, P. &
    Savage, A., 1974. Ap.J., **192**, L65.

McCray, R. & Hatchett, S., 1974. Ap.J., **199**, 196.

Nagase, F., Sato, N., Makishima, K., Kawi, N. & Mitani, K.,
    1983. Talk presented at the Summer Workshop in
    Astronomy, Santa Cruz, University of California.

Nagase, F. et al., 1984. Ap.J., in press.

Pallavicini, R., Golub, L., Rosner, R., Viana, G.S., Ayres,
    T. & Linsky, J.L., 1981. Ap.J., **248**, 279.

Parks, G.E., Mason, K.O., Murdin, P.G., Culhane, J.E., 1976.
    M.N.R.A.S., **191**, 547.

Petterson, J.A., 1978. Ap.J., **224**, 625.

Pounds, K.A., Cooke, B.A., Ricketts, M.J., Turner, M.J. &
    Elvis, M., 1976. M.N.R.A.S., **172**, 473.

Pravdo, S.H., White, N.W., Kondo, Y., Becker, R.H., Boldt,
    E.A., Holt, S.S., Serelmitsos, P.J., McCluskey, G.E.,
    1980. Ap.J., **237**, L71.

Rappaport, S. & Joss, P.C., 1977. Nature, **266**, 683.

Rappaport, S. & van den Heuvel, E.P.J., 1982. In IAU
    Symp. 98, Bestors, ed. M. Jaschek and H.-G. Groth
    (Reidel: Dordrecht), p. 327.

Sanford, P.W., Mason, K.P., Hawkins, F.J., Murdin, P. &
    Savage, A., 1974. Ap.J., **190**, L55.

Savonije, G.J., 1978. Astr.Ap., **62**, 317.

Savonije, G.J., 1980. Astr.Ap., **83**, 375.

Schreirer, E., Swarz, K., Giacconi, R., Fabbiano, G.,
    Morin, J., 1976. Ap.J., **204**, 539.

Shapiro, S.L. & Lightman, Ap.J., 1976. Ap.J., **204**, 555.

Swank, J.H., Becker, R.H., Boldt, E.A., Holt, S.S., Pravdo,
    S.H., Rothschild, R.E., Serlemitsos, P.J., 1976. Ap.J.,
    **209**, L57.

Tarter, C.B., Tucker, W.H. & Salpeter, E.E., 1969. Ap.J.,
    **156**, 943.

Tuohy, I.R. & Cruise, A.M., 1975. M.N.R.A.S., **171**, 33p.

Underhill, A.B., 1983. Ap.J., **268**, L127.

Wang, Y.-M., 1981. Astr.Ap., **102**, 36.

Wang, Y.-M. & Welter, G.L., 1982. Astr.Ap.J., **113**, 113.

Watson, M.G. & Griffiths, R.E., 1977. M.N.R.A.S., **178**, 513.

Watson, M.G., Warwick, R.S., Corbett, R.H.D., 1982.
    M.N.R.A.S., **199**, 915.

White, N.E., Kallman, T.R. & Swank, J.H., 1983. Ap.J.,
    **269**, 264.

White, N.E., Mason, K.O. & Sanford, P.W., 1978.  M.N.R.A.S.,
    **184**, 67p.
White, N.E. & Swank, J.H., 1984.  Ap.J., submitted.
White, N.W., Swank, J.H., Holt, S.S., 1983.  Ap.J., **270**, 711.
White, N.W., Swank, J.H., Holt, S.S. & Parmar, A.N., 1982.
    Ap.J., **263**, 277.

# CATACLYSMIC VARIABLES AS INTERACTING BINARY STARS

Richard A. Wade

Dept. of Physics and Astronomy
Northwestern University
Evanston  IL 60201
U.S.A.

## 1.  SCOPE OF THE REVIEW

The periodical literature on cataclysmic variables (CVs) and closely related topics has grown rapidly in recent years. In the Astrophysical Journal alone, the number of papers published on this subject in 1982 was about 80, the rate having doubled in three years. The other major journals taken together yield a similar annual total. In view of this enormous burst of activity it is impossible to review the whole field in a few pages. Fortunately it is also not necessary to do so, for several excellent recent reviews of specific topics, as well as published conference proceedings, are available.

After a brief description of the visual and ultraviolet properties of CVs, this review concentrates on two areas: (1) the basic binary star model of CVs and the evidence supporting it, and (2) "standard" accretion disks as the means of mass exchange between the component stars. The emphasis is on observations, particularly on assessing how reliably the physical quantities of interest can be or have been determined. I believe it is appropriate to have clearly in mind the limitations of our data, as well as to celebrate our partial successes in studying these stars.

Scientific subjects not covered in this review include: the novae (outbursts, abundances, production of dust, photoionisation models, chemical enrichment of the Galaxy), the nature of recurrent novae, and the AM Her and DQ Her stars (magnetic accretion, cyclotron emission, polarisation, etc,). X-radiation from CVs has been reviewed elsewhere, and the study

P. P. Eggleton and J. E. Pringle (eds.), Interacting Binaries, 289–326.
© 1985 by D. Reidel Publishing Company.

of far infrared and radio emission from CVs is in its infancy.
Outbursts of dwarf novae, high and low states, rapid
oscillations and superhumps are briefly described, but models
are not discussed. Finally the secular evolution of these
binary stars, including consideration of their space density
and pre- and post- cataclysmic phases, is omitted.

Important prior reviews of the basic model for CVs include
those by Robinson (1976) and Warner (1976). Gallagher &
Starrfield (1978) review the classical novae. Pringle (1981)
reviews accretion discs in general, with particular
consideration given to the active disks in CVs. Cordova &
Mason (1984) review all classes of CVs, giving emphasis to the
X-ray observations. Published conference proceedings include
those of the Rochester, Toronto, and Haifa meetings (van Horn &
Weidemann 1979, Plavec et al. 1980, and Livio & Shaviv
1983, respectively). The Cambridge (Mass.) meeting in January
1983 is in press at the time of this writing (Lamb & Patterson,
1984). In addition to these, the proceedings of the various
NASA and ESA symposia on ultraviolet astronomy with IUE contain
important reviews and/or recent results. Many other
conferences from recent years, especially on variable stars as
a whole, contain reviews of interest to students of CVs.

## 2.  OVERVIEW OF OPTICAL AND ULTRAVIOLET OBSERVATIONS OF CVs

### 2.1  Visible Light

The defining phenomenon common to all cataclysmic
variables is their violent optical variability. In the
classical novae this takes the form of an enormous outburst
whose decay takes typically weeks or months. The "novalike"
variables in their "high" states (UX UMa stars, VY Scl stars,
AM Her stars) resemble post-eruptive novae; some of these stars
hve been observed in "low" states, 4 to 6 magnitudes fainter.
Dwarf novae show frequent outbursts of shorter duration and
smaller amplitude. Various subclasses of dwarf novae also show
"standstills" or "superoutbursts". A light curve for the
well-studied dwarf nova SS Cygni is shown by Campbell (1934;
reprinted by Bath & van Paradijs 1983).

In addition to these long-timescale forms of variability
(weeks to centuries), many CVs show periodic light variations
on the timescale of the orbit (hours). These include eclipses
(which are sometimes "double" or otherwise asymmetric),
photometric "humps" or shoulders" just prior to or after
conjunction, and ellipsoidal variations due to the changing
aspect of the distorted secondary star. During
"superoutbursts" stars in the SU UMa subclass of dwarf novae

show " superhumps", similar  to the orbital humps but generally recurring with a period slightly longer than the orbital period.  CVs also vary on short timescales (seconds or minutes);  these phenomena are known variously as flares, flickering, or oscillations, depending upon their amplitudes and power spectrum characteristics.  See the reviews already mentioned and, for example, the papers by Robinson & Nather (1979), Patterson (1981), and Vogt (1980) for details.

The photovisual spectra of CVs in the "normal" states (post-eruptive novae, quiescent dwarf novae, novalike variables in the high state) are characterized by strong, broad emission lines of H, He I,  and Ca II. More rarely, weaker lines of He II and the CNO group are seen.  In the more luminous stages of the CV phenomenon (post-eruptive novae, erupting dwarf novae), broad shallow absorption lines of the same elements, often with weak emission cores, may be present instead; these objects tend to show the higher excitation He II and CNO lines more strongly.  In some quiescent dwarf novae, especially ones with orbital periods longer than about six hours, absorption lines characteristic of a late-type star can be found.  These absorption lines move in approximate anti-phase with the orbital variation of the emission lines.

The emission Balmer decrement is nearly always very shallow or reversed in CVs, suggesting electron densities in excess of $\sim 10^{13}$ cm$^{-3}$ and probably requiring an optical depth greater than $\sim 10$ at the centre of the H$\alpha$ line (Oke & Wade 1982).  Other means of estimating the gas density in the emission line region also suggest electron densities well above $10^{13}$.

The spectral energy distribution of faint CVs (mostly quiescent dwarf novae) is characterised by the Balmer jump in emission and a Paschen continuum that can be represented by the sum of a late-type star and a component that has $f_\nu$ approximately constant.  Brighter CVs often have the Balmer jump in absorption and have Paschen continua of the form $f_\nu \propto \nu^s$, $s > 0$. Examples of these types of spectra are given by, for example, Oke & Wade (1982) and Wade (1982).

The AM Her stars bear an overall similarity to the description given above for CVs that are not overtly magnetic, but there are some deviations.  The presence of circularly polarized light can be regarded as an important secondary defining characteristic for these systems.  See the review by Liebert & Stockman (1984) for details.

## 2.2 Ultraviolet Observations

In quiescent dwarf novae, the satellite ultraviolet (UV) spectra show strong emission lines of species ranging in excitation from Mg II through Si III/IV and He II up to C IV and N V. An exception is U Gem, the prototypical dwarf nova, where absorption lines are present instead. In the more luminous kinds or stages of CV, absorption in these same species is the rule. P Cygni profiles of these lines, with unequal fluxes in the absorption and emission components, are sometimes present; the absorption/emission ratio changes during the course of a dwarf nova outburst. The P Cyg profiles are discussed further below.

The ultraviolet lines are, of course, less well studied than the visual lines, but already they have shown behaviour that is hard to interpret. In UX UMa, for example, the UV lines changed from absorption to emission in observations separated by several months, while the continuum was almost unchanged (Holm et al. 1982). Similarly the presence of Mg II emission from one system to another seems to be correlated at best only weakly with the strength of other UV lines. These observations suggest that the UV resonance lines arise in gas which is in some sense a "trace constituent" of the binary system, not directly responsible for the bulk of the emitted continuum radiation.

Anticipating for a moment part of the physical model for CVs which will be elaborated in the next section, we can interpret the P Cygni profiles as arising in a "wind" from the neighbourhood of the white dwarf primary star. The terminal velocities that are indicated from the blue edge of the absorption component are in the range 3000 to 5000 km s$^{-1}$. For comparison the escape speed from the surface of a 1 $M_\odot$ white dwarf is about 5000 km s$^{-1}$. If the white dwarfs in CVs are massive, then either the material in the wind does not escape from the white dwarf or some modification to the simple picture needs to be made. This might be that the wind arises on the surface of the accretion disk relatively far from the white dwarf itself, or that the wind is accelerated slowly and for some reason (geometry? unfavourable ionization balance?) we do not observe the true terminal velocity of the wind. Finally the white dwarfs in these systems may be much less massive than 1 $M_\odot$. Estimates for the apparent mass flux in the observed winds are difficult to make and very uncertain, but numbers near $\dot{M}_{wind} \simeq 10^{-11}\ M_\odot yr^{-1}$ are typical. P Cygni profiles and wind characteristics for CVs are discussed further by, for example, Cordova & Mason (1982), Hassall et al. (1982), and Kallman (1983).

The ultraviolet energy distributions of CVs can be represented crudely by power laws, $f_\nu \propto \nu^s$. For faint systems (quiescent dwarf novae), $s < 0$ typically. More luminous CVs, including erupting dwarf novae, tend to have $s \gtrsim 0$. Energy distributions are discussed in more detail in a later section.

## 2.3  Absolute Magnitudes and Orbital Periods

Table 1 gives a schematic representation of our knowledge of the absolute magnitudes of CVs. The mean magnitudes quoted are uncertain by perhaps 1 mag. The ranges are at least as large as indicated.

Quiescent dwarf novae exist over the range $+6.2 \lesssim M_V \lesssim +10.3$ at least, partly as a result of aspect effects and partly as a result of intrinsic differences (Wade 1982). The mean value $M_V = +7.5$ of Kraft & Luyten (1965) or the more recent mean of $+8.5$ adopted by Duerbeck (1984) are, of course, statistically derived. It is not safe to apply them to individual objects. The range in $M_V$ indicated for erupting dwarf novae is my very conservative estimate. A mean value near $+4$ is often adopted.

Nova absolute magnitudes are derived from distance moduli estimated in various ways and corrected for interstellar extinction. There is a real spread in $M_V$ for post-eruptive novae, with HR Del being near the bright end of the range and V1500 Cyg being at the faint end. The mean $M_V$ is about $+4$ (see Warner 1976).

No direct measurements of $M_V$ for novalike variables in the high state exist. The mean is usually taken to be 4 or 4.5, from their spectroscopic and other similarities to the novae. The range in $M_V$ is uncertain. The luminosity in the low state is equally uncertain for these objects, being based usually on an assumed high state magnitude and a magnitude difference. For one or two AM Her stars, estimates of distances or absolute magnitude are available in the low state from spectrophotometric arguments (brightness temperature of the exposed white dwarf, inferred $M_V$ for the secondary star, etc.). These low state values of $M_V$ overlap the better known faint end of the quiescent dwarf nova range. By using magnitude differences between high and low states, we can with great uncertainty infer that in the high state AM Her stars are one or two magnitudes fainter than other luminous cataclysmic variables.

Clearly much work remains to be done on establishing $M_V$ for CVs, but each star needs to be treated individually. Statistical means are inappropriate since even the subclasses are inhomogeneous. Standard assumptions, such as the spectrum-absolute magnitude relation for isolated main-sequence stars need to be applied cautiously to the dis-equilibrated and/or possibly evolved secondary stars in some CVs (see, for example, Rappaport, Joss & Webbink 1982; Wade 1979). Newer relations, such as one between emission line equivalent widths and absolute magnitude (Patterson 1984), show promise but need to be very thoroughly evaluated before they can be applied with confidence to large numbers of stars. The danger is that we will enforce an artificial similarity on an inherently diverse population of objects.

Disregarding T CrB (P = 227 d), a recurrent nova containing a giant secondary, and nova GK Per (P = 2.0 d), all known orbital periods for CVs cluster below $\approx 15$ h. The two shortest periods belong to the peculiar objects AM CVn (= HZ 29, P = 18 m) and GP Com (P = 46 m), which are thought to be "twin degenerate" stars. These objects flicker but have not shown major outbursts. The shortest period for "ordinary" CVs is found near 81 m, and the range from 81 m to $\sim$6 h is densely occupied with the exception of a gap between 2.2 h and 2.8 h, where no systems are found. One detached binary with a period of 2.7 h is known: the nucleus of the planetary nebula Abell 41; Grauer & Bond 1983. This gap can get narrower with time, and it has done so over the years, but most observers believe the gap to be real. There are no obvious selection effects that would artificially create such a gap. See Robinson's (1983) review for further discussion of the period distribution.

## 3.  CATACLYSMIC VARIABLES AS BINARY STARS

In the 1960's work by a number of observers, most notably R.P. Kraft, established that cataclysmic variables are binary stars. Today we believe that a basic model, consisting of a near main-sequence star and a white dwarf star orbiting in a semi-detached configuration, along with the associated mass transfer, applies to all the short period CVs with the exception of the twin degenerates already mentioned. In Eggleton's notation, these binaries are of type (M,W;S), and this section reviews the evidence supporting such a classification.

### 3.1 Evidence for a White Dwarf

Most of the blue light from a non-magnetic CV arises from a "primary star" that is now understood to consist of a white dwarf and surrounding accreting disk. But may the accreting object be a hot subdwarf or a neutron star instead? There is, after all, enough room inside the Roche lobe for anything the size of a main sequence star or smaller. We can rule out the larger objects by requiring that the gas stream from the secondary star form a disk, rather than striking the primary star directly (Smak 1971; Warner & Nather 1971). Observations pointing specifically to a white dwarf primary include short period oscillations, coherent oscillations, rapid eclipses, and possible direct detection of the spectrum in a few cases. Evidence tending to rule against a neutron star includes the theory of the nova outburst and the existence of a separate class of close binaries with high X-ray luminosity. The argument is an important one, for it is usually assumed that the mass of the primary star is below the Chandrasekhar limiting mass corresponding to $\mu_e = 2$, and this assumption often strongly constrains mass solutions for these binary systems.

Short-period oscillations in white light, of amplitude typically less than one percent, have been observed in many erupting dwarf novae and in a few other luminous CVs. The periods are in the range 8 to $\sim 10^2$ s, the shorter period oscillations usually being termed "coherent" and the longer ones "quasi-periodic" (Robinson & Nather 1979). The periods are similar to calculated periods for non-radial pulsations of a white dwarf, but the oscillations could also be and have been attributed to motions of the inner accretion disk. The coherent oscillations of erupting dwarf novae change their periods with time during the course of the outburst (see, for example, Patterson 1981), so only a small fraction of the white dwarf's mass, if that is what causes these oscillations, could be involved.

There are a few CVs that show very stable oscillations, however, suggesting that rotation of a massive object is the driving mechanism. For DQ Her, the observed 71 s oscillations would imply a white dwarf mass greater than about 0.2 $M_\odot$, to avoid breakup. For AE Aqr's 33 s period, the corresponding limit is $M_{wd} \gtrsim 0.38\ M_\odot$. I have used Eggleton's representation of the white dwarf mass-radius relation to derive these values:

$$R_{wd} = 7.8 \times 10^8 \left[ \left( \frac{M}{M_{ch}} \right)^{-\frac{2}{3}} - \left( \frac{M}{M_{ch}} \right)^{+\frac{2}{3}} \right]^{\frac{1}{2}} cm.$$

(1)

Eclipses of the "primary star" have in a few quiescent dwarf novae been observed to be "double", showing two distinct ingress dips and egress rises (Figure 1). These are interpreted to be the eclipse and reappearance of the compact primary star itself and the "bright spot", a region near the outer edge of the accretion disk located at the point of impact of the gas stream. Accurate timings show that (a) the eclipse of the star is stable in phase and (b) the star is too small to be a subdwarf or other object with an extended optically thick envelope.

The spectrum of the primary star in AM Her has been observed directly in visible light during one of its low states, showing clearly the Zeeman-shifted absorption lines from a white dwarf photosphere and incidentally allowing the surface magnetic field strength to be measured directly (see the review by Liebert & Stockman 1984 and references therein for details). Other "bare" white dwarf primary stars may have been observed in the ultraviolet part of the spectrum. These claims are based either on the continuum slope or on the presence of a white dwarf-like Ly$\alpha$ absorption line (e.g. Robinson et al. 1981; Panek & Holm 1984). In order to confirm these claims and to rule out alternative interpretations, more observations of this kind need to be made, especially since they offer the possibility of a direct measurement of the effective temperature of the white dwarf.

Numerical hydrodynamic calculations of the nova outburst have shown that thermonuclear runaway can occur in the envelope of a massive white dwarf if about $10^{-5}$ or $10^{-4}$ $M_\odot$ of H-rich material is present, in approximate agreement with the observed masses of nova ejecta. Models of the early stages of the nova light curve are also in reasonable agreement with observed behaviour. Thus there is no need to suppose that objects even more compact than white dwarfs, i.e. neutron stars or black holes, are present as the primary stars in cataclysmic variables.

Indeed, it is thought that the bulge-type X-ray sources are the neutron-star analogues of the CVs. These objects have $L_x/L_{opt} \gg 1$, in contrast to $L_x/L_{opt} \lesssim 1$ for CVs (Bradt & McClintock 1983). An additional discriminant is the spin-up of the mass gainer by accretion torques; existing evidence, mostly upper limits to $P_{rot}$ for magnetic CVs, indicate that the primary stars are white dwarfs rather than neutron stars (Lamb & Patterson 1983).

Warner (1979) gives further discussion of the evidence for white dwarfs in cataclysmic variables.

## 3.2. Evidence for a "Main-Sequence" Star

There are some close binaries, usually classified as CVs, that certainly do not have main-sequence secondary stars. These include the twin-degenerate (W,W;S) variables AM CVn and GP Com and the recurrent nova T CrB. The last of these has a giant-type secondary and may have a main-sequence gainer rather than a white dwarf gainer (Webbink 1976). It may make more sense to classify T CrB with the symbiotic stars, rather than with the CVs.

The remaining CVs can be divided into three groups based on the orbital period. The long-period group ($10$ h $\lesssim$ P $\lesssim$ $2$ d) contains secondary stars that are thought to be slightly evolved, in the sense that the radius of the star is typically 1.2 to 1.5 times larger than that of a main-sequence star of the same mass (or temperature class). Examples include GK Per, BV Cen, and AE Aqr (Gallagher & Oinas 1974, Gilliland 1982, and Patterson 1979, respectively).

There is little direct observational evidence concerning the nature of the secondary stars in the so-called "ultra-short period" CVs ($81$ m $\lesssim$ P $\lesssim$ $2$ h). It is at least possible to argue that a secondary star in one of these could be a normal main-sequence object, as follows: if it fills its Roche lobe (see next section), then the orbital period allows us to determine the star's mean interior density (Eggleton 1983); using a mass-radius relation for the main-sequence we can thus deduce its mass and hence its luminosity (Warner 1976); for the short-period systems, this luminosity is so small that normal spectroscopy at visible wavelengths will fail to detect the secondary star against the competing light from the accretion disk and primary star. This plausibility argument seems to work well whenever P $\lesssim$ $6$ h (Warner 1976), and holds a fortiori for P $\lesssim$ $2$ h. There are, however, theoretical reasons for thinking that these ultra-short period systems contain abnormal secondary stars. They are not evolved in a nuclear-burning sense, but rather they are out of thermal equilibrium because their mass-loss timescale is shorter than their Kelvin timescale (Rappport, Joss, & Webbink 1982). The $81$ m cutoff in the observed orbital period distribution is thought to arise ultimately from this dis-equilibrium and the simultaneous movement of the secondary star from a main-sequence structure to an electron-degenerate structure. Some models for the "period gap" require stars in systems with periods just longward of $3$ h to depart significantly from the normal mass-radius relation also.

There remains the group of CVs with periods in the range $3$ h $\lesssim$ P $\lesssim$ $10$ h. Secondary stars in this period range should have masses $\lesssim$ 1 M$_\odot$ (if normal) and, neglecting any prior mass loss, would not be expected to show the effects of nuclear evolution. More moderate deviation from the main sequence, as a consequence of the secondary star's abnormal environment, cannot be ruled out so readily, however, in view of the deviations that are thought to exist at both shorter and longer periods.

"Normal" main-sequence behaviour can be characterised in several ways, for example by pairing any of the following quantities: mass, radius, effective temperature, and luminosity or absolute magnitude. The most useful pair, to an observer who wishes to derive the mass of the secondary star, is the mass-radius law. Neither of these quantities is directly observable, however. A different pairing, that of spectral type and mean density (related straightforwardly to the orbital period), is in practice easier to use for a test of normality (see, for example, Wade 1979; Patterson 1979). In addition, this information is available for a large number of CVs. In the period range 3 h to 10 h, the spectrum - mean density diagram is consistent, within the rather large errors, with the empirical main sequence (H. Ritter, private communication). More and better data will in the future provide a more stringent test; for the present it is clear that claims and counter-claims about the main-sequence nature of this or that secondary star (e.g. U Gem: Wade 1981, Stover 1981) are concerned with one- or two- sigma deviations. A more interesting question is whether the moderate-period CVs deviate significantly as a group from the main sequence, and there is a tendency for the well-studied CVs to cluster on the cool side (or the low-density side) of the main-sequence band in the spectrum - density diagram.

Stover's (1981) work on U Gem clearly shows that a secondary star may fall near the main-sequence in the mass-radius plane, yet have the wrong spectral type (i.e., be underluminous) with respect to its mass. There is also not a monotonic relation between orbital period and spectral type, as Table 3 shows. For work on individual stars, therefore, it is very dangerous to assume a particular form for the main-sequence relations between mass and radius, mass and spectral type, mass and absolute magnitude, etc., for the purpose of finding particular values of mass, luminosity, distance, etc. It may also be dangerous to do statistical studies, for example, mass transfer rate as a function of orbital period, this way.

To summarize, there is no evidence that the secondary stars in most CVs are not core hydrogen-burning stars, but it cannot be assumed that in every respect they lie on the zero-age main sequence.

## 3.3  Evidence for a Semi-Detached Configuration

Mass transfer clearly takes place in all CVs, except possibly during the low states of the VY Scl and AM Her variables. The evidence includes short-timescale flickering and dwarf nova outbursts (interpreted as accretion events), as well as the "S-waves" and "bright spots" that refer specifically to the gas stream leading from the secondary star to the region of the white dwarf primary. Reviews already mentioned, especially the ones by Robinson (1976) and Warner (1976) give details of these observations.

In order to have mass transfer, it is sufficient for the secondary star to fill its Roche lobe, but perhaps it is not necessary. Chincarini and Walker (1981) have argued concerning AE Aqr that the secondary star underfills its Roche lobe, and that material is transferred through prominence activity on that star. The argument relies on the radius adopted for a normal K5 V star and on the observed spectral class and assumed normality of the secondary star in AE Aqr, all of which are perhaps still uncertain.

"Ellipsoidal variations" in brightness, due to the periodically changing aspect of the distorted secondary star, have been reported or suspected for several CVs. In the best studied case, that of U Gem, Berriman et al. (1983) fitted models in which the "fill-out factor" was allowed to vary along with the mass ratio, inclination, and irradiating luminosity. They concluded that to produce the observed orbital light curves at J, H, and K the secondary star in U Gem must fill or nearly fill its Roche lobe. In particular, the reflection effect is much less important than the ellipsoidal variation.

It is clearly impossible to establish whether a star exactly fills its Roche lobe. The bulk of the observational evidence, however, indicates that this approximation is an appropriate one to make for cataclysmic variables.

## 3.4  Evidence for an Accretion Disk

If the primary star in a CV is a white dwarf, then material transferred from the secondary star needs to be channeled on to a small target. In the magnetic CVs, this channeled flow ultimately occurs along magnetic field lines (see, for example, Liebert & Stockman 1984). In the absence of a strong field, material falling from the L1 point has sufficient angular momentum to form a disk around the white dwarf.

That disk formation actually occurs in cataclysmic variables follows from several lines of observation. "Doubled" emission lines are interpreted as arising from the surface of a prograde differentially rotating disk, the peaks corresponding to the maximum projected velocities of the outer edges of the disk and the wings corresponding to higher velocites nearer to the central star (Smak 1969; 1981). This interpretation also accounts for the following details. (a) The value of $\Delta\lambda/\lambda$, where $\Delta\lambda$ is for example the separation of the peaks, is often approximately constant with wavelength and atomic species. (b) The emission lines are generally broader and the peaks more separated in eclipsing systems. (c) The lines are more often doubled in eclipsing systems. (d) The eclipse "disturbance" of emission line velocities, as first one side of the disk and then the other is screened from view by the secondary star, is observed in favourable cases (Greenstein & Kraft 1959; Rayne & Whelan 1981; Young & Scneider 1980).

Primary "double eclipses" (Fig. 1) occur in some systems, being the eclipse of the white dwarf (or central region of the disk) and a "bright spot" (sometimes called a "hot" spot) at the disk edge. The bright spot is not an object fixed in position with respect to the white dwarf, since the eclipse shape for any one CV changes with time (e.g. Cook 1982). The reinterpretation of photometry of U Gem by Smak (1971) and Warner & Nather (1971) established that the bright spot is an entity well separated from the white dwarf, and that much of the light from quiescent dwarf novae comes from the disk and spot rather than the white dwarf itself (see also Bath et al. 1974).

The coherent oscillations seen in the light curve of DQ Her and sometimes UX UMa are observed to undergo a phase shift at the time of eclipse. The phase shifts have been interpreted in terms of the progressive occultation of parts of an accretion disk, which is periodically illuminated by a beam from the disk centre (see e.g. Petterson 1980).

Further evidence about accretion disks in CVs includes measurements of their size, either directly from eclipse timing data (Sulkanen, Brasure & Patterson 1981; also Ritter 1980b) or from simple dimensional arguments ($R^2 \propto L\ T^{-4}$). These show that a disk is large, typically $\gtrsim 70\%$ of its maximum possible size. Important recent work based on multi-colour photometry of eclipses has convincingly demonstrated that the disks are multi-temperature objects (Horne 1983; also in this volume).

Disk spectra will be discussed in connection with the determination of the mass accretion rate in Section 5.

## 4.   MASS DETERMINATIONS FOR CATACLYSMIC BINARIES

### 4.1   Statement of the Problem

In the following discussion it will be assumed that the orbital period of the CV is known. Given the separation of the stars, a, and the mass ratio, $q = M_1/M_2$, the masses follow directly from Kepler's third law. The separation cannot be measured directly, but is inferred from the radial velocity semi-amplitudes $K_1$ or $K_2$ at the expense of introducing the inclination, i, as an additional unknown. For example, a sin i $= P K_1 (1+q) / 2\pi$. Incorporating Kepler's law results in the usual formulation of the "mass function", $f(M_2) = P K_1^3 / 2\pi G = M_2^3 \sin^3 i / (M_1 + M_2)^2$ (subscripts can be interchanged). In general, then, three quantities, e.g., $K_1$, $q = K_2/K_1$, and i, are required to solve for the masses.

Usually for spectroscopic binaries the required three quantities can be measured only for double-lined systems that eclipse. For CVs, however, it is possible to use the properties of the basic model to introduce additional constraints, so that sometimes the masses, or at least bounds on the masses, can be inferred in less favourable cases. In this section I first outline many of the constraints that have been developed for use with CVs and then describe how they are applied to systems where differing degrees of information are available.

### 4.2   Constraints on Masses:   A Catalogue

### 4.2.1   Constraints Related to the Secondary Star

C1. The secondary star is assumed to fill its Roche lobe, i.e. the fill-out factor is unity. This means that the shape and size (relative to a) of the secondary star are fixed if q is known. In the absence of eclipses, this information is used to give an upper bound on i as a function of q. If eclipses occur, i can be treated as a function of q and the duration of the eclipse, $\Delta\phi$. ("$\Delta\phi$" could also represent other forms of timing information, for example the phase difference between eclipse of the white dwarf and the bright spot; see C11 below.) If ellipsoidal variations are observed, light curve modelling fixes i as a (weak) function of q.

C2. The secondary star is often assumed to lie on the main sequence. If the spectrum of the secondary star is visible, its mass follows directly. No information about $M_1$ is available unless q is known or can be inferred from other constraints.

C3. If the fill out factor is unity, the normalized volume-equivalent radius $R_2/a$ is a function only of q. Eggleton (1983) gives a recent tabulation, along with an accurate analytic fit that is valid over the whole range of q. The secondary star is often assumed to obey some mass-radius relation. Usually this is taken to be a simple parametrization of the lower main sequence, such as $(R_2/R_\odot) = \alpha (M_2/M_\odot)^\beta$, where $\alpha$ and $\beta$ are determined either empirically (e.g., Lacy 1977) or from a series of stellar models. Other possibilities include using models of evolved stars constrained by temperature class and mean density (e.g., Patterson 1979; Gilliland 1982), or degenerate or non-equilibrium stars. By equating $R_2(M_2)$ to $R_2 = (R_2/a) \cdot a$, a is found as a function of $M_2$, in a manner independent of radial velocity information. The second equation relating a and $M_2$ is, of course, Kepler's third law. The mass ratio q enters into both equations, but when $M_2$ is found as a function of P and q, the dependence on q nearly vanishes. The result is that $M_2$ is known as a function of P "alone", if the correct mass-radius relation has been chosen.

An alternative way of viewing this constraint is to note that $R_2/a = f(q)$ combined with Kepler's law fixes the mean density of the secondary star (if P is known). The mass-radius relation, derived independently from stellar structure calculations, fixes the star's mean density as a function of its mass and evolutionary state. Thus its mass is related directly to the orbital period of the binary system. (See Warner 1976 for a full exposition; an earlier statement of the method by Faulkner et al. [1972] is also valuable).

This method does not constrain $M_1$ unless q is known independently.

Note that C2 and C3 make use of assumed information about the secondary star, namely that it lies on the main sequence (or some other sequence). C2 posits a relation between mass and spectral type, while C3 relates mass and radius.

4.2.2 Constraints related to the white dwarf and its disk

C4. $M_1$ is assumed to be less than the Chandrasekhar limiting mass for a white dwarf with mean molecular weight $\mu_e = 2$ (per electron). This usually enters as a lower limit on the inclination.

C5. The full width at zero residual intensity of the emission lines is used to constrain the mass of the white dwarf: the relation

$$v_{circ,max} = \left(\frac{G M_1}{R_1}\right)^{\frac{1}{2}} \geqslant v_{obs} = \frac{1}{2} \cdot FWZI$$

(2)

combined with the white dwarf mass-radius relation imposes a lower limit on the white dwarf's mass. The assumption is that Keplerian motion in the disk is the only agent acting to broaden the lines.

C6. Similarly, the shortest observed period of any "coherent" oscillations can be used to place a lower limit on the white dwarf mass. The interpretation here is that the oscillation period corresponds to the orbital period of material around the white dwarf, or is related by some other definite prescription to the basic dynamical time $\sim (R_1^3 / G M_1)^{\frac{1}{2}}$ of the primary star.

C7. Another equation of constraint is available if the eclipse duration $\Delta\phi$ of the white dwarf primary itself is accurately known. In this case $(K_1, q, i)$ or a similar trial set of quantities determine trial values of a and $M_1$. They also yield an "observed" radius $R_1$ of the primary star. $R_1$ is a function of $\Delta\phi$, a, i, and q. This $R_1$ must match the value computed from the white dwarf mass-radius relation. Only those trial sets $(K_1, q, i)$ that give consistency are retained; this acts as a constraint, typically relating q and i. In practice, the extra equation becomes an inequality, $R_1$ (observed) $\geqslant R_1 (M_1)$, since one can never be sure that the observed short eclipse of the "white dwarf" is not contaminated by light from the surrounding inner disk or ring. An example of the method is given in Patterson's (1981) study of HT Cas.

C8. MacDonald (1983) has used the CNO abundance in nova ejecta, combined with the mass and the principal expansion velocity of the ejected material, to estimate the white dwarf mass in selected novae. These estimates are based on the thermonuclear runaway model for nova outbursts.

4.2.3 Constraints related to the mass-transfer model

C9. The radial velocity of the bright spot or some other point in the gas stream is a function of i, q, and orbital phase $\phi$. In some systems this emission-line velocity can be measured cleanly or deduced from the distortions that the extra light produces in the disk's emission line radial velocity curve (e.g., U Gem, Smak 1976). An admissible mass solution must correctly predict the behaviour of this extra component.

C10.   Warner (1973) argued that the ratio $K_1/(v_d \sin i)$, where $v_d \sin i$ is the projected velocity of gas orbiting at the outer edge of the accretion disk, might be a function of q alone.  By adopting FWZI/2 = $v_d \sin i$, Warner found a directly observable function of q.   The method has been criticized, but is still used.   Shafter  (1983a)  discusses an alternative way of using this relation.

C11.   The phase delay $\Delta\phi$ between  eclipses of the white dwarf and bright spot is a function of i,  q,  and  the  disk  radius. Here  the  spot  is  assumed  to be  on  the  single-particle trajectory  from L1, at  the  edge  of  the  disk.   Admissible solutions must correctly predict $\Delta\phi$  and  related quantities. The  use  of  this  constraint  for  double-eclipse  systems is exemplified by Ritter's (1980a) study of Z Cha.

C12.   The  disturbed  behaviour  of  the  emission lines during eclipse has been labelled the "Z-wave" by  Young,  Schneider, & Shectman  (1981b).   They  have  modelled  the  Z-wave  as  the selective  occultation  of parts of the disk surface, where the emission lines are  presumed  to  originate.   The  occultation pattern  is  a function of orbital phase, q, i, and the assumed emissivity pattern on  the  disk surface.  The amplitude of the disturbance measures the projected velocity  of  the gas in the disk, assumed to be in Keplerian orbit  around the white dwarf. Thus  an  additional  equation,  involving  no new unknowns but several  extra  assumptions,  is  available  if  time-resolved measurements  of sufficiently high signal-to-noise ratio can be obtained.  The  method  has  been  applied  to  DQ Her (Young & Schneider  1980),  LX  Ser (Young,  Schneider & Shectman 1981a), and HT Cas (Young, Schneider & Schectman 1981b).

In  the  use of all of these special  methods,  additional information is available because the model of a CV (white dwarf primary,  stream and  disk kinematics, origin of emission lines) is thought  to  be  firmly established.  A possible drawback is that  these methods are therefore  less  capable  of  revealing flaws  in  the  model.  They  should  therefore  be  used with caution.   A  second  drawback  is  that  the  extra  algebraic manipulation  or  numerical  modelling  required  to  extract a solution  from  the  observations  makes  error propagation and analaysis less transparent.

4.2.4  Empirical "rules"

C13.  From double-lined  systems  and  other  CVs for which the mass ratio q is known (perhaps by  using  constraint  C10),  an empirical  relation  q(P) between mass ratio and orbital period is developed.  The  mass  ratio  for additional systems is then inferred  from  the empirical curve (see,  for  example,  Bruch

1982).

C14.  In the same spirit as C13, $M_1$ is taken to be, say, 1 $M_\odot$.

C15.    Other  empirical  rules  are  derived,  e.g.,  $M_2(P)$,
$M_2$(spectrum), or $M_2$(colour), and then applied  to  new systems.
These  particular  examples  are  similar  to  constraint  C2
described  above,  but  they  have  an  inductive rather than a
deductive basis.

     Such  constraints, which  have  actually  been  used  when
other  information  was  unavailable,  span  a  wide  range  of
sophistication and are easily challenged.  Their use in any but
a very  crude statistical sense requires careful discussion and
justification of the assumptions implicit in each one.

4.3  Applications of the Constraints

     It is clear  how  the  constraints enumerated above can be
used  to  help  determine masses for  the  component  stars  in
cataclysmic variables.  For example, double-lined spectroscopic
binaries (SB2) that eclipse  provide direct measurements of $K_1$,
q, and "$\Delta\phi$".  Constraint  C1  then provides i as a function of
$\Delta\phi$  and  q,  allowing the mass  solution  to  be  carried  out
directly.  Of course, some  numerical  modelling of the eclipse
of a complicated light distribution (the disk) may be required,
and additional assumptions or measurable unknowns may enter the
problem at this stage.  To date  only two eclipsing SB2 systems
are known among the CVs.  These are  the dwarf novae EM Cyg and
U Gem; in neither case is the white dwarf itself eclipsed.

     In the case of SB2 systems that do  not eclipse, i remains
unknown.  Constraints 2, 4, 5, 6, 10, 14, and  15  can  all  be
applied,  but  the one most often used is C3, the "mean density
constraint".  Usually a  main-sequence mass-radius relation for
the secondary star is  adopted.   This method has the advantage
of yielding definite values for both  $M_2$  and  $M_1$, from which i
can  be deduced using measured quantities.  This in turn allows
consistency  checks,  e.g.  C5,  to  be  applied.   Well-known
examples of  SB2  systems are the longer-period systems SS Cyg,
RU Peg, AE Aqr, Z Cam, BV Cen and GK Per.

     Single-lined CVs do not show the spectrum of the secondary
star.  For eclipsing SB1  systems,  constraint  C1 gives i as a
function of $\Delta\phi$ and the unknown q.  Constraint C3 can yield $M_2$,
but one of  the  special  techniques  C9,  10,  11,  or 12, often
supplemented by one or more of the white  dwarf constraints (C4
through C8) is usually required to obtain $M_1$.  Examples include
OY Car, Z Cha, HT Cas, DQ Her, and WZ Sge.

Non-eclipsing SB1 systems leave both q and i undetermined. Here constraint C3 is used heavily, often supplemented by white dwarf constraints or empirical rules. Consistency checking is more difficult, since in general only wide limits can be placed on q and i (method C3 is not sensitive to them).

When only the orbital period is known (e.g. from superhumps in SU UMa systems), constraint C3 can still be used to estimate $M_2$. Only limits are available on $M_1$, unless an empirical mass is assumed (C13 or C14). When not even the orbital period is known, rule C14 can, of course, still be used. This procedure is not recommended.

## 4.4   Remarks on Methods of Mass Determination

The mean density constraint C3 (which combines the Roche geometry with a stellar mass-radius relation) is of central importance in the great majority of mass determinations for CVs. It therefore requires careful examination. The assumption that the fill-out factor is unity, has been discussed in Section 3.3; if the secondary star underfills its lobe by, say, 10%, the mass derived by method C3 will be in error by a similar amount. More dubious, perhaps, is the assumption of a mass-radius relation for the secondary star. Section 3.2 discussed evidence for· and against the main-sequence nature of the star. In addition to the basic assumption, there is the problem of adopting a specific form of the mass-radius relation: shall it be theoretical or empirical? Shall it be ZAMS, TAMS, evolved, or non-equilibrium? The use of constraint C3 needs to be tested whenever possible, i.e. by using other constraints as consistency checks.

The assumption that the primary star is a white dwarf is almost certainly justified, but it is an assumption. The constraints resulting from it are powerful. For example, Stover (1981) concluded that the secondary star in RU Peg must be evolved, because the use of C3 with a main-sequence mass-radius law led to $M_1 > 1.4\ M_\odot$ for this SB2 system.

The lower limits imposed on the white dwarf's mass due to the velocity width of emission lines (C5) are likewise very powerful (see Figure 2, from the study of T Leo by Shafter & Szkody 1984). If pressure broadening or electron scattering, for instance, are important line broadening agents, this constraint will be less meaningful. Likewise the use of oscillation periods to constrain $M_1$ (C6) requires a knowledge of the mode of oscillation which we do not possess. The dwarf nova EM Cyg has $M_1 = 0.55 \pm .05\ M_\odot$ measured spectroscopically (Stover et al. 1981), but oscillations with a period of

14.6 s have been observed (Stiening et al. 1982). If these reflect orbital motion around the white dwarf, the inferred lower limit to $M_1$ is 0.68 $M_\odot$.

The use of eclipse timing or phase offsets (C7, C11) relies on measurements that are difficult to make, in view of the flickering that is common among CVs. It is possible to suppress the flickering "noise" averaging many eclipses, with the risk that the system structure (e.g. disk size or bright spot location) has changed during the interval covered. See Cook 1982 and Smak 1971 for examples of change. The numerical modelling required to extract i and q from such data is also hard, and error propagation is not straightforward.

## 4.5 Measuring $K_1$ from Emission Lines

Because $K_1$ enters the mass function in the third power, the error in mass determinations is often dominated by the error in determining $K_1$. The motion of the white dwarf is not measured from its own lines, but from the velocity variations of the emission lines arising in the gas around the white dwarf. It is worthwhile to consider how well $K_{em}$, the velocity semi-amplitude of the gas, represents $K_1$, the dynamical value which is used in mass solutions.

The model for the formation of the emission lines in an accretion disk was described in Section 3.4 above. The confidence that was built up as a result of the generally successful comparison of model and observation has led observers to equate $K_{em}$, the velocity semi-amplitude of the emission lines, with $K_1$, the dynamical velocity semi-amplitude of the white dwarf.

There are, however, many anomalous observations that call into question the automatic identification of the measured $K_{em}$ with the dynamical $K_1$. The emission lines are not doubled in every case, even in centrally eclipsing systems such as LX Ser. The model does not predict that the "valley" between the emission peaks should extend below the continuum, as is observed in the higher Balmer lines of, for example, Z Cha and OY Car (see Fig. 3, from Bailey & Ward 1981). In these same centrally eclipsing systems the velocity separation of the peaks changes along the Balmer series, showing that Doppler broadening is not the only agent acting to shape the lines. The $\gamma$ (zero-crossing) velocity changes from one emission line to another (V436 Cen, Gilliland 1982) or between emission and absorption (i.e. secondary star) line measurements (U Gem, Wade 1981). Emission line zero-crossing occurs at the wrong phase in some systems, measured relative to eclipses or absorption line velocity curves (Stover 1981). Strong asymmetries sometimes appear, such as the anomalous blue wing on the H$\beta$ line in RX And (Hutchings & Thomas 1982). Finally, the emission line velocity curves of some systems are suggestive of gas streams in non-standard locations, or additional emission components (T Leo, Shafter & Szkody 1984; RW Tri, Kaitchuck et al. 1983).

Even if these disturbing exceptions to the customary simple picture can be incorporated in a systematic and quantitative enlargement of the model, so that $K_{em}$ or some function of $K_{em}$ can with confidence be identified with $K_1$, there remains the more technical problem of how best to measure $K_{em}$. Of course, distortions of the basic emission line profile due to underlying absorption lines or emission "S-waves", etc., should be removed before the line can be measured by any technique; this is not always done. Since in the model the broad wings of the emission lines arise from gas orbiting close to the white dwarf, it is usual to measure $K_{em}$ from these wings and to exclude the "core" of the line. The core is the region including the valley and the peaks, where emission from the bright spot, or distortions of the circular orbits in the outer disk, can perturb the velocity measurement. The questions are, how much core should be excluded, and how far into the wings should the measurement extend? Many modern attempts at measuring $K_{em}$ adopt thoroughly quantitative approaches to this problem. This allows the measurement to be reproduced by other observers, but the criteria used are still arbitrary, i.e. not fully understood.

As an example of the importance of choosing a measuring technique, consider the study of T Leo by Shafter and Szkody (1984). They choose the size of the excluded core region (parametrized by "a", the separation of two gaussian passbands in the wings of the line) by considering the standard error $\sigma$ of a sinusoidal fit of the phase-binned velocity data. Where $\sigma(K_{em})/K_{em}$ is plotted as a function of $\underline{a}$, it is at first roughly constant and then rises steeply beyond a critical value of $\underline{a}$ (see Figure 4). The adopted value of $K_{em}$ corresponds to this $\underline{a}_{crit}$; coincidentally (or perhaps not?) this $K_{em}$ is the largest value found over the range of $\underline{a}$ considered by the authors. The criterion chosen for $\underline{a}$ is primarily a signal-to-noise ratio criterion: moving further out into the line wings increases the relative importance of noise in the continuum, so that after some point $\sigma/K$ rises. The authors choose to measure as far as possible into the wings of the line, consistent with moderate values of $\sigma/K$. it makes sense to wonder whether original data of better or worse signal-to-noise ratio, if treated the same way, would result in a significantly different value for $K_{em}$.

Other authors have adopted similar criteria, although the details of the numerical treatment differ. For example, it is possible to use $\sigma$ rather than $\sigma/K$ as the discriminant for $a_{crit}$ (Shafter 1983b). By singling out the study of T Leo I do not mean to criticize that work in particular. Indeed, I have no suggestion at present for a demonstrably better procedure. The problem of how to measure $K_{em}$ is serious and difficult. Certainly observers should measure as many different emission lines as possible, with a view to detecting any systematic shifts in $\gamma$ or $K_{em}$ with wavelength or atomic species. More complex models of line formation in a disk, including radiative transfer effects and the hydrodynamics of the disk-stream interaction, may eventually show us how better to measure $K_1$ from emission lines.

## 4.6   Results of Mass Determination

Ritter's (1983) compilation of data for cataclysmic variables is the source for this brief summary of present knowledge of masses, and further details and references can be found there. It should be mentioned that in cases where more than one study has been made of a system, the resultant mass "determinations" sometimes differ markedly. Ritter's choices are reflected in the following discussion.

Fig 5 plots $M_1$ against $M_2$ for 31 CVs, some of which are identified by their two-letter first names. Error bars are omitted from the figure; the upper limits for RU Peg result from Stover's (1981) work already mentioned. In all but three cases, the reported mass of the primary star exceeds that of the secondary star.

There is a range of white dwarf masses extending from about 0.3 ($\pm 0.2$) $M_\odot$ to the Chandrasekhar limit near 1.4 $M_\odot$. No strong correlation with orbital period is evident. In particular, there is no difference in average mass between the short and long sides of the "period gap". As expected, the masses of the secondary stars show a tight correlation with orbital period over a range from 0.1 $M_\odot$ to 1.1 $M_\odot$. This reflects the use of the mean-density method for estimating $M_2$. The secondary star in the twin degenerate system AM CVn is thought to have a mass of 0.04 $M_\odot$. The most significant outliers to this nearly linear trend are the longest period systems such as AE Aqr and BV Cen, which are thought to contain evolved secondary stars.

For eighteen of the $M_1$ values that I collected, authors' estimates of the error are available. For all but one case, $\sigma(M_1)/M_1$ is reported as less than 0.3, often below 0.2. I wonder whether these estimates are in every case reasonable.

Except for SB2 systems, where a measurement of q together with constraint C3 can give $M_1$ with only linear dependence on $K_1$ and $K_2$, one expects that $\sigma(M)/M \gtrsim 3 \, \sigma(K)/K$. Thus the values quoted above imply that $K_1$ is thought to be known to better than 10% in most cases, and that other sources of random error (such as from eclipse timing data, or the finite width of the main-sequence mass-radius relation) are unimportant. Of more concern than the random errors, which $\sigma(M)$ represents, is the possibility of systematic errors whose origin we are not sure of, and whose presence we may fail to detect. Stover (1981) points out that "variations of a least 10% [in $K_1$] have been seen in our models for phase shifts" in the emission line velocity curve. The delicate question of how much or what part of the emission lines to measure has already been discussed.

Similar, possibly optimistic, estimates of $\sigma(M_2)/M_2$ exist. Many of the reported values for $\sigma(M_2)$ are 0.1 $M_\odot$ exactly, suggesting that the authors have used a notional value in lieu of a careful assessment of the error.

In view of the importance of mass determinations for CVs and the difficulties in measuring such quantities as K and q, it is necessary for observers to report accurately and in full detail how their studies were carried out. This must include realistic estimates of the errors of measurement, both from internal and external checks, and correct propagation of those errors. Just as important, they need to discuss all of the model assumptions and whenever possible to assess the impact of changing these assumptions.


5.    ACCRETION DISKS IN CATACLYSMIC VARIABLES

The idea that accretion disks are present in non-magnetic CVs is so attractive, and the evidence is so strong, that an accretion disk has to be regarded as a fundamental part of the basic model for CVs. There is still an enormous amount of work to be done on the details, both from the theoretical and observational sides. The recent reviews by Pringle (1981) of accretion disks in general and the spate of recent theoretical work on time-dependent disks in particular (Faulkner et al. 1983, and references to other studies therein) should be consulted for a more complete picture than can be presented here.

This section limits itself to a brief description of some features of steady, optically thick disks, from the point of view of using observations to deduce the mass accretion rate through the disk.

## 5.1  Simple Disk Physics

Consider a disk revolving about the white dwarf primary star of a CV. The disk is assumed to be physically thin and optically thick. That is, the local vertical thickness, $2h(r)$, of the disk is much smaller than the local radius r, and energy is radiated through the two faces of the disk. If the disk is in a steady state so that the radial mass flux through the disk is independent of radius, and energy dissipation is radiated locally, then there is a local effective temperture $T_{eff}(r)$ that characterizes the radiative flux from the disk faces. $T_{eff}(r)$ is determined by the mass and radius of the accreting star $(M_1, R_1)$, the mass flux $(\dot{M})$, and r. Ignoring "edge effects", such as an inner boundary layer and a bright spot or tidal heating at the outer edge of the disk, $T_{eff}(r)$ varies approximately as $r^{-3/4}$. The disk thickness is determined by local hydrostatic equilibrium between pressure support on the one hand and the vertical component of the gravity from the accreting star on the other. Again ignoring details, $h(r)$ varies roughly as r, so the effective gravity at the disk "surface" varies as $g(r) \propto h \, r^{-3} \propto r^{-2}$.

Given $T_{eff}(r)$ and $g(r)$ it should be possible to compute the local emergent spectrum of radiation, and summing over annuli the spectrum of the disk as a whole. This integrated spectrum will depend on, among other things, $M_1$, $R_1$, $\dot{M}$, the outer disk radius, the inclination of the disk to the observer's line of sight, and details of the vertical disk structure (temperature and density gradients, etc.). In practice, real disks may not be optically thick everywhere, neither are they likely to satisfy the steady-state assumption, nor will the local spectrum radiated from each surface element of the disk be easy to compute. Then, of course, the (largely uncertain) edge effects must be taken into account, along with radiation from other parts of the binary system, such as the two stars and the gas stream that feeds the disk.

It is no wonder that, even for disks that are in fact expected to be steady and optically thick, few real attempts to test the disk model have been made. The far more usual approach is to assume that the model is correct and try to derive $\dot{M}$ and other disk parameters from the observations. Usually extra light sources are neglected and simplifying assumptions about disk physics and local spectra are made. What few careful tests there are (e.g. work by Kiplinger 1979, 1980; Horne 1983; Hassall 1983) have had only mixed success in confirming our ideas about disks in CVs: the simplest steady-state models do not adequately describe the observations. If the model used to predict accretion disk spectra is not secure, however, neither are the estimates of $\dot{M}$

resulting from its use. In published work, mass accretion rate estimates range from $\sim 10^{-11}$ $M_\odot$ $yr^{-1}$ up to $10^{-6}$ $M_\odot$ $yr^{-1}$ for CVs. While some of this wide scatter is undoubtedly real, some of it is due to different modelling assumptions and reflects the uncertainties mentioned above. Certainly some, possibly many, of the published values of $\dot{M}$ will turn out to be in error by more than a factor of 10.

## 5.2 Integrated Spectra of Accretion Disks

### 5.2.1 Further Disk Physics

In the models I describe below I have used the expressions for $T_{eff}$ and g adopted by Herter et al. (1979; see also references therein). The expression for $T_{eff}$ is standard:

$$T_{eff}^4(x) = T_*^4 x^{-3}\left(1 - x^{-\frac{1}{2}}\right),$$

(3)

where $x = r/R_1$, $R_1$ = inner radius of the disk, usually taken to be the radius of the accreting star. In this expression,

$$T_* = 4.1 \times 10^4 \dot{M}_{16}^{\frac{1}{4}} M_1^{\frac{1}{4}} R_{1,9}^{-\frac{3}{4}} K,$$

(4)

where $\dot{M}_{16}$ is the (steady) mass accretion rate in units of $10^{16}$ g $s^{-1}$, $M_1$ is the stellar mass in solar units, and $R_{1,9}$ = $R_1$ in units of $10^9$ cm. The maximum value of $T_{eff}$ is $T_{max} \simeq T_*/2$, and the ratio $T_{max}/T_{min} = T_{max}/T_{out} \simeq x_{out}^{-3/4}/2$, where $x_{out} = r_{out}/R_1$, $r_{out}$ being the outer radius of the disk. Typical values for $\dot{M}$ lead to values of $T_{max}$ of a few times $10^4$ K. Typical values of g in the inner disk are $10^6$ cm $s^{-2}$.

For $T_{max}$ in the range just indicated, most of the radiation from the disk will emerge at ultraviolet wavelengths. Thus one of the common methods for estimating $\dot{M}$ is to try to measure $T_{max}$ from the shape of the ultraviolet continuum. Then $\dot{M} \propto T_{max}^4 (M_1 R_1^{-3})^{-1}$, and small fractional errors in $T_{max}$ are magnified enormously. Factors affecting estimates of $T_{max}$ include interstellar reddening, accretion disk inclination (i.e. "limb-darkening"), and, if $T_{max}$ is estimated in part from the continuum slope over a long baseline, non-simultaneous observations at different wavelengths. Note also that for a white dwarf mass-radius relation, the combination $M_1 R_1^{-3}$ also varies rapidly with $M_1$, so that incorrect mass estimates can seriously bias estimates of $\dot{M}$. A second method of estimating $\dot{M}$

is to match the observed flux at a specific wavelength with the flux from a model. This technique is less sensitive to $M_1$, but is highly sensitive to the angular size of the disk (i.e., its distance and outer radius), its inclination, possible edge effects, and other sources of light.

### 5.2.2. Blackbody Models

If Nature were so kind as to allow her accretion disks to radiate locally as blackbodies, the life of the model-builder would be simpler than it is. Alas, emission lines, absorption lines, and Balmer jumps all testify that life is not simple. Blackbody models remain popular nevertheless, so it is worth pointing out how they differ from other models and how they may lead us astray.

The spectrum from a blackbody disk can be computed by integrating the local emissivity over the surface of the disk,

$$f_\nu \propto \int_{x_1}^{x_{out}} 2\pi x \cdot B_\nu \left( T_{eff}(x) \right) dx,$$

(5)

where $B_\nu$ is the Planck function. This is equivalent to an integral over $T_{eff}$, and it can be shown that if $T_{max}/T_{min}$ is large enough, then $f_\nu \propto \nu^{1/3}$ over some range of $\nu$ (see e.g., Pringle 1981). Thus a $\nu^{1/3}$ spectrum is often taken as the signature of a steady-state accretion disk and ultraviolet spectra with logarithmic slopes close to 1/3 (or -7/3 in $\lambda$, $f_\lambda$ terms) have in fact been observed many times among the CVs.

There can, however, be no direct connection between these observations and the inference that the disks in these CVs are steady-state and radiate as blackbodies according to the formulas given above. This is because the necessary condition on $T_{max}/T_{min}$ is rarely obtained in CVs, which are truncated by the tidal action of the secondary star. Fig. 6 shows $x_{out}$ and $T_{max}/T_{min}$ as functions of orbital period and white dwarf mass. These are derived using Eggleton's (1983) expression for the Roche lobe size, Lacy's (1977) mass-radius relation for main-sequence stars (his equation 13) along with constraint C3 from Section 4.2 above, and Eggleton's white dwarf mass-radius relation (given in Section 3.1). The line $q = M_1/M_2 = 1$ is also shown. Certainly for moderate white dwarf masses and short orbital periods, $T_{max}/T_{min}$ will not be larger than about 20. Fig. 7, which plots log f for three values of $T_{max}/T_{min}$ (units arbitrary, but same $T_{max}$ in each case), shows that a logarithmic slope of 1/3 is attained only at a tangent point for these modest values of $T_{max}/T_{min}$. Note also that the "turnover" at short wavelengths is not abrupt and does not coincide with the turnover in $B_\nu$ $(T_{max})$.

### 5.2.3 "Stellar Atmosphere" Disk Models

Model spectra for accretion disks can of course be constructed in many different ways. Williams & Ferguson (1982) for example, construct disks which are allowed to be optically thin according to a definite prescription. The spectrum radiated by a surface element is then

$$ f_\nu = \left[1 - exp(-\tau_\nu)\right] . B_\nu , \tag{6}$$

where the optical depth $\tau_\nu$, as well as $T_{eff}$, is a function of radius. The integrated spectrum is computed as before; for low values of $\dot{M}$, the Balmer jump and Balmer lines are predicted to be in emission. Extensive use of these models, which incidentally were computed for $M_1 = 1\,M_\odot$ and a fixed outer radius, has been made by e.g. Szkody (1984) and Guinan & Sion (1982).

For those of us who are too lazy to do the atomic physics necessary to calculate $\tau_\nu$, or who believe that vertical temperature gradients in the disk modify the emergent spectrum even where the disk is optically thick, an alternative procedure is to obtain the elemental disk spectra from published stellar atmosphere computations. Kiplinger (1979, 1980) pioneered this approach, and more recent work has been done with real stellar spectra by Wu & Panek (1982). In my study I specifically compare the blackbody and model stellar atmosphere approaches (Wade 1984). Here I note some of the results of that study.

- The Lyman discontinuity is strongly present in hot stellar-atmosphere models. As a consequence, it is not necessarily the case that a large part of the power from a disk is emitted in the extreme ultraviolet ($\lambda < 912$ Å). Optical depth effects, combined with temperature gradients, also cause the Balmer continuum to behave approximately as a power law. In the stellar-atmosphere disk models, there is no gradual turnover of $f_\nu$ in the wavelength range observable by IUE, although a very gradual turnover is present in corresponding blackbody models.

- Because individual hot stellar atmospheres have approximate power law spectra in the Balmer continuum, it follows that the corresponding disk spectra do as well. Any moderately positive spectral index s, where $f_\nu \propto \nu^s$, including s = 1/3, can be obtained with a suitable choice of $T_{max}$ and $T_{min}$. The range of $T_{eff}$ required for s = 1/3 is much smaller than in the blackbody case. Even a single-temperature "disk", with $T_{eff} \simeq$ 18000 K, has a Balmer continuum spectral index near 1/3. The small range needed may explain why s = 1/3 spectra are actually observed. For fixed $T_{max}/T_{min}$, higher values of $T_{max}$ result in higher values of s.

   - The slope of the Balmer continuum and the slope of the Paschen continuum (e.g. V-R) can be used as two constraints in determining disk parameters such as $T_{max}$ and $x_{out}$. Neither ultraviolet nor optical data alone is sufficient to point to a unique set of model parameters.

   - The Balmer jump is in absorption in the "atmosphere" models, as is observed in the more luminous (hence presumably optically thick) CV disks. The Balmer jump is generally too large in the models, however. Obviously the Balmer jump is not present in blackbody models.

   Differences in the inferred values of $T_{max}$, $\dot{M}$, and disk luminosity will occur when the same observed disk spectrum (with s $\simeq$ 1/3 in the Balmer continuum) is modelled by "atmosphere" disks and blackbody disks. The atmosphere model will typically be smaller and cooler. Differences in $\dot{M}$ of order 10 to 100, depending on which type of model is used to interpret the observations, can arise.

## 5.2.4 Real Disks

   The important differences that exist between the energy distributions that are predicted for "blackbody" disks and "atmosphere" disks lead to the question, which type of model fits the data better and is more likely to be correct? Comparison with observations of cataclysmic variables shows that blackbody disks are less successful than atmosphere disks in matching details of the spectra, such as the Balmer jump and the power law slope of the Balmer continuum. Moreover, the temperature range required by the atmosphere disks is generally more compatible with the size constraints imposed by the binary system.

   Nevertheless, the quantitative agreement between these models and the data is less convincing than it ought to be. Part of the reason must be that parts of the basic model have been neglected. These include additional light sources and also such modifications as tidal dissipation at the outer edge of the disk (which would alter the expression for $T_{eff}$ (r)) or irradiation of the disk surface (which would alter vertical temperature gradients).

Probably the largest modifications to the ultraviolet part of model disk spectra will come from a more appropriate treatment of radiative transfer. Neither blackbodies nor classical stellar atmospheres are likely to provide a close match to the physics of a disk of finite optical and physical depth, where energy production and radiation happen together. Careful work on this topic, including a consistent treatment of both emission and absorption features, may very well result in models that match the observations much more closely than anything tried so far. Confidence in the mass accretion rates thus derived would be very high.

On the other hand, any effect that impedes our ability to predict $T_{eff}$ (r) would diminish confidence in $\dot{M}$ determinations. Kiplinger (1979) showed that without the constraint provided by something like equations (3) and (4) in Section 5.2.1, it is much easier to find some combination of local spectra (specified for example by a set of temperatures, gravities, and optical depths) that will reproduce the observations. The difficulty is that such a combination may not be unique, and in any case there would be no simple prescription to relate such a set of local spectra to a value for $\dot{M}$. Time-dependent disks, or "clumpy" disks which mix hot and cool regions, are examples of disks with a less definite prescription for $T_{eff}$ .

## 6.  FUTURE WORK

Just as it is impossible to review everything that has been done in the field of CVs, it is impossible to make an exhaustive list of things that ought to be done in the future. Some of the more obvious topics are presented below.

Emission lines are major sources of information for most CVs, and they need to be understood better. Carefully calculated line profiles due to winds from accretion disks should be compared with observations, so that mass loss rates can be reliably estimated. More predictions of line strengths need to be made, so that ionic and elemental abundances can be measured. The possibility of an emission line spectrum arising near the secondary star, perhaps as a reult of irradiation, needs to be looked into. The optical emission lines in particular, from which we measure $K_1$, need to be very carefully modelled. Special efforts to understand the origin of line asymmetries and phase shifts be valuable.

Our analysis of observations needs to be done with as much

care as we can give. Details should be (and increasingly are being) reported in full. Attention should be paid to the correct assessment and propagation of errors. Outstanding discrepancies, such as different values for $K_1$, $K_2$, or $\gamma$ resulting from different investigations ought to be tracked down and eliminated. This is not "somebody else's problem".

Flickering, oscillations and other high-speed photometric phenomena did not figure prominently in this review, but such variability must be the manifestation of very interesting physics in accretion discs. Recent studies have shown how new insight can be gained by obtaining simultaneous data in more than one waveband.

Observations of dwarf nova outbursts need to be made in greater detail, so that high quality data are available to test new theoretical models of the outburst process. The recent CN Ori monitoring campaign (Schoembs 1983) is a good example of what can be tried, as are the various simultaneous ground-based and IUE observations of outbursts. Such short-term and telescope-intensive observations complement but do not supplant the continuous enthusiastic work done by the amateur organizations such as AAVSO, VSS/RASNZ, and AFOEV.

There is hardly any need to suggest more extensive theoretical modelling of outbursts. Both non-steady disks and the secondary-star instability mechanism (e.g. Bath 1975) need to be given fair trials.

Finally, new instruments, techniques, and passbands should be included in the CV observer's bag of tools. Voyager and Space Telescope provide more space platforms for ultraviolet studies and possible astrometry. New infrared array detectors and near-infrared CCD instruments are making additional important emission lines available for study. Radio and extreme-ultraviolet observations, while difficult, are also very valuable in furthering our understanding of cataclysmic variables.

Table 1

Absolute Visual Magnitudes of Cataclysmic Variables

| Type | Mean | Range |
|------|------|-------|
| Dwarf Novae | | |
| Quiescent | +7.5, +8.5 | 6.2 - 10.3 |
| Erupting | +4 | 3.5: - 5.5: |
| Novae (at minimum) | +4 | 1.5 - 9.: |
| Novalike | | |
| Low State | ? | 8.5: - ? |
| High State | +4.5 | 3. : - 6.: |
| AM Her Stars | | |
| Low State | ? | 8.5: - ? |
| High State | +6? | 5.:: - 8.: |

TABLE 2

Some Orbital Periods and Spectral Types

| Object | Period | Spectrum | References |
|--------|--------|----------|------------|
| U Gem | 4.2 h | M4.5 | Wade 1979; Stauffer et al., 1979. |
| "YY Dra"* | 3.9 h | M3 | Patterson et al., 1982; Green et al., 1982. |
| AM Her | 3.1 h | M4+ | Young & Scneider 1981. |

* Suggested optical counterpart of 3A1148+719
(Patterson et al., 1982).  Wenzel (1983) has concluded
that this star is not YY Dra.

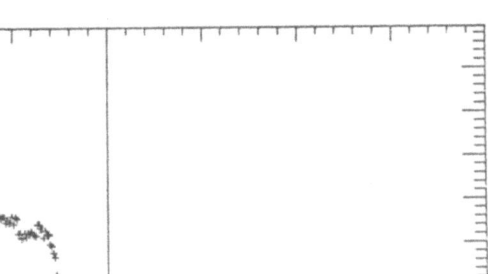

Figure 1.   An average light curve for the dwarf nova Z Cha in quiescence.   Observations by B. Warner.   From Cook 1982 (by permission).

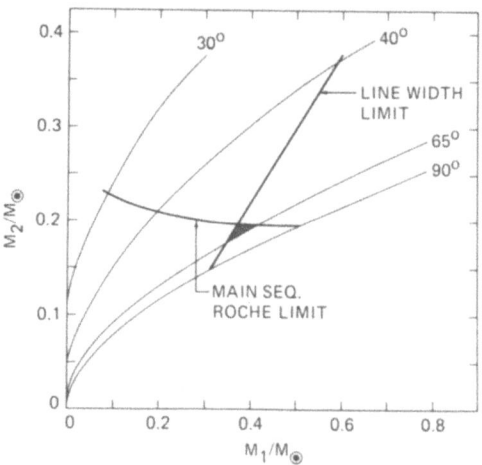

Figure 2. The $(M_1, M_2)$ plane for the dwarf nova T Leo. Note how use of the line width limit (constraint C5), along with constraint C3 and the absence of eclipses, restricts the range of possible mass solutions for this system. From Shafter & Szkody 1984 (by permission).

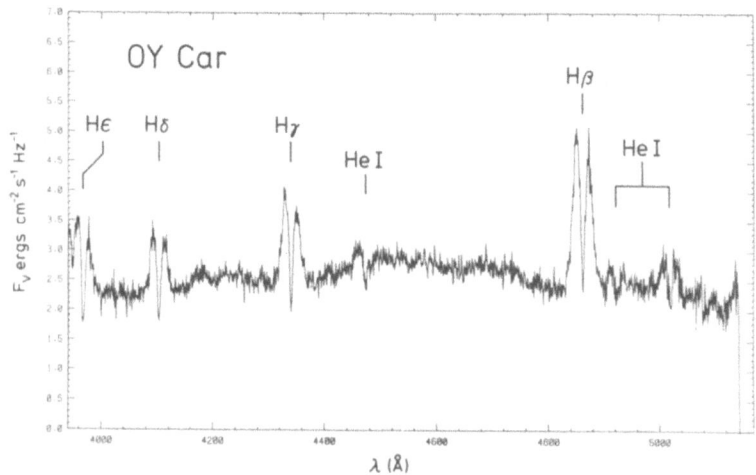

Figure 3. The visible spectrum of the dwarf nova OY Car, showing how the "valley" between the emission peaks extends below the continuum. From Bailey & Ward 1981 (by permission).

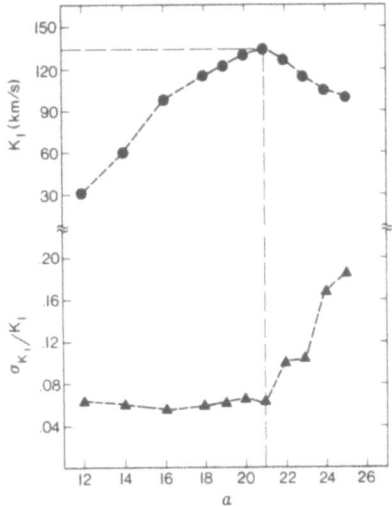

Figure 4. A figure showing the dependence of $K_1$ on measurement technique, for the dwarf nova T Leo. From Shafter & Szkody 1984 (by permission).

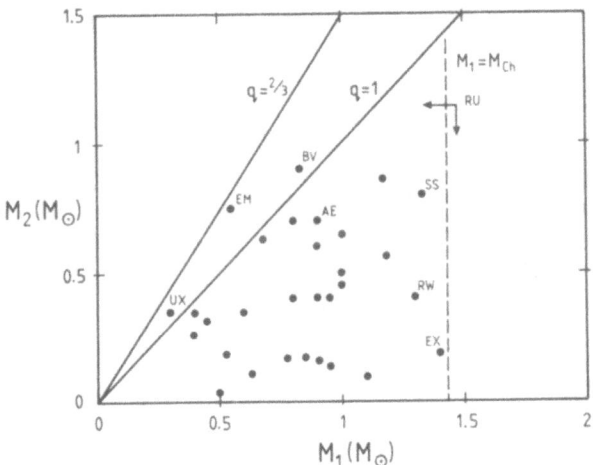

Figure 5. Masses for cataclysmic variables, from Ritter 1983. The letters stand for UX UMa, EM Cyg, BV Cen, AE Aqr, SS Cyg, RW Tri, EX Hya, and RU Peg.

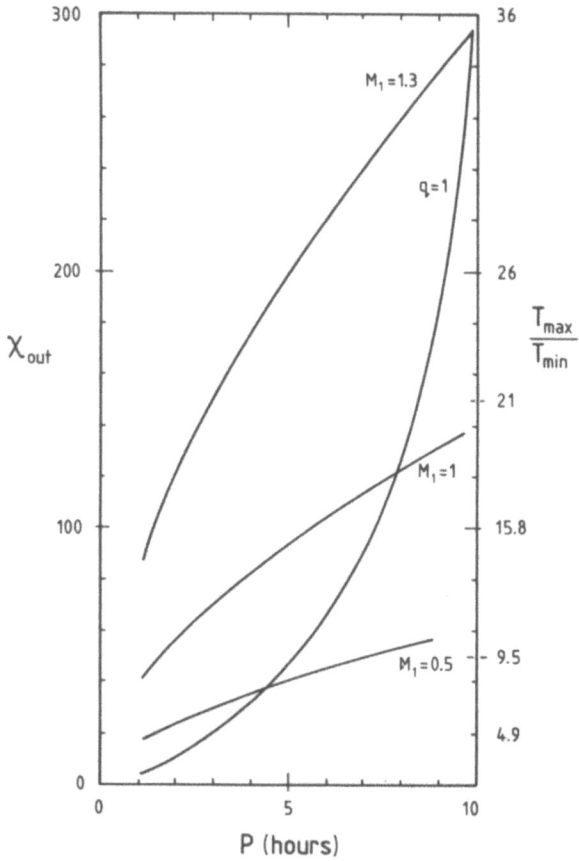

Figure 6. $T_{max}/T_{min}$ and $x_{out} = r_{out}/R_1$ as functions of orbital period and white dwarf mass, for accretion discs in cataclysmic variables. In the figure the outer radius of the disc is taken to be equal to the volume-equivalent radius of the Roche lobe around the white dwarf. This is an upper limit to the actual value of $r_{out}$, so $T_{max}/T_{min}$ in the figure is also an upper limit.

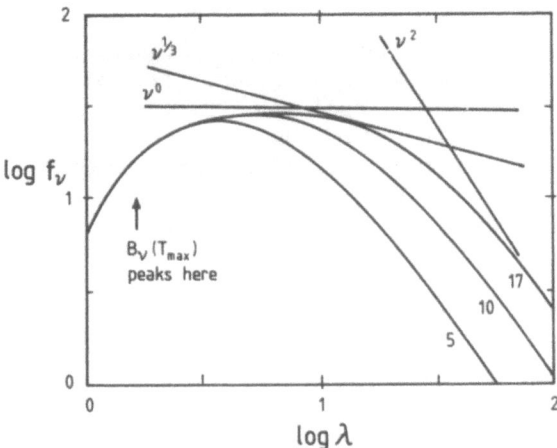

Figure 7. Spectra of blackbody discs for $T_{max}/T_{min}$ = 5, 10, and 17. Representative power laws are also shown. The units on the axes are arbitrary.

REFERENCES

Bailey, J. & Ward M.J., 1981. Mon.Not.R.astr.Soc., **194**, 17P.
Bath, G.T., 1975. Mon.Not.R.astr.Soc., **171**, 311.
Bath, G.T., Evans, W.D., Papaloizou, J. & Pringle, J.E., 1974.
     Mon.Not.R.astr.Soc., **169**, 447.
Bath, G.T. & van Paradijs, J., 1983. Nature **305**, 33.
Berriman, G., Beattie, D.H., Gatley, I., Lee, T.J., Mochnacki,
     S.W. & Szkody, P., 1983. Mon.Not.R.astr.Soc., **204**, 1105.
Bradt, H.V.D. & McClintock, J.E., 1983. Ann.Rev.Astr.Astrophys.
     **21**, 13.
Bruch, A., 1982. Publ.Astr.Soc.Pacific, **94**, 916.
Campbell, L., 1934. Annals Harvard coll.Obs. **90**, 93.
Chincarini, G. & Walker, M.F., 1981. Astr.Astrophys., **104**, 24.
Cook, M.C., 1982. Ph.D. thesis, Univ. of Cambridge.
Cordova, F.A. & Mason, K.O., 1982. Astrophys.J., **260**, 716.
Cordova, F.A. & Mason, K.O., 1984. In W.H.G. Lewin & E.P.J.
     van den Heuvel (eds.) "Accretion-Driven Stellar X-Ray
     Sources" (Cambridge: Cambridge Univ. Press).
Duerbeck, H.W., 1984. To appear in proceedings of IAU Coll.
     80, "Double Stars: Physical Properties and Generic
     Relations" (Lembang, June 1983).
Eggleton, P.P., 1983. Astrophys.J., **268**, 368.
Faulkner, J., Flannery, B.P. & Warner, B., 1972. Astrophys.J.,
     **175**, L79.
Faulkner, J., Lin, D.N.C. & Papaloizou, J., 1983. Mon.Not.R.
     astr.Soc., **205**, 359.
Gallagher, J.S. & Oinas, V., 1974. Publ.astr.Soc.Pacific,
     **86**, 952.
Gallagher, J.S. & Starrfield, S., 1978. Ann.Rev.Astr.Astrophys.
     **16**, 171.
Gilliland, R.L., 1982. Astrophys.J., **263**, 302.
Grauer, A.d. & Bond, H.E., 1983. Astrophys.J., **271**, 259.
Green, R.F., Ferguson, d.H., Liebert, J. & Schmidt, M., 1982.
     Publ.Astr.Soc.Pacifc, **94**, 560.
Greenstein, J.L. & Kraft, R.P., 1959. Astrophys.J., **130**, 99.
Guinan, E.F. & Soin, E.M., 1982. astrophys.J., **258**, 217.
Hassall, B.J.M., 1983. Ph.D. thesis, Univ. of Cambridge.
Hassall, B.J.M., Pringle, J.E., Wade, R.A. & Whelan, J.A.J.,
     1982. Proc. Third European IUE Conf. (Paris: European
     Space Agency), 179.
Herter, T., LaCasse, M.G., Wesemael, F. & Winget, D.E., 1979.
     Astrophys.J.Suppl., **39**, 513.
Holm, A.V., Panek, R.J. & Sciffer, III, F.H., 1982.
     Astrophys.J., **252**, L35.
Horne, K.D., 1983. Ph.D. thesis, California Institute of
     Technology.
Hutchings, J.B. & Thomas, B., 1982. Publ.Astr.Soc.Pacific
     **94**, 102.

Kaitchuck, R.H., Honeycutt, R.K. & Schlegel, E.M., 1983. Astrophys.J., **267**, 239.

Kallman, T.R., 1983. Astrophys.J., **272**, 238.

Kiplinger, A.L., 1979. Astrophys.J., **234**, 997.

Kiplinger, A.L., 1980. Astrophys.J., **236**, 839.

Kraft, R.P. & Luyten, W.J., 1965. astrophys.J., **142**, 1041.

Lacey, C.H., 1977. Astrophys.J.Suppl., **34**, 479.

Lamb, D.Q. & Patterson, J., 1983. In M.Livio & G. Shaviv (eds.) "Cataclysmic Variables and Related Objects" (Dordrecht: Reidel), 229.

Lamb, D.Q. & Patterson, J. (eds.) 1985. "Cataclysmic Variables and Low-Mass X-Ray Binaries" (Dordrecht: Reidel).

Liebert, J. & Stockman, H.S., 1985. In D.Q. Lamb & J. Patterson (eds.) "Cataclysmic Variables and Low-Mass X-Ray Binaries" (Dordrecht: Reidel).

Livio, M. & Shaviv, G. (eds.) 1983. "Cataclysmic Variables and Related Objects" (Dordrecht: Reidel).

MacDonald, J., 1983. Astrophys.J., **267**, 732.

Oke, J.B. & Wade, R.A., 1982. Astrophys.J., **87**, 670.

Panek, R.J. & Holm, A.V., 1984. Astrophys.J., **277**, 700.

Patterson, J., 1979. Astrophys.J., **234**, 978.

Patterson, J., 1981. Astrophys.J.Suppl., **45**, 517.

Patterson, J., 1985. In D.Q. Lamb & J. Patterson (eds.) "Cataclysmic Variables and Low-Mass X-Ray Binaries" (Dordrecht: Reidel).

Patterson, J. & 8 others, 1982. Bull.Amer.Astr.Soc., **14**, 618.

Petterson, J.A., 1980. Astrophys.J., **241**, 247.

Plavec, M.J., Popper, D.M. & Ulrich, R.K. (eds.) 1980. "Close Binary Stars: Observation and Interpretation" (Dordrecht: Reidel).

Pringle, J.E., 1981. Ann.Rev.Astr.Astrophys., **19**, 137.

Rappaport, S., Joss, P.C. & Webbink, R.F., 1982. Astrophys.J., **254**, 616.

Rayne, M.W. & Whelan, J.A., 1981. Mon.Not.R.astr.Soc., **196**, 73.

Ritter, H., 1980a. Astr.Astrophys., **86**, 204.

Ritter, H., 1980b. Astr.Astrophys., **91**, 161.

Ritter, H., 1983. "Catalogue of Cataclysmic Binaries, Low-Mass X-Ray Binaries and Related Objects," 2nd edition. (Garching: Max-Planck Inst. Phys. Astrophys.).

Robinson, E.L., 1976. Ann.Rev.Astr.Astrophys., **14**, 119.

Robinson, E.L., 1983. In M. Livio & G. Shaviv, (eds.), "Cataclysmic Variables and Related Objects (Dordrecht: Reidel), 1.

Robinson, E.L. & Nather, R.E., 1979. Astrophys.J.Suppl., **39**, 461.

Robinson, E.L., Barker, E.S., Cochran, A.L., Cochran, W.D. & Nather, R.E., 1981. Astrophys.J., **251**, 611.

Schoembs, R., 1983. The Messenger (ESO) **32**, 6.

Shafter, A.W., 1983a.  Ph.D. thesis, Univ. of California at
    Los Angelas.
Shafter, A.W., 1983b.  Astrophys.J., **267**, 222.
Shafter, A.W. & Szkody, P., 1984.  Astrophys.J., **276**, 305.
Smak, J., 1969.  Acta Astr., **19**, 155.
Smak, J., 1971.  Acta Astr., **21**, 15.
Smak, J., 1976.  Acta Astr., **26**, 277.
Smak, J., 1981.  Acta Astr., **31**, 395.
Stauffer, J., Spinrad, H. & Thorstensen, J., 1979.
    Publ.Astr.Soc.Pacific, **91**, 59.
Stiening, R.F., Dragovan, M. & Hilibrand, R.H., 1982.
    Publ.Astr.Soc.Pacific, **94**, 672.
Stover, R.J., 1981.  Astrophys.J., **249**, 673.
Stover, R.J., Robinson, E.L. & Nather, R.E., 1981.
    Astrophys.J., **248**, 696.
Sulkanen, M.E., Brasure, L.W. & Patterson, J., 1981.
    Astrophys.J., **244**, 579.
Szkody, P., 1985.  In D.Q. Lamb & J. Patterson (eds.),
    "Cataclysmic Variables and Low-Mass X-Ray Binaries"
    (Dordrecht: Reidel).
van Horn, H.M. & Weidemann, V. (eds.) 1979.  "White Dwarfs
    and Variable Degenerate Stars" (Rochester: Univ. of
    Rochester).
Vogt, N., 1980.  Astr.Astrophys., **88**, 66.
Wade, R.A., 1979.  Astr.J., **84**, 562.
Wade, R.A., 1981.  Astrophys.J., **246**, 215.
Wade, R.A., 1982.  Astr.J., **87**, 1558.
Wade, R.A., 1984.  Mon.Not.R.astr.Soc.,**208**, 381.
Warner, B., 1973.  Mon.Not.R.astr.Soc., **162**, 189.
Warner, B., 1976.  In P.P. Eggleton, S. Mitton, & J.A.J.
    Whelan (eds.), "Structure and Evolution of Close Binary
    Systems" (Dordrecht:  Reidel), 85.
Warner, B., 1979.  In H.M. van Horn & V. Weidemann (eds.),
    "White Dwarfs and Variable Degenerate Stars" (Rochester:
    Univ. of Rochester), 417.
Warner, B. & Nather, R.E., 1971.  Mon.Not.R.astr.Soc., **152**,
    219.
Webbink, R.F., 1976.  Nature **262**, 271.
Wenzel, W., 1983.  Mitt.Ver.Sterne **9**, 141.
Williams, R.E. & Ferguson, D.H., 1982.  Astrophys.J.,
    **257**, 672.
Wu, C-C. & Panek, R.J., 1982.  Astrophys.J., **262**, 244.
Young, P. & Schneider, D.P., 1980.  Astrophys.J., **238**, 955.
Young, P. & Schneider, D.P., 1981.  Astrophys.J., **247**, 960.
Young, P., Schneider,  D.P. & Shectman, S.A. 1981a.
    Astrophys.J.2 **244**, 259.
Young, P., Schneider,  D.P. & Shectman, S.A. 1981b.
    Astrophys.J., **245**, 1035.

# MAXIMUM ENTROPY RECONSTRUCTION OF ACCRETION DISK IMAGES FROM ECLIPSE DATA

Keith Horne

Institute of Astronomy
Madingley Road
Cambridge CB3 0HA
U.K.

## 1. INTRODUCTION

The eclipsing cataclysmic variables provide a unique opportunity for the observational testing of accretion disk theory. Cataclysmic variables (CVs) are close binary systems which consist of an accreting white dwarf primary, and a late-type secondary star that fills its Roche-lobe. The accretion disk around the white dwarf is fed by a gas stream which emerges from the secondary star near the inner Lagrangian point. A compact bright region is frequently present where the stream injects fresh material into the accretion disk. In most cases the disk is the dominant source of light in the system, and the luminosity of the secondary star is comparatively small. Eclipses by the secondary star thus act as precise probes of the spatial structure of the accretion flow.

This article treats the eclipse of an accretion disk as an image reconstruction problem. The numerical methods used to synthesize light curves for accretion disk eclipses are reviewed in Section 2. Section 3 then develops techniques for reconstructing the intensity distribution on the face of the disk (an image of the disk) from an observed eclipse light curve. Results of simulation tests in which disk images were reconstructed from synthetic light curve data are discussed in Section 4. Section 5 summarizes the present status and future prospects of the eclipse mapping method. The analysis of observed light curves will be reported elsewhere (for example Horne & Stiening 1984).

*P. P. Eggleton and J. E. Pringle (eds.), Interacting Binaries, 327–348.*
© *1985 by D. Reidel Publishing Company.*

## 2.   LIGHT CURVE SYNTHESIS

A basic  prerequisite  for  any  quantitative  analysis of
eclipse  observations  is  the  capability  to synthesize light
curves.  The light curve synthesis program developed  for  this
work  makes  several  approximations.  The  tidally  distorted
surface  of the occulting secondary star is approximated by the
critical  equipotential  of the  dimensionless  Roche potential

$$ \Psi = \frac{1}{1+q} \cdot \frac{1}{R_w} + \frac{q}{1+q} \cdot \frac{1}{R_+} - \frac{P^2}{2} , $$

(1)

in  which $R_w$ and  $R_+$ are  distances  (in  units  of  the  binary
separation a) from the white dwarf and red secondary star, P is
the  distance  from  the axis through the centre of  mass,  and
$q = M_+/M_w$  is  the  binary  mass  ratio.  The  accretion disk is
assumed to be flat and to lie in the plane of the binary orbit.
This  flat  disk  approximation  is  not  too  restrictive,
however,  because  deep  eclipses occur in cataclysmic binaries
with inclinations as small  as  70  deg,  while theory predicts
disk opening  angles  to  be  only  a few degrees.  The eclipse
model will break down, of course, for systems with inclinations
very close to 90 deg.

The  disk  surface  is  divided into  $N^2$  discrete  elements
covering a  square  region  centered  on  the  white  dwarf  and
bisected  along  one  edge  by the inner Lagrangian point.  The
preferred unit of distance is  $R_{L1}$,  the distance from the disk
centre to the inner Lagrangian point, because  in  these  units
the  mean radius of the white dwarf Roche lobe is approximately
independent of the mass ratio q (Figure 1).  The cartesian grid
is  computationally  efficient,  since  its  surface  elements
subtend equal solid angles, but was chosen  primarily  for  the
convenience  of  displaying  the  accretion  disk  brightness
distribution with conventional image display hardware.

The spatially integrated accretion  disk flux $f_v$ at binary
phase $\phi$ is calculated by summing contributions from all visible
surface elements

$$ f_v(\phi) = \frac{\theta^2}{4N^2} \sum_{J=1}^{N^2} I_v(J) . V(J,\phi) . $$

(2)

Here $I_v(J)$ is the intensity  along the line of sight to surface
element J, and $V(J,\phi)$ is the  fraction  of  element  J  that is

visible   at   phase $\phi$ .   The   angular   scale   of   the system is
measured by the parameter

$$\theta^2 = \left(\frac{R_{LI}}{D}\right)^2 \cos i ,$$

(3)

where i is the binary inclination and D is  the distance of the
system  from earth.  Convenient units for $\theta$ are solar radii per
kiloparsec (4.56 x $10^{-6}$ arcseconds).

Occultations of  disk  elements  by the secondary star are
taken into account in (2)  by  the  visibility function V(J,$\phi$),
which normally is assigned a value of 1  or  0  depending  upon
whether or not a point at the centre of element J is visible at
phase $\phi$.  When an element straddles the boundary of an occulted
region,   a   more   accurate   value  is calculated by refining the
location  of  the  boundary along  two  edges  of  the  surface
element.  To determine whether  a  given  point  on the disk is
visible, a ray is traced from the point  in  question along the
path of a photon toward the earth.  A series of  analytic tests
first  decides  whether the ray intersects a cylindrical volume
enclosing the companion  star, and hence whether an occultation
is possible.  If the  ray  intersects  the  cylinder, the Roche
potential is evaluated at small intervals along the  segment of
the ray interior to the cylinder.  The location is  occulted if
the  Roche  potential  on this segment falls below the critical
value defining the surface  of the companion star.  This method
was originally used by Mochnacki  (1971)  to  synthesize  light
curves of contact binary stars.

The  three parameters that specify  the  geometry  of  the
eclipse are  (1)  the  binary  mass  ratio  q  =  $M_r/M_v$, which
determines  the  size  of  the occulting Roche lobe relative to
that occupied by the disk, (2) the phase of conjunction $\phi_o$, and
(3) the phase width of  the  eclipse  at  disk centre $\Delta\phi$, which
determines the inclination of the binary when the mass ratio is
given (see Figure 2).  The third parameter is taken  to  be $\Delta\phi$,
rather  than  the  inclination,  because  the  former is rather
better  constrained  by  the  eclipse  observations.   The
inclination  would  of  course  be used for shallow eclipses in
which  the  centre of the  disk  is  not  occulted.   Figure  3
illustrates  the  eclipse  geometry  for  the  case  q  =  0.9,
$\Delta\phi$ =  0.081,  which corresponds to an inclination near 75 deg.
The two Roche  lobes are projected on to the orbital plane, and
a network of ingress/egress  arches  is  drawn on the accretion
disk.  The arches outline regions of the disk that are occulted
at different binary phases.  A small section  of  the accretion
disk at the back of the Roche lobe escapes the eclipse.

The synthetic eclipse light curves shown in Figures 4 and 5 resemble the two basic forms found among the observed light curves of CVs. To generate these light curves we adopted the eclipse geometry of Figure 3, and assumed simple models for the intensity distribution on the face of the disc. Figure 4 shows a deep but round-bottomed eclipse with broad, asymmetric wings produced by a bright spot at the intersection of the gas stream and the outer rim of the disk. Old novae, nova-like variables, and dwarf novae in outburst show such eclipses. In quiescence, the eclipses of dwarf novae resemble the one shown in Figure 5. Sudden changes in brightness are caused by the ingress and egress of two compact bright sources, one at disc centre and one at the bright spot. These light curves will serve as benchmarks to demonstrate the image reconstruction method developed in the next section.


## 3. IMAGE RECONSTRUCTION FROM ECLIPSE DATA

### 3.1 Model-fitting vs Image Reconstruction

Having considered the synthesis of an eclipse light curve from a given intensity distribution, we now turn the problem around and ask what may be learned about the disk intensity distribution from a given light curve. Model-fitting and image reconstruction are two different approaches to this problem. In the conventional model-fitting approach, the parameters of an accretion disk model are adjusted to find the model which best fits the eclipse observations. The model-fitting approach can only succeed if the underlying model is essentially correct. Model-fitting has been enormously successful in the analysis of light curves of e.g. contact binary stars. The current models of accretion disks, on the other hand, contain very basic uncertainties which prevent their realistic comparison with observations.

The accretion disk theory is incomplete in at least two important ways. One is the theoretical idealization of azimuthal symmetry. The observed light curves of CV eclipses are generally asymmetric. Good fits to high quality eclipse observations thus require an asymmetric brightness distribution, which must be included in the model in an ad hoc way. The asymmetric light curve obtained by adding a circular Gaussian bright spot to a symmetric disk model produces less-than-satisfying light curve fits (see for example Frank et al. 1981). Any further increase in the number of nuisance parameters makes the parameter estimation problem indeterminate (unpublished experiments by the author). Another limitation of the theory is that it considers only vertically-integrated properties of the disk. The vertical structure depends sensitively on the nature of the viscosity mechanism, which is known to be anomalously efficient in CV accretion disks but has never been identified. Thus although the run of effective temperature with disk radius is predicted from very basic energy balance considerations, the form of the emerging spectrum, which depends on the vertical structure, is unknown.

These basic uncertainties in accretion disk theory motivate interest in the image reconstruction approach, which provides a framework for tests of the validity of existing models, and for the construction of more realistic ones. Image reconstruction is in one sense a limiting case of model-fitting, in which the intensity of each image pixel is an independent parameter. With this enormous flexibility, light curve fits as good as those shown in Figures 4 and 5 are typical. Image reconstruction is, however, an ill-posed problem. The one-dimensional eclipse light curve contains insufficient information to completely specify a two-dimensional accretion disk image. Thus in general infinitely many different images of the disk will equally well reproduce a given eclipse light curve. To select a single image, some form of additional constraints must be imposed. Positivity, for example, is clearly a desirable but generally not sufficient requirement. Model-fitting imposes constraints by forcing the image to lie within the parameter space of the model. The form of constraint used by the reconstruction method will be reflected in the resulting images, and so the reconstruction method must be carefully chosen to reflect the aims of the investigator.

## 3.2  A Linear Image Reconstruction Method

We first briefly consider a linear reconstruction method used routinely by geophysicists in their investigation of conditions within the body of the earth from measurements which are necessarily confined to its exterior (see review by Parker 1977). Geophysical inverse theory can be used in any reconstruction problem in which predicted data are given by linear combinations of the image values

$$F(K) = \sum_{J=1}^{N^2} B(K,J) . I(J) . \tag{4}$$

Inspection of (1) shows that the eclipse mapping problem has the required form. The method restricts attention to images I that are linear combinations of the observations,

$$\hat{I}(J) = \sum_{k=1}^{M} A(J,k).F(k).$$

$$(5)$$

The reconstructed image value $\hat{I}(J)$ is thus a convolution of the true image I with a point spread function given by

$$\Delta(J,I) = \sum_{k=1}^{M} A(J,k).B(k,I).$$

$$(6)$$

The B coefficients are specified by the eclipse model, but the freedom to select the A coefficients can be used to some extent to tune the point spread function, for example to optimize spatial resolution, or to limit sensitivity to light curve noise.

Linear reconstruction of disk images from eclipse data has not been investigated in detail. We expect that negative intensities and "ringing" will occur because the method does not incorporate a positivity constraint. It also remains to be seen whether the A coefficients offer sufficient control to produce a useful point spread function. The advantage of the linear reconstruction method is that the propagation of errors from the observed light curve to the reconstructed image is well understood. The method deserves further study in the context of eclipse mapping.

### 3.3 Maximum Entropy Image Reconstruction

We now describe a reconstruction method based on the principle of maximum entropy. The maximum entropy principle has been used in image reconstruction problems from several branches of astronomy (for example Gull & Daniell 1978; Bryan & Skilling 1980, Willingale 1981). This method selects the unique positive image which maximizes a scalar function on image space, the image "entropy", subject to constraints provided by data. A robust and efficient algorithm for the constrained maximization of entropy in the large dimensional image space was developed by Skilling (1981). The method can be applied to any problem in which a model has been devised to calculate predicted data from a given image.

A consistency statistic C is used to measure the "goodness-of-fit" between the observed data d(K), and the predicted data f(K) associated with the image. The most frequently used consistency statistic, and the one employed here, is

$$\chi^2 = \frac{1}{M} \sum_{K=1}^{M} \left[ \frac{f(K) - d(K)}{\sigma(K)} \right]^2$$

(7)

in which $\sigma(K)$ are estimates of the error in $d(K)$. Data constraints are imposed on the image by requiring the consistency statistic to have some target value CAIM.

The image entropy is defined relative to a default image. Its specific form is

$$S = - \sum_{J=1}^{N^2} I(J) \left[ \ln\left(\frac{I(J)}{D(J)}\right) - 1 \right],$$

(8)

where $I(J)$ is the intensity in pixel $J$, and $D(J)$ is the corresponding default intensity. The logarithmic term in the entropy ensures that the image is strictly positive. By differentiating (8) we obtain

$$\frac{\partial S}{\partial I(J)} = - \ln\left[ \frac{I(J)}{D(J)} \right],$$

(9)

and see that the entropy is maximized, in the absence of data constraints, when the image is identical to the default image. For small departures from the default image, the image entropy decreases quadratically from its maximum value. In loose terms, the entropy of an image measures its similarity to the default image.

Images produced by maximum entropy reconstruction are the result of a balance between two opposing influences. Data constraints on the one hand influence the image through the consistency statistic C. The default image, on the other hand, exerts its influence through the entropy S. If the data constraints are weak, the reconstructed image will depart little from the default image. If the data constraints are strong, the image will depart greatly from the default image, (unless of course the default image is itself consistent with the data). When insufficient information is present in the data to form an image, the missing information is supplied by the default image. The default image may thus be regarded as containing prior information about the image, which will be modified only if the observations require it.

The use of a default image in the definition of the image entropy lends the maximum entropy method a useful flexibility that proves to be essential for reconstructions of accretion disk images from eclipse data. Two forms of default image are considered here. The first is the one used in most previous maximum entropy applications, a uniform image whose value is the average image intensity. With a uniform default image, maximizing the entropy produces the most uniform image consistent with the observations. The second form of default image is an azimuthally symmetric image derived by averaging the image values around lines of constant radius. With this default, the entropy is sensitive only to departures from axi-symmetry, and its maximization leads to the most nearly axi-symmetric accretion disk image that is consistent with the eclipse data. In this case the entropy contains no information about the radial profile of the disk; this aspect of the image must be determined entirely by the eclipse data.

## 4.    SIMULATION TESTS

Simulation tests were used 'to evaluate the performance of the maximum entropy eclipse mapping method, since the nonlinearity inherent in the maximum entropy principle makes it difficult to evaluate analytically. In each simulation, a known intensity distribution was used to compute a synthetic eclipse light curve, to which noise was added. A reconstructed image of the disk was then formed and compared to the original model. Such tests lead to a clear understanding of the way the eclipse data and the default image conspire to form the maximum entropy image of the disk. The results presented here will be useful in the evaluation of images formed from observed eclipse light curves.

The disk models used to produce the synthetic light curves in Figures 4 and 5 are shown in Figures 6 and 7, along with reconstructed images made with the uniform and azimuthally-averaged forms of default image. Figure 6a is the 5500 Å intensity distribution of a steady-state blackbody accretion disk model. The model was truncated abruptly at radius 0.7, and a gaussian bright spot was added to make the light curve asymmetrc. The model in Figure 7a is a uniform intensity disk, again truncated at radius 0.7, to which gaussian bright spots were added at the disk centre and near the rim of the disk.

Reconstructed images were formed with the light curve synthesis programme described in Section 2, and the software package MEMSYS (Burch, Gull & Skilling 1983), which implements the Skilling (1981) algorithm for a general image reconstruction problem. MEMSYS makes iterative corrections to an initial image. During each iteration, a quadratic approximation to the constrained maximization problem is solved on a small-dimensional subspace of the image space. The subspace is spanned by from 3 to 6 images, including the image space gradients $\nabla C$ and $\nabla S$, which are evaluated anew on each iteration. The early stage of the image adjustment, which requires about 30 iterations, seeks to decrease the value of C until the target value CAIM is reached. Subsequent iterations aim to hold C fixed while increasing the entropy S. Eventually, the $\nabla C$ and $\nabla S$ become parallel, signalling that a solution of the constrained maximization problem has been found, and the iteration is halted.

The initial image used to start the MEMSYS iteration was either a uniform or a circular gaussian image chosen to roughly reproduce the depth and width of the eclipse light curve. The final reconstructed images were not sensitive to the choice of initial image, althought the number of iterations required to produce them did depend on the initial image. The final images were also stable to perturbations; spurious bright spots added at several different places on the image were always eliminated after about 20 additional iterations.

The intensity in the reconstructed images was forced to vanish at radii larger than 0.85, a physically justifiable requirement since this radius is outside the white dwarf Roche-lobe. The eclipse geometry, namely q = 0.9, $\Delta\phi$ = 0.081, (Figure 3), remains fixed as the image is adjusted. Correct values for q, $\Delta\phi$ and $\phi_o$ were used in the reconstruction; the explicit effect of uncertain eclipse geometry is taken up in Section 4.4.

## 4.1 Image Distortions Produced by a Uniform Default

The reconstructed images shown in Figures 6b and 7b were formed with a uniform default image, and hence are the most uniform images of the disk which produce the correct eclipse light curve. These images are severely distorted. Striking artifacts stretch along the two ingress/egress arches that cross at the position of each compact bright spot. "Ghost" spots are present at positions corresponding for example to the ingress of one spot and the egress of another.

The formation of these image artifacts can be understood in terms of the interplay between the eclipse constraints and

the uniform default image. Consider first the nature of the
constraints placed on an image by eclipse data. The slope of
the eclipse light curve at any given phase can be expressed as
a weighted line integral of the disk intensity along the edge
of the region occulted at that phase. This line integral is
negatively weighted on the ingress branch of the ingress/egress
arch, where a strip of disk surface is entering eclipse, and
positively weighted along the ingress branch. Eclipse
observations thus specify the total flux of the disk image, and
the values of these weighted line integrals of the disk
intensity along each of the ingress/egress arches. The effect
of a uniform default image is to suppress extreme image values.
In the eclipse mapping problem, the flux removed from a compact
bright spot to increase the entropy must be re-distributed over
the image in such a way as to preserve the weighted line
integrals along the ingress/egress arches. Flux thus spreads
primarily along the two arches which pass through the bright
spot.

Eclipse constraints closely resemble those encountered in
tomography, where for example the density on a two-dimensional
slice of a three-dimensional object is to be reconstructed from
integrals of the density along straight lines at various angles
in the plane of the slice. The eclipse mapping problem, apart
from its more complicated geometry, is very similar to a
two-angle tomography problem, with one angle corresponding to
ingress and the other to egress. Kemp (1980) found for
tomographic reconstruction that progressively larger artifacts
appear along the viewing directions as the number of viewing
angles is decreased. These artifacts are a general feature of
maximum entropy reconstruction whenever observations constrain
line integrals of the image. They arise from the non-local
nature of the data constraints, rather than from noise in the
data.

## 4.2 Eclipse Mapping with a Radial Profile Default

Maximum entropy reconstruction using the azimuthally-
averaged radial profile of the image as the default produces
the most nearly circularly-symmetric image that is consistent
with the eclipse data. This form of default image was
originally motivated by a desire to reduce the image
distortions produced by a uniform default. A preference for
axi-symmetric disk images may also be justified on physical
grounds because the strong Keplerian shear in the accretion
disk should rapidly eliminate any azimuthal structure that is
not actively maintained (Pringle 1981).

The images shown in Figures 6c and 7c are reconstructions
made with a radial profile default image. The action of the

radial profile default image is to re-distribute flux in the azimuthal direction by as much as the eclipse data will permit. Thus the bright spot near the outer rim of the disk is distended in azimuth, and a bright annulus encircles the disk at its radius. Because the radial profile default image more closely resembles the reconstructed image, the tendency to suppress extreme values, and hence to form artifacts along the ingress/egress, is greatly reduced. The radial profile of the reconstructed image is determined entirely by the eclipse observations. Figure 8 compares the reconstructed brightness temperature profile from the light curve of Figure 4 to the $T \propto R^{-3/4}$ law from which the light curve was formed. The agreement suggests that useful temperature maps for accretion disks can be reconstructed from observed eclipse light curves of sufficient quality.

## 4.3 The Trade-Off between Resolution and Noise

Image reconstruction always entails a compromise between spatial resolution and noise; a high signal-to-noise light curve will allow greater resolution of features in the disk image. In maximum entropy reconstructions, this trade-off can be controlled by artificially adjusting the strength of the eclipse constraints through the value of CAIM. If a high value of CAIM is chosen, the predicted data are not forced to match the observed data very closely, and the image relaxes toward the default image, degrading the resolution. If too low a value of CAIM is chosen, the predicted data is forced to follow data noise, and spurious features develop in the image. A network of fine "grooves" along the ingress/egress arches appears in the reconstructed image when the value of CAIM is too low. The appearance of these features is helpful in choosing the value of CAIM when the error estimates are uncertain.

## 4.4 The Effect of Uncertain Eclipse Geometry

In the analysis of an observed light curve, the eclipse geometry will not be precisely known. It is therefore important to consider what effect errors in the assumed geometry may have on the reconstructed disk image. The changes produced in the reconstructed image as the geometric parameters are changed may be understood by considering the affect these parameters have on the eclipse geometry. The distortion of the eclipse geometry produced by separate changes in $\phi_0$, $\Delta\phi$ and q are illustrated in Figure 9.

Small changes in $\phi_0$ and $\Delta\phi$ produce a trivial translation of the image. A change in $\phi_0$ rotates the occulted region about the centre of the secondary star (Figure 9a), and an error in

$\phi_o$ will therefore rotate the disk image, but will not otherwise affect the surface brightness distribution.    A  change  in $\Delta\phi$ corresponds  to  a change in the inclination of the system, and hence in the  distance  from  the  secondary  star to which the occulted region extends (Figure 9b).  The first order affect of changes in $\phi_o$ and $\Delta\phi$ near the disk  centre is thus to translate the  image.   In  practice,  the  parameters $\Delta\phi$ and $\phi_o$ may  be determined  from the phases of steepest slope  in  the  eclipse light curve,  and errors in the adopted values may be diagnosed and corrected, provided that  the  disk intensity peaks at its centre.

The eclipse geometry is  only weakly sensitive to the mass ratio q.  A change in  q  alters  the  width  of  the  occulted region.  The result is a second-order distortion of the eclipse geometry  whose affect vanishes at the disk centre (Figure 9c). Thus the  position  of a bright spot in the outer region of the disk is sensitive to  errors  in  q, but the surface brightness near  the  disk centre, and the azimuthally-averaged properties of the disk are little affected.

## 4.5  Neglected Background Light

There is always uncertainty as to  what  part of the light observed at the bottom of an eclipse is  due  to  the companion star  and  other  non-disk  sources of light in the system.  To test the effect of this  neglected  background  light,  various constants  were  added to synthetic eclipse light curves before reconstructing images of  the  disk.   The  extra flux appeared primarily in the small section at the  back  of  the  disk that escapes  the  eclipse.  The reason for this is that the eclipse light  curve  strongly  constrains  the  total  flux  in  the complementary region of  the  disk that does participate in the eclipse.   Neglected  background  light  is  therefore  not  an essential  problem.   It  becomes  important  only  when  a substantial fraction of the total  light  of the system is from non-disk sources.  In that case the direct  contribution of the secondary  star may be taken into account  in  the  light  curve synthesis programme at the cost  of  one  additional  parameter (for example its surface temperature).

## 5.   SUMMARY AND PROSPECTS

We  have used maximum entropy  techniques  to  reconstruct images of  accretion disks from eclipse light curve data.  This image reconstruction approach  provides a broader framework for testing  accretion  disk  theories  than  do  the  conventional model-fitting methods, in which  the parameters of an accretion disk model are adjusted to  find  the model which best fits the eclipse observations. Image reconstruction is in one sense a limiting case of model-fitting, in which the intensity of each image pixel is an independent parameter. This enormous flexibility makes it possible to achieve satisfying fits to observed eclipse curves that are only approximately reproduced by parameterized disk models.

The limited constraints provided by eclipse data generally cannot specify an image completely, and reconstruction techniques must compensate for this in some way. Maximum entropy techniques supply the missing information in the form of a default image. A uniform default image, which is used in most applications of maximum entropy, leads in the eclipse mapping problem to severe image distortions along the edges of the occulted regions at different phases. In an effort to reduce these distortions the azimuthally-averaged radial profile of the image was used as the default. Images with approximate circularly symmetry are then satisfactorily recovered from eclipse data. Other forms of default might also be useful. For example, a standard accretion disk model could be used for the default image. The difference between the maximum entropy image and the model default would then guide the construction of a more realistic model.

The application of image reconstruction techniques to high-quality multi-wavelength eclipse data provides stringent tests of our current ideas about disk accretion. Images of disks at several different wavelengths in effect show us the spectrum of the light emitted at each place on the disk. The form of this local spectrum depends sensitively on the vertical structure of the disk. Current disk theory treats only the vertically-integrated properties of disks. Radial temperature profiles derived from eclipse observations directly tests the $T_{eff} \propto \dot{M}.R^{-3/4}$ law, which is predicted from very basic energy balance considerations, and provides a reliable means of measuring the mass transfer rate $\dot{M}$. Eclipses may also be used to form images of the emission line regions associated with the disk. In this case the velocity information provided by line profiles may obviate the need for an axisymmetrized image.

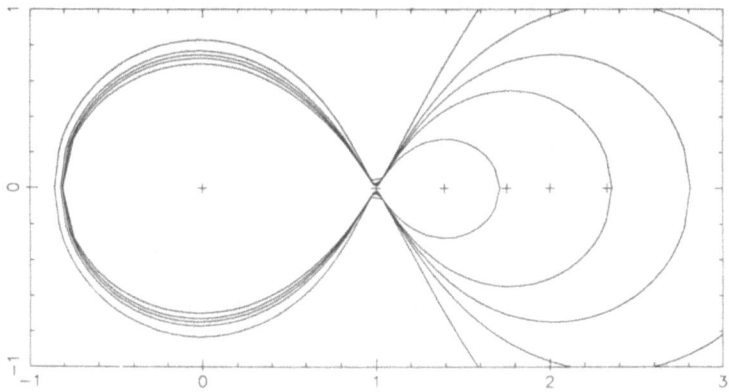

Figure 1.  Roche-lobes are shown in projection on to the binary
orbital plane for mass ratios 0.1, 0.5, 1, 2 and 10.

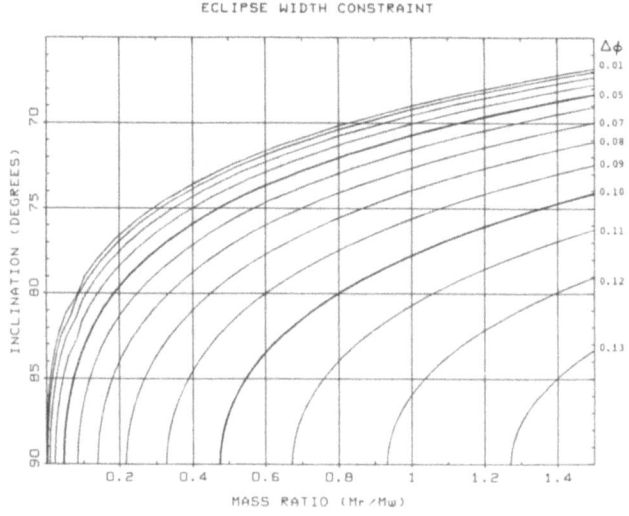

Figure 2.  The relationship between the mass ratio q and the inclination i is shown for different values of the eclipse phase width
$\Delta\phi$ at disk center.

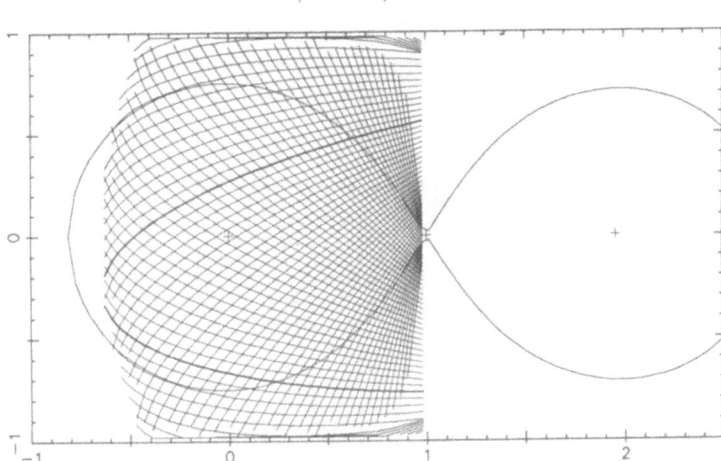

$q = 0.9 \quad \Delta\phi = 0.081$

Figure 3. The eclipse geometry is illustrated here by a network of ingress/egress arches crossing the face of the accretion disk. Each arch is the boundary of a region of the disk that is occulted by the secondary star at a particular binary phase. Orbital motion is in a counter-clockwise direction.

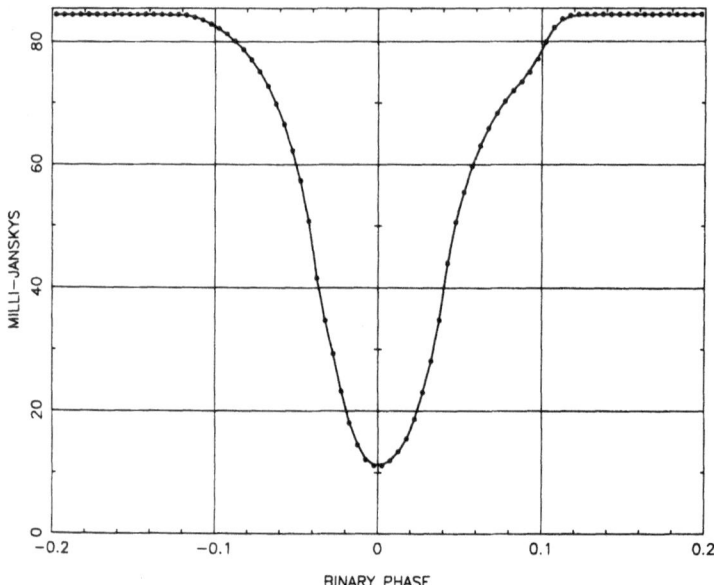

Figure 4.  Points give a synthetic eclipse· light curve that
resembles the observed light curves of many old novae, nova-like
variables, and dwarf novae in outburst.  The continuous curve
gives the light curve fit produced by a maximum entropy reconstruction of the disk intensity distribution.

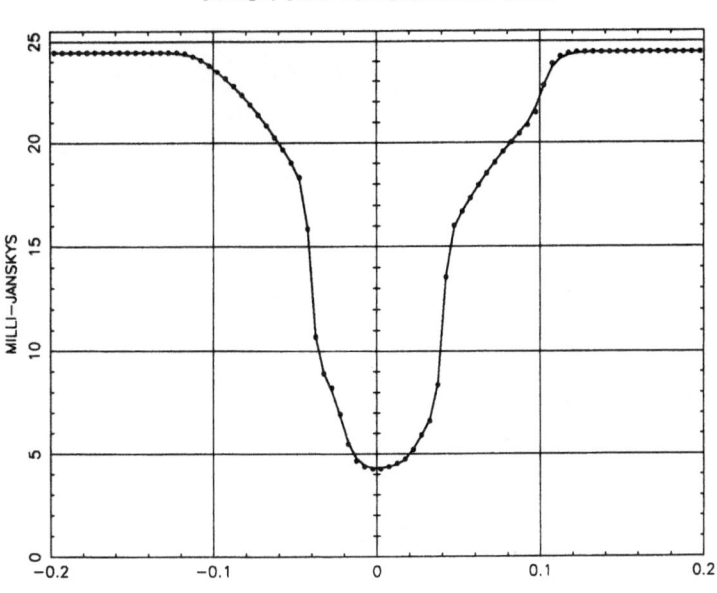

Figure 5. Points show a synthetic eclipse light curve that resembles the observed light curves of dwarf novae in quiescence. The continuous curve is the fit produced by a maximum entropy reconstruction of the disk.

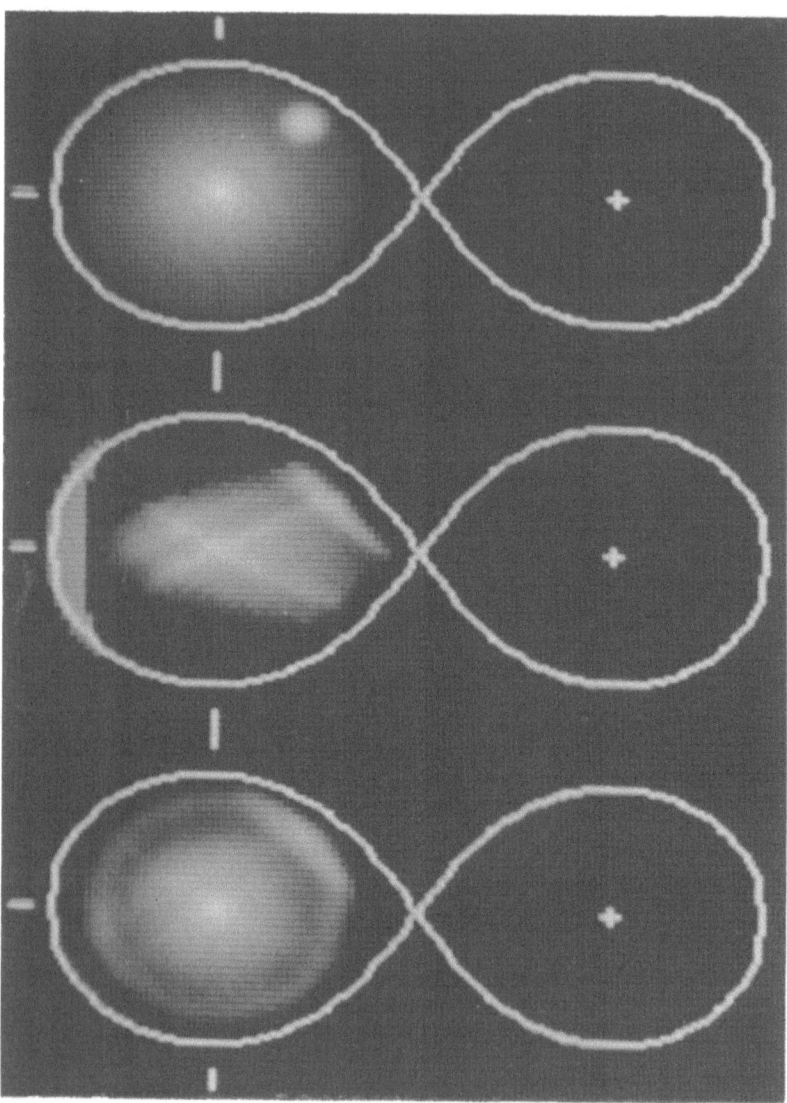

Figure 6. (a) The accretion disk model used to generate the light curve shown in Figure 4. (b) Maximum entropy reconstruction of the image in (a) from the synthetic light curve of Figure 4, and using a uniform default image. (c) Maximum entropy reconstruction using the radial profile of the image as the default image.

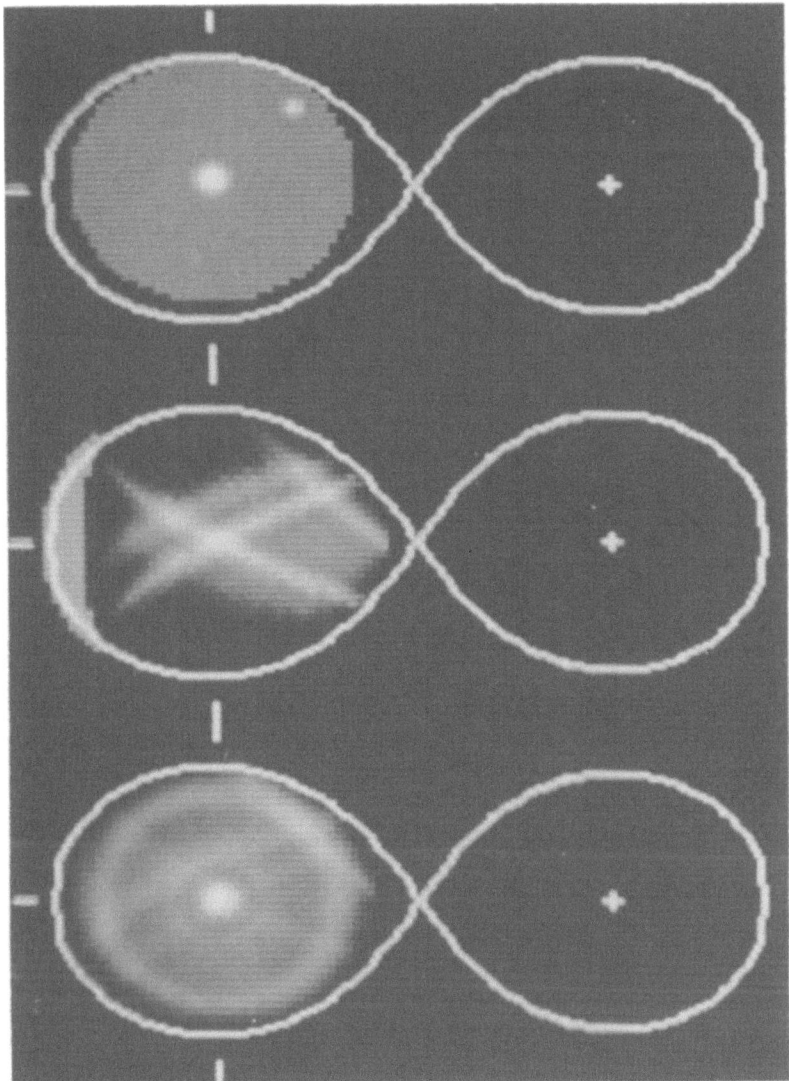

Figure 7.  Same as Figure 6 except for the light curve of Figure 5

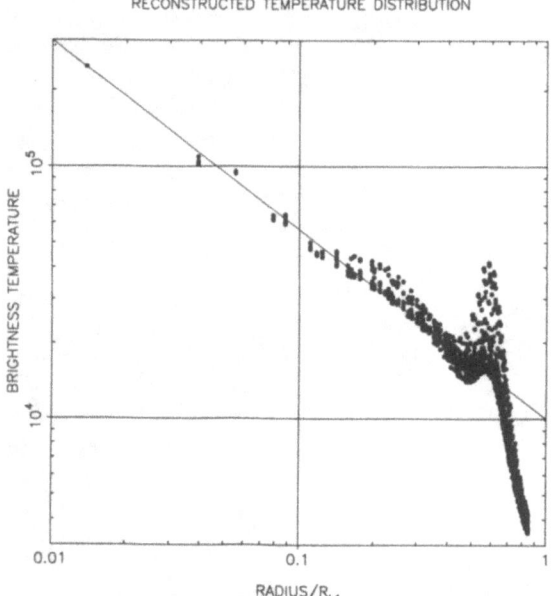

Figure 8.  The radial temperature profile based on the reconstructed image in Figure 6c has the corréct $T \propto R^{-1/4}$ dependence on disk
radius R.

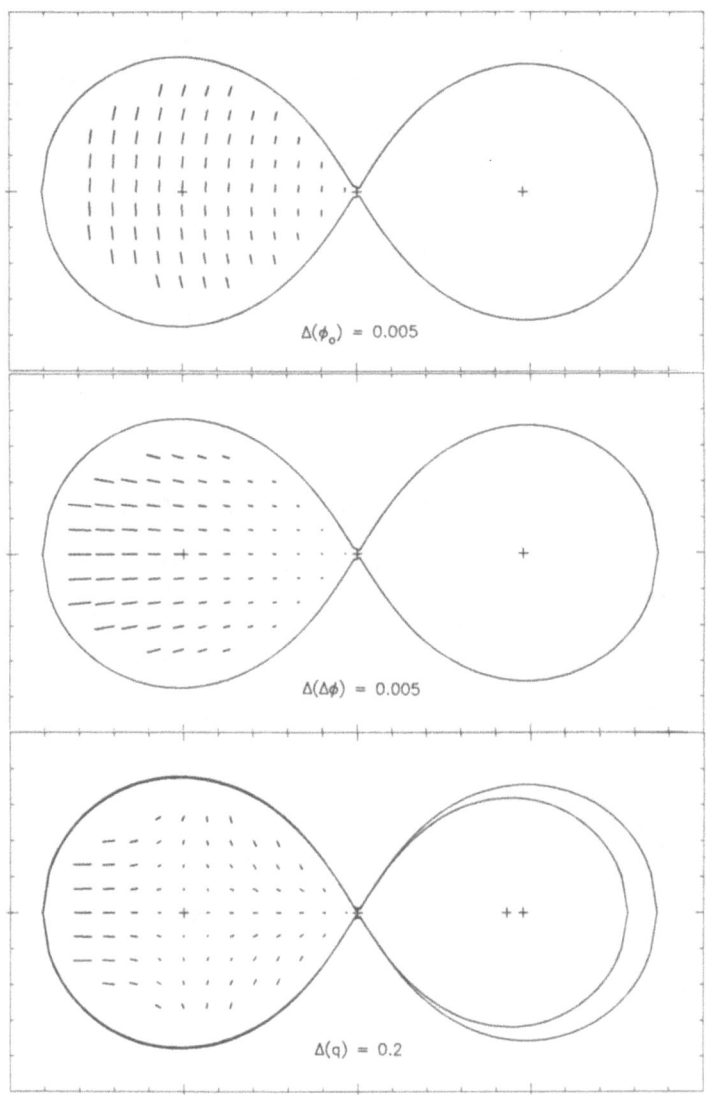

$\Delta(\phi_o) = 0.005$

$\Delta(\Delta\phi) = 0.005$

$\Delta(q) = 0.2$

Figure 9. Vectors showing the change in the eclipse geometry produced by the parameters $\phi_o$, $\Delta\phi$, and q.

ACKNOWLEDGEMENTS

I thank A.C.S. Readhead, S.W. Mochnacki, R.P. Lindfield, S.F. Gull, and K.A Whaler for conversations providing insight at critical stages in the development of the eclipse mapping method. Access to the maximum entropy data analysis package MEMSYS, developed by J. Skilling and made available to me by S.F. Gull, was essential. The computations were carried out with VAX 11/780 computers of the Astronomy Department at Caltech, and the Cambridge node of the SERC Starlink Network. The support of a Graduate Research Assistantship at Caltech, and an SERC Postdoctoral Research Assistantship at Cambridge is gratefully acknowledged.

REFERENCES

Bryan, R.K. & Skilling, J., 1980.  Mon.Not.Roy.Astr.Soc., **191**, 69.

Burch, S.F., Gull, S.F. & Skilling, J., 1983.  Comp. Vision Graphics Image Processing, **23**, 113.

Frank, J., King, A.R., Sherrington, M.R., Jameson, R.F. & Axon, D.J., 1981.  Mon.Not.Roy.Astr.Soc., **195**, 505.

Gull, S.F. & Daniell, G.J., 1978.  Nature, **272**, 686.

Horne, K. & Stiening, R.F., 1984.  Ap.J., submitted.

Kemp, M.C., 1980.  Medical Radionuclide Imaging, **1**, 313.

Mochnacki, S.W., 1971.  M.S.Thesis, University of Canterbury, New Zealand.

Parker, R.L., 1977.  Ann.Rev.Earth Planet Sci., **5**, 35.

Pringle, J.E., 1981.  Ann.Rev.Astron.Ap., **19**, 137.

Skilling, J., 1981, Algorithms and Applications, presented at Workshop on Maximum Entropy, Laramie, Wyoming.

Skilling, J., Strong, A.W. & Bennett, K., 1979.  Mon.Not.Roy.Astr.Soc. **187**, 145.

Willingale, R., 1981.  Mon.Not.Roy.Astr.Soc., **194**, 359.

# THE LATE STAGES OF INTERACTIVE STELLAR EVOLUTION

R.E. Nather

McDonald Observatory
and the
University of Texas at Austin
U.S.A.

## 1. INTRODUCTION

In his review of the cataclysmic variabiles in 1976, Robinson (1976) observed that we could point with assurance to neither the progenitors of these objects nor to their descendents. Things are a bit better now: we can point with modest confidence to the interacting binary stars with orbital periods measured in days as likely progenitors; in many cases the original mass ratio has been reversed by mass transfer, and will be followed by the death and collapse of the mass-losing component. Subsequent evolution can lead to a common envelope phase in which enough mass and orbital momentum are lost to yield final orbital periods of a few hours. The recent discovery by Grauer & Bond (Grauer 1983) of a binary system with an orbital period of 2.7 hours inside the planetary nebula Abell 41 makes such a scenario reasonably plausible.

With somewhat less confidence we can point to likely descendents as the small handful of known interacting objects with orbital periods so short that only two collapsed objects can be involved. This paper will concentrate on these Interacting Collapsed Objects (ICOs) already known, and will speculate on those not yet known observationally but which ought to exist in any reasonably designed Universe.

*P. P. Eggleton and J. E. Pringle (eds.), Interacting Binaries, 349–366.*
© *1985 by D. Reidel Publishing Company.*

## 2.   INTERACTING WHITE DWARF BINARIES

There are now three examples known of systems which exhibit direct evidence that they are composed of a pair of interacting white dwarfs: AM CVn (= HZ-29), PG1346 + 082, and GP Com (= G61-29). All three systems exhibit the following characteristics, which might be thought of as defining this class of objects:

- .  No detectable hydrogen in the spectrum

- .  Broad spectral lines characteristic of an accretion disk

- .  Orbital periods well under the minima expected from models of hydrogen-rich mass donor systems

Of the three systems, only GP Com exhibits periodic variations that arise unambiguously from orbital motion; an orbital origin for the variations seen in the other two is likely, but not yet proven.

Observational proof of orbital motion is tough: the objects are intrinsically faint, and collecting spectra with very short exposure times is essential. The result is a lot of noise and very little signal. The mass ratios are extreme: the mass donor in AM CVn retains only .04 solar masses of its original material if current models are correct, and the donors in the other systems are even less massive. The mass accretor just sits there while the donor orbits around it. Since all of the spectral lines come from the immediate vicinity of the more massive component, even a carefully designed search for periodic radial velocity modulation, using a large telescope and state-of-the-art detectors, can fail to detect it. (Robinson & Faulkner 1975). The "S-wave" modulation found in the spectral lines of GP Com, the direct evidence for orbital motion in that object, involved less than 5% of the light in the lines, and much less than 1% of the total light of the system (Nather et al. 1981).

### AM CVn

The first of these objects discovered, and the one with the shortest orbital period (17.5 min), is AM CVn (Smak 1967). Smak's original suggestion that it might be a very short-period binary was expanded by Faulkner et al. (1972) into a reasonably plausible scenario: they proposed that the basic driving mechanism for the system was the loss of orbital momentum due to gravitational radiation. Mass loss from the degenerate donor, the less massive of the pair of white dwarfs, would cause the orbital separation to increase to conserve angular momentum (Kuiper 1941); the objects would tend to spiral out, increasing the size of the Roche lobes surrounding them. The mass donor's radius would also increase as a result of the mass loss, so subsequent mass loss could occur at a slightly greater orbital separation than before. Thus the increased separation would be preserved, and subsequent mass transfer would continue the process.

Observational confirmation of the prediction that the orbital period should be increasing followed quickly - somewhat too quickly, as it turns out. Patterson et al. (1979) reported that the observed period in AM CVn was increasing at about 1000 times the predicted rate. The theoretical consternation that followed was taken (by the observers) as evidence that further theoretical refinements were needed.

It is now clear that the reported rapid increase in the orbital period is wrong. Because of the obvious importance of the problem, we decided to mount an independent observational effort to prove, beyond any doubt, that the original conclusions were correct. We learned, instead, that the light curve of AM CVn is sufficiently variable that the "primary" minimum cannot be distinguished from the "secondary" minimum by inspection alone, an assumption made in all prior analyses. The subsequent possible error of 1/2 cycle can propagate into a convincing description of a changing orbital period. We are not the first to learn this, of course, only the most recently embarrassed. Details of the most recent analysis were reported at this conference by Jan-Eric Solheim.

## PG1346 + 082

Following a suggestion by Jim Liebert, we explored the faint blue object PG1346 + 082 for rapid variability. Initial observations showed the rapid flickering characteristic of mass exchange, and some evidence for periodicity on a somewhat longer time scale. Later observations found the object somewhat fainter, displaying far less flickering but much more prominent long-term periodic behaviour. Figure 1 shows a portion of the light curve obtained by Don Winget & Suchitra Balachandran using the McDonald 36-inch telescope. The repetitive character of the variations is readily apparent, showing a period of 25.3 minutes. We identify this as the orbital period of the system. Time-resolved spectroscopic data now in hand, by Liebert at Arizona and Ed Barker at Texas, should confirm or deny this assertion. In what follows, I will assume it is correct.

Figure 2 shows the average spectrum of the object, together with the spectrum of a DB white dwarf (GD 190, both spectra obtained by Liebert) and a portion of the average spectrum of AM CVn from Robinson & Faulkner (1975). It is clear that the spectrum of this new object is a dead ringer for that of AM CVn, and very different from that of a single DB white dwarf.

Robinson and Faulkner suggested that the curious line ratios evident in the lines of He I in AM CVn could be explained as a combination of an absorption spectrum and an underlying emission spectrum, which completely fills in some of the absorption lines and partially fills in others. The same explanation can clearly be applied to PG1346 + 082 as well. Except for the slightly different orbital periods, the two objects are spectroscopic and photometric twins. In particular, there is notable absence of the Balmer lines of hydrogen in both objects.

## GP Com

The remarkably detailed optical spectrum of GP Com is shown in Figure 3. As first pointed out by Smak (1975), all of the emission lines seem to be composed of two components: a broad, double-humped component shaped very much like the line profiles obtained from disk models, (the "D" component in Smak's notation) with a single, narrow "S" component superimposed. For both components in all lines $\Delta\lambda/\lambda$ is constant, as expected from line broadening due to the Doppler effect. Curiously, the narrow component is always located at a gamma velocity about 70 km/sec lower than that of the broad component.

We first thought this redshift might indicate ejecta aimed preferentially away from us, and sought confirmation of that idea in area scans of the neighbourhood of GP Com on the Palomar Sky Survey plates. There is, indeed, a faint luminosity surrounding the image, but Stover (1933) showed that the extra luminosity comes from three faint stars that surrounded the image of GP Com in 1950; its large proper motion has carried it away from those stars, and its image is now seen as point-like.

The "S" component is clearly resolved in all the lines and displays an apparent Doppler broadening of about 400 km/sec FWHM. It might arise from circumbinary material, photoionized by photons from the hot accretor - but then why is it red-shifted with respect to the lines from the disk? This is one of the many puzzles to solve in trying to model the system from the data.

Figure 4 shows two different (separated) portions of the spectrum on a larger scale. Note the apparent absence of emission at the locations of H$\alpha$ and H$\beta$ , illustrating the purity of the Helium being transferred. The two He emission lines shown are located on each side of the blend that includes He II 4686, and illustrate how well the fairly complex line shape is repeated. If we make the assumption that the emission line blended with $\lambda$4686, located half-way between the two lines shown in Figure 4, has the same basic shape as its red and blue brethren, we can deconvolve the blended pair and see what the He II emission line looks like; the result of this exercise is shown in Figure 5.

The resulting shape of the $\lambda$4686 emission region is plausible: it clearly contains both "S" and "D" line components, but the "S" component is considerably stronger relative to the "D" component than in any of the lines of He I. The "D" component is also obviously broader, suggesting it arises in a region of higher velocity, nearer the "accreti object", than do the emission lines of He I. The width of the He II "S" component is not measurably different from those of the He I "S" components. Table 1 shows the "radial velocity profile" of the system - the various velocities which any detailed model of the system must explain.

Note also, in Figure 5, the apparent absence of N III $\lambda$4640, whose excitation potential lies between those of He I and He II. This is most curious. Virtually all cataclysmic variables that show $\lambda$4686 in emission also show a blend of N III and C around $\lambda$4640-4650. Carbon may well be detectably absent, but nitrogen shouldn't be - N V $\lambda$1240 is the strongest emission line in the IUE spectrum of GP Com obtained by Lambert & Slovak (1981). The only other line clearly present in that spectrum is He II $\lambda$1640, also in emission, and both lines show a width consistent with that expected for the "D" component. The spectrum is not of sufficient quality to determine whether or not the "S" component is also present.

## What it all means

While there are clearly puzzles about the detailed model for these systems, and therefore much to be learned from them, there seems little doubt about the overall morphology: mass is being stripped from the remnant core of an evolved star, forming an accretion disk with a "hot spot" around another compact object, probably a far more massive white dwarf. The absorption/emission lines in the systems of shorter period (AM CVn, PG1346 + 082) come about because the mass transfer rate is high and the disk largely opaque to its own radiation. As the orbital separation increases and the effect of gravitational radiation is reduced the mass transfer rate falls, until the disk becomes thin enough for the lines to be seen entirely in emission (GP Com). The orbital period of 46.5 minutes in GP Com translates, based on the model of Faulkner et al., into a residual mass of 0.02 solar masses left in the donor.

The evolutionary path leading to these systems is somewhat puzzling. As modeled by Rappaport et al. (1982), if the mass-losing component in a cataclysmic variable is a hydrogen-rich star on the main sequence, the components will exhibit a minimum orbital period corresponding to the condition that the mass-losing star becomes fully convective; the orbital separation should increase thereafter, and the mass donor will never detach from its Roche lobe. These objects are not likely progenitors of the ICOs anyway, since they lack the evolved central core so much in evidence in the more compact systems.

Warner (1978) suggested that some of the cataclysmic variables might contain a mass donor with an evolved core, its giant status disguised by the absence of an extended envelope. In this case further mass removal might lower the internal temperature enough to extinguish the stellar furnace, allowing the donor to detach from its lobe and collapse to a white dwarf (Paczinski & Seinkiewicz 1983). An extended period of inactivity would then ensue, putting the objects into a kind of cosmic deep-freeze, until gravitational radiation could remove enough orbital momentum to bring material from the less massive degenerate into contact with its inner Langrangian point, at an orbital period of one or two minutes. Following this "kiss of death" mass transfer will begin again and the orbital period should increase, passing through the periods seen in the interacting white dwarf objects.

This scenario predicts that there should exist a small group of double white dwarf binary systems with very short orbital periods. If close, non-interacting white dwarf pairs really exist they are unlikely to be found by accident: the long exposure times required to observe their spectra would smear out any short-period Doppler effects, and the pressure-broadened lines normal to objects would make radial velocity variations difficult to detect. Eclipses would be too rare to support a purely photometric search programme. But the interacting white dwarf binaries did come from somewhere, and the most likely immediate progenitors are pairs of detached white dwarfs in close orbit. Such a system should exhibit a readily detectable evolution in its orbital period, due entirely to the effects of gravitational radiation, and measurable to good accuracy with a baseline of only a few years. Somebody really ought to look for them.

The future evolution of the interacting white dwarf
binaries is less problematical: the process of mass loss will
continue until the donor is completely consumed. At some point
it will no longer have enough mass to be degenerate, and will
then not expand as mass is removed so the orbital period will
stop increasing, but gravitational radiation will continue to
remove orbital momentum until there is nothing left in orbit.
The final result will be a white dwarf with an outer atmosphere
of helium (unless there is an inner core of carbon not yet
reached in the stripping process). A considerable portion of
the original orbital angular mometum should still be present as
rotational momentum: the accretor should be spinning very
fast, and should do so for a very long time unless it can find
some method of slowing down.

Whether or not all of the helium-atmosphere white dwarfs
were manufactured in this way is an open question. Occam would
insist on it, but arguments based on a sample of three objects
are not very convincing. We have no real way of knowing how
many of these objects there are, nor how long the process
takes. We are encouraged, however, by the fact that the number
of known objects has increased very recently by 50%.


## 3. OTHER INTERACTING COLLAPSED OBJECTS

4U 1626-67

The enormously deep gravitational well produced by a
neutron star will have dramatic but predictable effects on any
material falling into it. The many known X-ray objects can be
explained as interacting binary systems, cousins of the
symbiotic and cataclysmic binaries, with a neutron star as the
mass-accreting component. It should be no surprise, then, that
two X-ray binary systems have been identified which seem to be
cousins of the interacting white dwarf pairs, with accretion
taking place onto a neutron star. The mass donor is most
likely a low mass white dwarf.

The X-ray object exhibits rapid pulsations at a period of
7.7 seconds, readily explained as the rotation rate of an
accreting neutron star in a close binary system. Orbital
modulation of this basic periodicity would be expected, similar
to that found in many other X-ray binaries, but none has been
detected in 4U 1626-67 to impressively stringent limits (Joss &
Rappaport 1984). This implies either that the model is wrong,
or the mass ratio is so extreme that orbital modulation of the
central object is below the limit of detectability.

This latter interpretation was materially strengthened by
the discovery of a low-power sideband in the (optical) power
spectrum of the objects's visual counterpart by Middleditch et
al. (1981), which they explain as the difference between the
synodic and sidereal periods of the binary system. Apparently
a tiny fraction of the X-radiation is reprocessed by the
mass-losing companion and therefore betrays its Doppler
velocity. The orbital period is deduced as 41.5 minutes,
neatly within the range of periods expected for Interacting
Collapsed Objects.

Unfortunately the system is at too high a temperature for
the spectroscopic presence of hydrogen to be ruled out (or
confirmed), so the arguments ascribing core-evolved status to
the mass donor are indirect. The implied orbital separation of
about 1 light-second constrains its mass to be less than $\sim 0.1$
$M_\odot$, making a hydrogen-rich composition possible but very
unlikely.

## 4U 1915-05

Another strong candidate for identification as an ICO with
a neutron star accretor and a white dwarf donor is the object
4U 1915-05, studied by White & Swank (1982) and by Walter et
al. (1982). The X-ray "light curve" of this object exhibits
periodic reductions in intensity with a period of 49.8 minutes,
ascribed by both studies to obscuration of the accreting
object, assumed to be a neutron star, by the turbulent material
surrounding the hot spot in an accretion disk. Although the
absorption "dips" do not occur at exactly the same time for
each rotation, apparently due to erratic turbulence around the
hot spot, the stability of the underlying period is retained
over a baseline of several years.

White & Swank have studied the changes in the spectrum of
X-ray absorption during the regular "dips" and conclude that
the absorbing material cannot be of normal solar composition.
They propose that the material must be metal-deficient by a
factor of 17 with respect to the sun, and suggest it belongs to
Pop II if it is a normal star. The optical counterpart has not
been positively identified, although Walter et al. have
located a candidate object of magnitude 22. We are not likely
to learn much about the optical spectrum of so faint an object,
so it is fairly safe to predict that accreting material should
be composed of pure helium, as seen in the interacting white
dwarf objects; such a composition is consistent with the dearth
the high-Z absorption seen in the "dips" (along with a lot of
other possible compositions).

## Interacting Binary Neutron Stars (IPNS)

Although there is far less constraining evidence for the interaction of WD-NS star pairs than for the interacting WD pairs, what exists is completely consistent with simple models involving mass stripping of an evolved stellar core; the only real difference lies with the accreting object. It is therefore appropriate to ask whether NS-NS interactions are possible, and what they might look like.

This question is given some currency by the discovery of a pulsar with the remarkably short period of 1.5 milliseconds (Backer et al. 1982) and by suggestions that its extremely rapid rotation may be the results of mass accretion and subsequent spin-up. Henrichs & van den Heuvel (1983) express well-reasoned doubts about models involving massive stellar donors, and propose that the coalescence of two neutron stars could do the job, based on earlier theoretical work by Clark & Eardley (1977).

Whether such an exotic origin is required to explain the rapid rotation rate of PSR1937 + 214 is an open question, but many authors have pointed out that the two components of the binary pulsar PSR1913 + 16, given a time far less than the age of the Universe, will inevitably drift together in orbit until some kind of interaction takes place. The unseen member of that pair is also though to be a neutron star, although a moderate-mass black hole cannot be ruled out. Tidal disruption of a neutron star spiraling into a black hole was examined theoretically by Lattimer & Schramm (1976). Their model indicates that disruption will occur on the dynamical time scale of a neutron star, and would therefore be essentially unobservable. They simply assumed that such disruption would occur, however, and did not demonstrate that it would.

The coalescence of two neutron stars was described by Clark & Eardley as an extension of the work by Lattimer & Schramm, and is based on a number of simplifying assumptions, at least some of which may be open to question. They assume, for example, that the neutron star will encounter its Roche radius for tidal disruption at the same time it encounters the inner Langrangian point of the Roche lobe structure. For any possible mass ratio involving neutron stars the inner Lagrangian point will be encountered first, so whether disruption will occur at all depends on the effects that take place there. They further assume that the less massive member of the pair, destined by its larger radius to be the one consumed, will retain its character as a gravitationally collapsed neutron star even though the gravitational potential along the line of centres between the two objects will be progressively reduced as they drift together in orbit. This seems unlikely indeed.

Figure 6 shows the classical picture of the Roche lobe structure surrounding two point masses. The centre of mass of the system is located where the two axes cross, and the horizontal axis, as shown, holds three of the five Lagrangian points. The inner Lagrangian point, of course, represents the point of zero gravitational potential in the rotating frame, through which mass can flow when binary systems interact. A second point of zero potential exists at the outer Lagrangian point somewhat farther from the object of lesser mass. Although the gravitational potential rises rapidly from zero inward of these two points it will not provide the needed gravitational compression to hold a neutron star together if the two orbiting objects drift very close to each other. Expansion along the line of centres would seem to be inevitable.

The details of such an effect clearly depend on the equation of state chosen to describe the neutron star's construction, as well as on the (asymmetric) properties of the gravitational lobe system. How far can a neutron star distort along the line of centres before something drastic happens? It seems very likely that mass will find its way through the inner Lagrangian point long before the striking lobe structure will contact the surface of the less massive member of the pair. Such mass transfer will redistribute the system's angular momentum in such a way as to oppose the drift to shorter orbital periods, just as it does in models involving less exotic objects. White dwarf donors would, of course, be subject to the same distorting effects, but current models do not include them.

What is far from clear is how a neutron star will behave in response to a slowly changing gravitational environment. If we assume, for argument, that it will manage to adjust itself fairly gently, without invoking its very short and potentially explosive dynamical time scale, then a prolonged interval of mass transfer might well be possible, much as envisioned for the NS-WD and WD-WD interacting pairs. The phase might even last long enough for the process to be observable. Detailed models will be required to see whether such slow adjustment is possible, but it seems far more likely than sudden tidal disruption, based on the assumption that the neutron star remains unchanged until it encounters the critical Roche radius.

If we assume for the moment that controlled mass transfer is possible between two neutron stars, or between a neutron star and a moderate-mass black hole, what will be the material transferred? It will certainly not be free neutrons, as described in the Clark and Eardley model, but what will it be? Iron?

Whatever gravitational potential remains at the putative neutron star surface will be insufficient to force electron capture, but may be enough to provide nuclear stability to neutron-rich isotopes near the maximum in the curve of binding energy; the subsequent reduction in the potential should result in iron-peak elements mixed with whatever decay products could exist in that environment. Carefully constructed models may tell us what to look for.

Another point of interest involves the ultimate fate of the mass-losing object. If it is a degenerate white dwarf, it seems likely that slow mass erosion can be accommodated without a sudden change in structure. Erosion of material from a neutron star is another matter entirely. A neutron star cannot, according to current theory, retain its collapsed structure if it is spherical and its mass falls below about 0.2 $M_\odot$. Under the conditions encountered in the tidal lobe of a binary system, it would seem that explosive expansion might well take place when the gravitational compression at any point on the surface of the (distorted) neutron star drops below that on a spherical neutron star of this minimum mass. The part of the surface nearest the inner Lagrangian point is clearly in the greatest danger. Directed along the line of centres joining the two stars and aimed at the more massive member of the pair such an event might approximate a cosmic jet engine with tremendous thrust, driving some fraction of the lighter object free of the gravitational potential well surrounding the system.

## Back to reality

The advance of scientific understanding, as Hermann Bondi has observed, is like a man walking on two legs, one labeled "theory" and the other "experiment". An advance in one comes to a halt until there is a corresponding advance in the other. The increasing store of evidence pointing to the existence of Interacting Collapsed Objects clearly needs help from theoretical models, just as the esoteric models of black holes need constraining observations. Neither job is very easy. Whether, in fact, Interacting Binary Neutron Stars exist for long enough to be observed, and whether they would be recognized as such even if they do, is presently only a matter for entertaining speculation. Interacting WD-WD and WD-NS pairs almost certainly do, although detailed models have yet to be constructed. I hope the existing observational evidence reviewed here is convincing enough to get them underway.

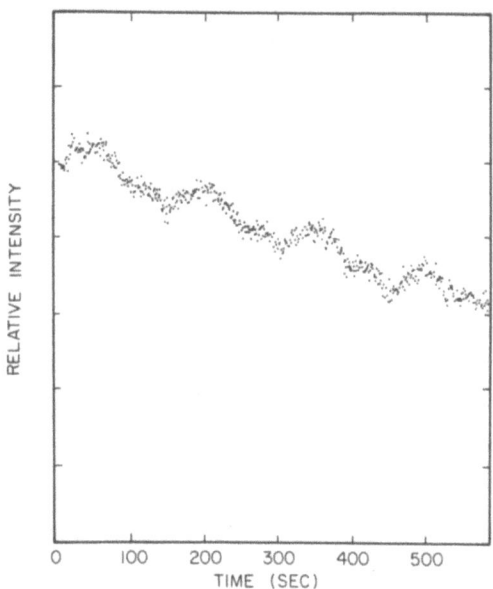

Figure 1.  The light curve of PG 1346 +082 in white light.  The deeper minima recur at intervals of 25.3 minutes; the overall shape is much like that of the light curve of AM CVn.

## Table 1

### A Radial Velocity Profile of GP Com

---

- $\Delta\lambda/\lambda \simeq$ constant for all He I emission lines

- "S" components are systematically redshifted with respect to the "D" components by $\simeq$ 70 km/sec

- The full Width at Half Maximum (FWHM) of the "S" component lines corresponds to $\simeq$ 400 km/sec

- The "S-wave" K velocity (velocity of the disk material at the location of the hot spot) $\simeq$ 600 km/sec

- The FWHM of the "D" component lines of He I $\simeq$ 2500 km/sec

- The FWHM of the "D" component line of He II $\simeq$ 3600 km/sec

---

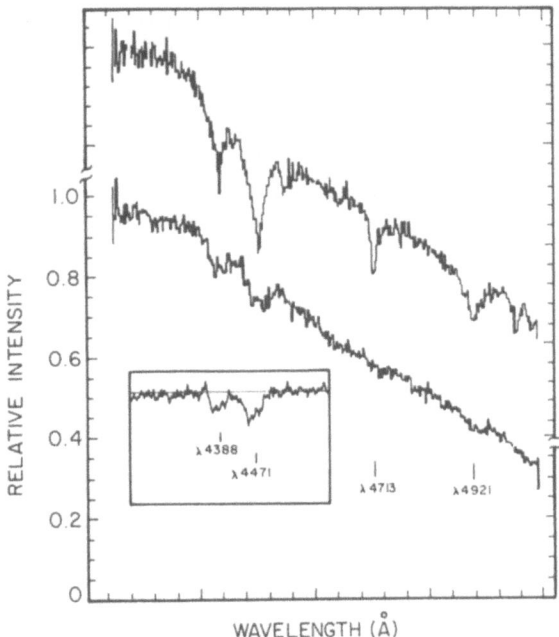

Figure 2.    The spectrum of the DB white dwarf GD 190 (top curve),
PG 1346 +082 (middle curve) and AM CVn (box).

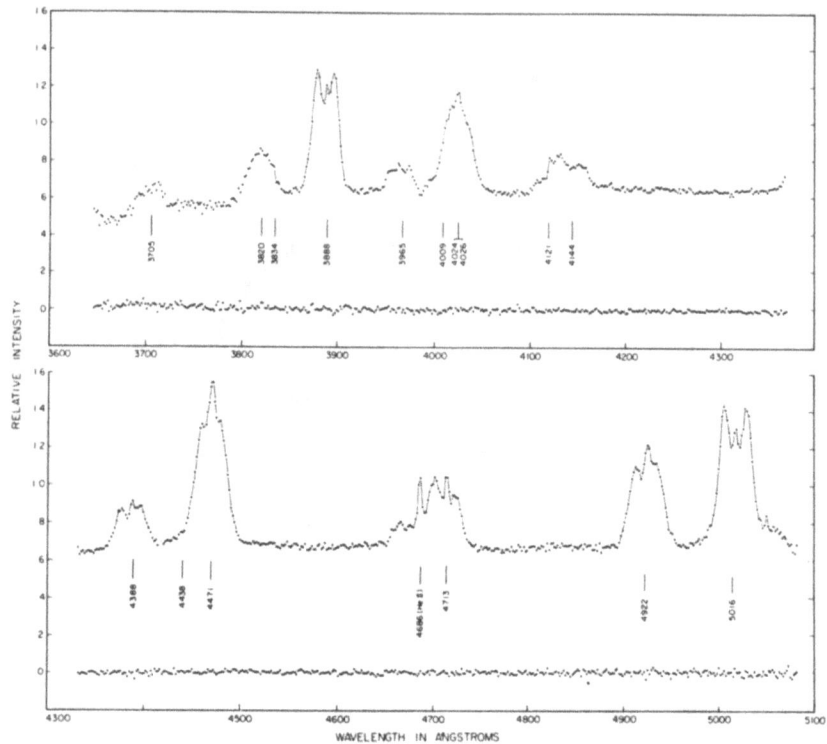

Figure 3.  The optical spectrum of GP Com.  All of the line identifications refer to lines of He I unless otherwise marked.  The upper curve in each box is the sum of two time-interlaced spectra; the lower curve in each is their difference, showing only the spectral noise of the measurement.

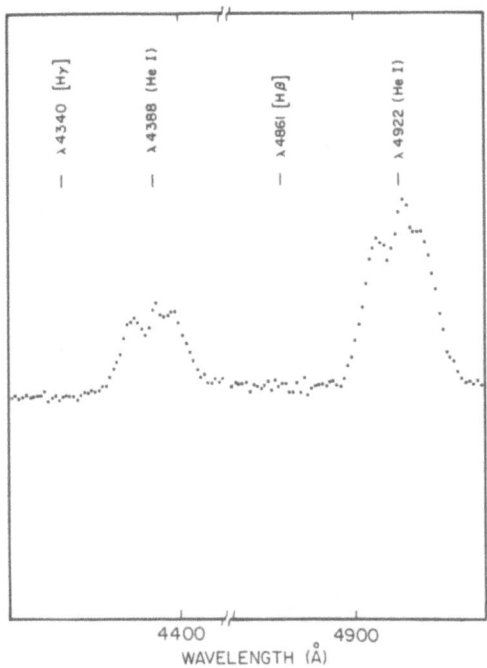

Figure 4. Two (separated) He I lines from the spectrum of GP Com, illustrating how well the complex line shape is reproduced. Note the break in the wavelength scale at the bottom. Hydrogen lines identified in square brackets are notable by their absence.

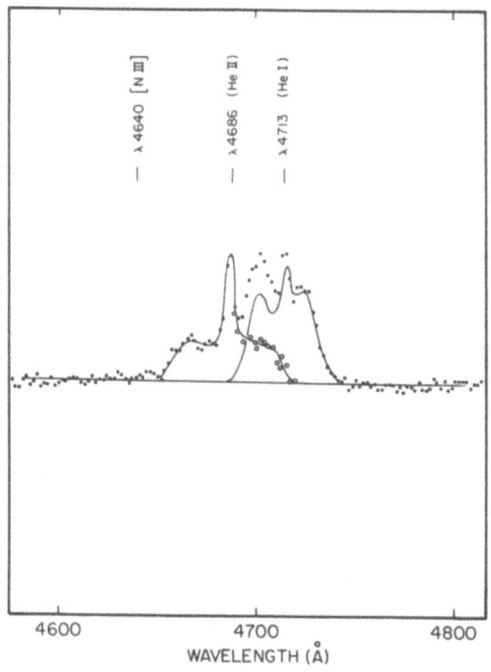

Figure 5. The region of He II λ4686 deblended. Filled dots are the original data points, open circles those obtained after subtracting the He I line with the shape shown. The region where N III would be expected is also indicated.

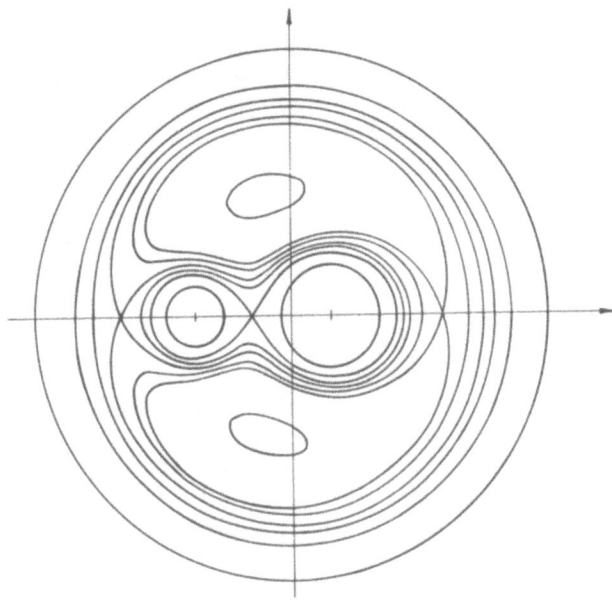

Figure 6.  The classical structure of the Roche equipotential
surfaces surrounding two point masses; the axes intersect at the
centre of mass for the system.

ACKNOWLEDGEMENTS

I thank Ethan Vishniac, Don Winget and Hugh van Horn for very helpful discussions on theoretical matters, and Jim Liebert, Winget and Suchi Balachandran for providing observational data in advance of publication. Much of the work reported here was supported by a grant from the National Science Foundation.

REFERENCES

Backer, D.C., Kulkarni, S.R. & Taylor, J.H., 1982. Nature **300**, 615.

Clark, J.P.A. & Eardley, D.M., 1977. Ap.J., **215**, 311.

Faulkner, J., Flannery, B.P. & Warner, B., 1972. Ap.J., (Letters) **175**, L79.

Grauer, A.D., 1983. In "Cataclysmic Variables and Low Mass Binaries," ed. J. Patterson and D.Q. Lamb (Dordrecht: Reidel).

Henrichs, H.F. & van den Heuvel, E.P.J., 1983. Nature, **303**, 213.

Joss, P.C. & Rappaport, S.A., 1984. Ann. Review Astron. Astrophys., 22, in press.

Kuiper, G.P., 1941. Ap.J., **93**, 133.

Lambert, D.L. & Slovak, M.H., 1981. P.A.S.P., **93**, 477.

Lattimer, J.M. & Schramm, D.N., 1976. Ap.J., **210**, 549.

Middleditch, J., Mason, K.O., Nelson, J.E. & White, N.E., 1981. Ap.J., **244**, 1001.

Nather, R.E., Robinson, e.L. & Stover, R.J., 1981. Ap.J., **244**, 269.

Paczynski, B. & Seinkiewicz, R., 1983. Ap.J., **268**, 825.

Patterson, J., Nather, R.E., Robinson, E.L., & Handler, F., 1979. Ap.J., **232**, 819.

Rappaport, S.A., Joss, P.C. & Webbink, R.F., 1982. Ap.J., **254**, 616.

Robinson, E.L., 1976. Ann.Rev.Astron.Astrophys., **14**, 119.

Robinson, E.L. & Faulkner, J., 1975. Ap.J., (Letters) **200**, L23.

Smak, J., 1967. Acta Astr., **17**, 255.

Smak, J., 1975. Acta Astr., **25**, 227.

Stover, R.J., 1983. P.A.S.P. (in press).

Walter, F.M., Bowyer, S., Mason, K.O. & Clark, J.T., 1982. Ap.J., (Letters) **253**, L67.

Warner, B., 1978. Acta Astron., **28**, 303.

White, N.E. & Swank, J.H., 1982. Ap.J., (Letters) **253**, L61.

# STARS THAT GO HUMP IN THE NIGHT : THE SU UMa STARS

Brian Warner

Department of Astronomy
University of Cape Town
S. Africa

## 1.    INTRODUCTION

The  gross  structure  underlying the cataclysmic variable
stars  was  worked out  in  the  1950's,  60's  and  70's  from
photometric  and spectroscopic  observations  (see  reviews  by
Robinson (1976) and  Warner  (1976)).   With  the discovery of
X-rays from cataclysmic variables, and the wealth  of data from
low mass X-ray binaries, there has been a  continuing  interest
in  understanding  the  more  exotic  phenomena associated with
these  stars.  In particular, dwarf nova  outbursts  provide  a
means of  studying accetion tori or disks in non-steady states.
The   abundance  of  curious  properties of  outbursting  dwarf
novae, which include rapid coherent  and  quasi-periodic  light
variations,    soft    X-ray    enhancements,    supermaxima   and
superhumps, demands every effort of understanding, not only per
se  but  also in order to model with confidence their energetic
cousins the X-ray  novae, and accreting neutron stars and black
holes in general.

## 2.    THE SU UMa STARS

### 2.1   Outbursts of Dwarf Novae:   General Background

Among the dwarf novae several classes  have  been  defined
according to the nature of their outburst light curves.   We are
concerned  here  with the U Gem stars, or ordinary dwarf novae,
and the SU  UMa  stars, or supermaxima dwarf novae.  The former
have non-periodic outbursts, recurring typically  on timescales
~20 - 100 days, with durations of a few days and amplitudes 2-5

*P. P. Eggleton and J. E. Pringle (eds.), Interacting Binaries, 367–392.*

magnitudes; the latter have, in addition, super-outbursts, often at regular intervals, in which the stars reach ~0.5 magnitude brighter than in normal outbursts and remain in the outburst state for ~15 days.

The reality of the distinction between the U Gem and SU UMa classes has been questioned by van Paradijs (1983). He points out that almost all dwarf novae have a bimodal distribution of outburst durations, i.e. they possess "narrow" and "wide" outbursts. The width of the narrow outbursts is strongly correlated with orbital period $P_o$, being ~12 days at $P_o$ = 9 hrs and ~3 days at $P_o$ = 1 1/2 hrs. The wide outbursts are less markedly correlated with $P_o$, showing only a slight decrease (from ~16 to ~11 days) between the above values of $P_o$. The SU UMa characteristic appears therefore to be more the result of reduction in length of the narrow outbursts than the appearance of distinct type of super-maximum in the stars of short $P_o$.

However, the picture is actually more complicated than these correlations would imply. Above the "period gap" (Warner 1976) ($0^d.090 < P_o < 0^d.115$) narrow and wide outbursts tend to be present at roughly equal frequencies, often alternating between the two. Below the period gap almost all dwarf novae are in the SU UMa class in which several (4-8) narrow outbursts occur between each super-maximum. In some of these systems (YZ Cnc, AY Lyr, VW Hyi, Z Cha) the super-maxima appear at very regular intervals; this is certainly not the case for wide outbursts in systems above the period gap. Other SU UMa stars have irregular and infrequent super-maxima: SU UMa itself had no super-maxima at all for three years; HT Cas may have been in this state for several years.

The period gap appears to represent also a transition from the SU UMa type of outburst light curves to that of U Gem stars. In this connection it is noteworthy that TU Men, which lies at the longward edge of the period gap is tri-modal: its outburst durations cluster around $D \sim 1^d$, $D \sim 8^d$ and $D \sim 20^d$ (Bateson 1979a, 1981). It therefore combines the outburst properties of the SU UMa and the U Gem stars. Only two of the $20^d$ duration outbursts (separated by 6 1/2 years) have been observed in TU Men; it was during the second of these that Stolz & Schoembs (1981) discovered the superhump characteristic of SU UMa supermaxima (see below). It will be interesting to see if the wide ordinary outbursts in TU Men also show superhumps. That they may do so is suggested by observations of CN Ori, which is the dwarf nova with the next longest orbital period after TU Men. The outburst light curves of CN Ori are described by Bateson (1979b) as having "a slight tendency for wide and narrow maxima to alternate, but this sequence is often broken by successive pairs of either wide or narrow maxima". Throughout two consecutive wide maxima (between which CN Ori hardly reached quiescence) Schoembs (1982) found two simultaneous photometric periods of $0\overset{d}{.}1595$ and $0\overset{d}{.}1631$; this is suggestive of the superhump phenomenon (see Table 1) and urges the question whether both periods will be found to exist during narrow maxima and quiescence.

The most enigmatic feature of the SU UMa stars is the appearance of the "superhumps" during super-maxima; it is to a discussion of the properties of these that the majority of this paper is devoted.

At quiescence several of the SU UMa stars show an orbital hump in their light curve, resulting from variations in aspect of the bright spot at the outer edge of the accretion disk. During most normal eruptions the orbital hump is reduced in relative amplitude, usually to undetectability, by the increase in background luminosity. In all supermaxima, however, a periodic hump appears irrespective of whether an orbital hump is seen at quiescence. As was discovered independently by Vogt (1974) and Warner (1975) from the December 1972 outburst of VW Hyi, these "superhumps" have a mean period $P_s$ a few percent longer than $P_o$. The seven stars for which both $P_s$ and $P_o$ are known are listed in Table 1; the beat period $P_{beat} = P_o P_s/(P_s - P_o)$ between them is also given.

Table 1

Comparison of Orbital and Superhump Periods

| Star | $P_o$ (days) | $P_s$ (days) | $P_{beat}$ (days) | Reference |
|------|------|------|------|------|
| WZ Sge | 0.05669 | 0.05714 | 7.2 | Patterson et al. (1981) |
| V436 Cen | 0.06250 | 0.06378 | 3.1 | Warner (1983c) |
| OY Car | 0.06312 | 0.06463 | 2.7 | Vogt (1982) |
| VW Hyi | 0.07427 | 0.07669 | 2.4 | Haefner et al. (1979) |
| Z Cha | 0.07449 | 0.07725 | 2.1 | Vogt (1981) |
| WX Hyi | 0.07481 | 0.07735 | 2.3 | Bailey (1979), Schoembs (1979) |
| TU Men | 0.1176 | 0.1262 | 1.7 | Stolz & Schoembs (1981) |

2.2 Spectroscopic Observations during Outbursts

Spectra of dwarf novae during ordinary outbursts in general show broad shallow absorption lines ascribed to Doppler broadening in the optically thick luminous accretion disk (Warner 1976). Super-maxima of the non-eclipsing systems V436 Cen and VW Hyi show similar spectra (Whelan et al. 1979). However, in the high inclination system Z Cha, much narrower

absorption lines are observed (Vogt 1981). In this star, and
in the similar system OY Car (Vogt et al. 1981), narrow
absorption lines, strongest in the higher members of the Balmer
series, are present at quiescence; their radial velocity
variations are in phase with, and of the same amplitude as, the
emission lines from the quiescent accretion disk. Their source
is evidently cool gas at the outer rim of the accretion disk,
seen against the bright centre of the disk. From details of
the eclipses in Z Cha (Cook & Warner 1984) the central region
is of white dwarf dimensions, so the material in the disk seen
projected against this source is moving tangentially to the
line of sight and has a small velocity dispersion.

During super-maxima of Z Cha these narrow, single
absorption lines are still present, but with the following
enigmatic properties (Vogt 1981):

S1. The strengths of the lines are correlated with phase
$\phi_s$ in the superhump cycle ( $\phi_s = 0$ at superhump maximum): very
strong at hump maximum, otherwise weak. They are absent during
eclipse (see Figure 6 of Vogt 1981).

S2. Vogt's observations of March 28/29, 1978, at which time
the superhump maximum occurred at orbital phase $\phi_o = 0.86$ (i.e.
$\phi_s = 0$ at  $\phi_o = 0.86$; $\phi_o = 0$ at mid-eclipse), provide a radial
velocity curve from the absorption lines which has the same
phase and amplitude as at quiescence, but with a mean ($\gamma$)
velocity of +267 km s$^{-1}$. The previous night, in which $\phi_s = 0$
occurred at $\phi_o = 0.38$, two radial velocity measurements imply $\gamma$
is correlated with the orbital phase at which superhump maximum
occurs; i.e. $\gamma$ probably varies periodically with period $P_{beat}$ .
The Balmer emission lines in Z Cha during super-maximum, in
contrast, do not show any significant departures of $\gamma$ from that
measured at quiescence.

2.3  Photometric Phenomena during Supermaxima

Vogt (1983) has given a comprehensive review of the
photometric behaviour during outbursts of VW Hyi, the best
studied of the SU UMa stars. From this and observations of
other SU UMa stars we select the following phenomena which have
a direct bearing on the underlying structure of these systems
during supermaxima.

P1. During the rise to both ordinary and supermaxima the
orbital hump retains the same intensity amplitude, period and
phase as at quiescence. Only VW Hyi has so far been thoroughly
studied during this important phase (Vogt 1974, Warner 1975;
Hefner et al. 1979).

P2.    Through    ordinary    outbursts    well    separated    from
supermaxima   the   orbital   hump   has   approximately   the   same
intensity   amplitude   as   at   quiescence.    However,   for those
outbursts occurring shortly before a supermaximum, enhancements
of the orbital hump by factors of  2-15  have  been  seen in VW
Hyi,   with   the   greatest enhancement occurring   when   the   two
outbursts  were  separated by  only  2  days  (Haefner  et  al.
1979, Vogt 1983).   In  WZ Sge the orbital hump was enhanced in
intensity by a factor ~100  for  the  first  ten  days  of  its
outburst,  with a phase shift $\Delta\phi \sim -0.38$, decaying in amplitude
later with $\Delta\phi \sim -0.23$ (Patterson et al. 1981).

P3.    Superhumps   first   appear   near   the   maximum   of   a
super-outburst, usually about one day  after  the  beginning of
outburst.   Occasionally superhumps  appear  late  but this is
attributable   to   a   supermaximum having developed   out   of   an
ordinary outburst (e.g. V436 Cen, Semeniuk 1980).

P4.    Superhumps   start   with   a   sharply   peaked   profile   and
amplitude  30-40  percent.   Their  amplitude  decays  somewhat
faster than the background  brightness of the system and at the
same   time   they   broaden   in   profile   (Haefner   et   al.   1979,
Patterson 1979, Section 2.4).

P5.    Superhumps   appear   during   all   supermaxima,   even   in
systems   of intermediate   or   low   inclination.   Their   initial
amplitudes do  not  depend noticeably on inclination:  they are
similar   in  Z Cha   and   OY Car  (eclipsing   systems),   VW   Hyi
(non-eclipsing but with a prominent quiescent orbital hump) and
the group  YZ  Cnc,  WX Hyi and V436 Cen (no detectable orbital
hump).

P6.   The mean period of  the  superhumps  (Table  1)  is  a few
percent   longer   than   the   orbital   period.   In the well-studied
systems there is an  apparent, statistically highly significant
decrease in the superhump period  during  outburst. Expressing
times of superhump maxima as

$$T_{max} = T_0 + P.E + C.E^2$$

(1)

where E is the number of elapsed  cycles, we find the following
values of P and C

Table 2

| Star | $P(d)$ | $C(d)$ | |
|------|--------|--------|---|
| YZ Cnc | 0.0920 | $-8(\pm 4)\times 10^{-6}$ | Patterson (1979) |
| V436 Cen | 0.06378 | $-1.0\times 10^{-6}$ | Warner (1983c) |
| Z Cha | 0.07749 | $-4(\pm 2)\times 10^{-6}$ | Warner (unpublished) |
| VW Hyi | 0.07712 | $-2.85(\pm 0.13)\times 10^{-6}$ | Haefner et al. (1979) |
| TU Men | 0.12625 | $-6.1\times 10^{-6}$ | Stolz & Schoembs (1981) |
| TY Psc | 0.0703 | $-4\times 10^{-6}$ | Bond et al. (1982) |

The superhump timings in WX Hyi (Bailey 1979) also indicate $C < 0$.

P7. There is no strong modulation of the superhump profile at the beat period. For example, the superhump profiles on 8 and 9 December 1974 (top two profiles in Figure 9 of Haefner et al. 1979) are very alike but occur respectively at orbital phases when at quiescence the orbital hump is invisible and when it is prominent. Similar independence of superhump profile on $\phi_s - \phi_o$ can be seen in Figure 8 of Warner 1975. We conclude that if, as is usually believed, the orbital hump at quiescence is produced from a combination of obscuration by the optically thick accretion disk and (when observed) varying geometric aspect of the bright spot, then the source of luminosity responsible for the superhumps is not strongly affected by such a modulation at period $P_o$.

P8. The presence of deep eclipses in OY Car and Z Cha during both normal and supermaxima demonstrates that at least the majority of the additional luminosity in the outbursts arises in the primary or accretion disk. In both stars the depth of eclipse increases during the course of the outburst, except when $\phi_s \sim 0$ at $\phi_o = 0$, when eclipses are always relatively shallow (Vogt 1982, Section 2.4).

P9. After the rapid decrease in brightness at the end of supermaximum, by which time the superhumps have disappeared, VW Hyi has an interval of several days of slow decrease in brightness during which orbital humps (at approximately their quiescent intensity amplitude) are present together with "late superhumps" which have amplitudes similar to that of the orbital hump but a period near that of the outburst superhumps. However, the late superhumps are shifted in phase by 180 deg. with respect to the outburst superhumps. Semeniuk's (1980) observations can be interpreted to show that the same phenomenon is present in V436 Cen (Warner 1983c). Other systems have not been observed sufficiently in the late stages of supermaxima for the detection of late superhumps.

<u>P10</u>. Both outburst superhumps and late superhumps frequently show structures in their profiles which roughly repeat over several cycles and even from night to night (Figure 1; Figures 2a and 2b of Schoembs & Vogt 1980). An alternative way of stating this is that the superhumps consist of multiple components all of which repeat with period $P_s$.

<u>P11</u>. The existence of rapid coherent or quasiperiodic oscillations (Warner 1976), with comparable periods (10-100 secs) and behaviour in both normal and superoutbursts, demonstrates a basic similarity in the two types of outburst, and that the primary or material in the disk near the primary is intimately involved in the eruptions.

2.4  New Photometric Observations of Z Cha

Over the past decade the writer has obtained high speed photometric observations of Z Cha whenever outbursts coincided with his observing runs. Only the data for January 1973 have so far been discussed (Warner 1974). The material now available, from outbursts in January 1973, April 1979, February 1980, February 1981 and December 1982, includes one ordinary and four supermaxima. Most stages of development of the supermaximum are covered and light curves are available with superhumps appearing at a representative sample of orbital phases. A selection of these light curves, greatly condensed, appears in Figure 2.

The following points emerge from study of these light curves

<u>Z1</u>. Eclipses near maxima of ordinary outbursts are symmetrical, indicate a disk radius similar to that at quiescence $(r_q)$, and are relatively broad and shallow suggesting that the outer parts of the disk are brighter than the inner.

<u>Z2</u>. The profile of the superhump is more peaked when it occurs near $\phi_o = 0.5$ than when near $\phi_o = 0$; in the latter case the hump lasts for almost all of the orbital cycle, whereas a hump occurring near $\phi_o = 0.5$ may last for less than half an orbital cycle in the early stages of a supermaximum. This conclusion - that there is a variation in superhump profile with period $P_{beat}$ - differs from (P7) above, but presumably only applies to high inclination systems.

<u>Z3</u>. All supermaxima eclipses are asymmetrical in the sense of having slower egress than ingress - the differences arising principally in protraction of the final stages of egress. This indicates that during supermaxima the following lune of the accretion disk at the time of eclipse has an extended bright

area.   The   slow   recovery in egress resembles (and   at   times
involves   a   similar   fractional   intensity   as)   that   seen   in
quiescence.   In quiescence, however, an orbital hump is present
with a range $\sim$50   percent   of   the mean light.   Although it is
difficult to disentangle the effects of the superhump, there is
no   clear evidence during supermaxima for any   modulation   like
the quiescent   orbital   hump.   This   implies   that   the excess
luminosity present in the following lune at eclipse   is   always
visible (whatever its location at other phases).

     Eclipse   first   contacts   give   disk radii $r_d$ in the range
$1/2 \; r_q \lesssim r_d \lesssim r_q$ for the preceeding lune; however last contacts
show that there is always   luminosity in the following lune out
to radii $\geqslant r_q$ ($r_q$ = quiescent disk radius).

Z4.   The behaviour of eclipse depth,   stated   in (P9) above, is
illustrated   for Z Cha in Figure 3, where   the   intensities   at
mid-eclipse, divided   by   mean intensity of Z Cha in regions of
the light curve free   from effects of eclipse or superhump, are
given.   Only when superhump maximum   occurs   in   the range 0.35
$< \; \phi_o \; <$   0.65   is   the   intensity   at   mid-eclipse   free   from
additional light.

Z5.   The decay with time since the beginning· of   outburst   in
the relative amplitude of the superhump (P4) is illustrated for
Z Cha in Figure 4.

Z6.   Observed   times   of   minima   during supermaxima show large
positive values of O-C if   $\phi_s$ = 0 occurs in the range $-0.2 < \phi_o$
$< 0.02$; this is illustrated in Figure 5.   (In order to make the
O-C   values   for   8   January   1973 consistent   with   the   other
observations,   it   has   been   necessary to   conclude   that   the
observer, in his excitement at finding for the first time Z Cha
at outburst, recorded the time   incorrectly at the start of the
run:   the required adjustment is exactly   one minute.)   As well
as the writer's observations, times of minima   during the March
1978   supermaximum   have   been   obtained from Table 4   of   Vogt
(1981), which include an observation determined visually by the
New Zealand amateurs.

     Figure 5 shows that whenever there is a significant amount
of the superhump present at   eclipse,   the   eclipse is delayed.
This   implies   that   at such times the following   lune   of   the
accretion disk is brighter than the preceeding time.

Z7.   Exceptions to the   statements   in (Z3) occur at the end of
both ordinary and supermaxima:   for at least a day the luminous
part of the disk shrinks in   radius   to $\sim 1/5 \; r_q$, i.e. to only
two   or   three   times   larger   than   the   radius of   the   primary.
Patterson (1981) has observed a similar phenomenon in HT Cas.

## 2.5 Comparitive Anatomy of two Z Cha Eclipses

We examine the eclipses observed on 16 and 17 December 1982, which occurred with superhump maxima at $\phi_0 = 0.46$ and $\phi_0 = 0.89$ respectively, i.e. in similar circumstances to those of Vogt's (1981) radial velocity measurements of March 27/28 and March 28/29, 1978. Between 16 December and 17 December, Z Cha decreased in brightness by $\sim$25 percent; in order to facilitate comparison, the two light curves have been scaled so that those regions free from superhump or eclipse have the same mean brightness. The light curves in the region of the eclipses are shown in Figure 6.

Consider first the eclipse of 16 December (eclipse A in Figure 6). Its asymmetry is made evident by the reflection of ingress about mid-eclipse. The latter corresponds closely to $\phi_0 = 0$ as predicted from the emphemeris given by Cook & Warner (1981) for mid-eclipse of the primary. From $\phi_0 = 0.04$ to 0.09 the egress curve lies below the reflected ingress (see dashed difference curve above). Before $\phi_0 = 0.04$ it is not possible to define an uncontaminated ingress curve: whatever is eclipsed and emerging from eclipse in $0.04 < \phi_0 < 0.09$ is likely to be immerging in $-0.04 < \phi_0 < 0$. We note, however, that emergence takes place in $0.075 < \phi_0 < 0.095$ and this is the same phaserange in which the bright spot emerges from eclipse (for example Smak 1979). It would appear that the asymmetrical egress is caused by a region on the following lune which is in approximately the position of the quiescent bright spot and which contributes 6 percent of the brightness of the system at $\phi_0 = 0.05$. As pointed out in Z3 above, this is typical of eclipses throughout supermaxima of Z Cha.

Eclipse B is remarkable in being much shallower than A and having mid-eclipse greatly displaced from $\phi_0 = 0$. Because of the pre-scaling, we can make a direct comparison of B with eclipse A and its reflected ingress (which approximates to an axially symmetric intensity distribution on the disk). From the difference curve (inset in Figure 6) the excess light is present almost unobscured up to $\phi_0 = 0$ (we interpret the slow variations in excess light in $-0.13 < \phi_0 < 0$ as intrinsic variations in brightness always present in the superhumps. In the region $-0.01 < \phi_0 < 0.03$ the excess slowly diminishes and from $\phi_0 = 0.03$ to 0.07 is completely absent (relative to the eclipse egress of the symmetrical disk). In $0.07 < \phi_0 < 0.1$ the excess reappears but at lower intensity as by now the system is $0.2P_S$ away from superhump maximum.

The interpretation of this behaviour appears straightforward: until $\phi_0 = -0.01$ the region responsible for

the superhump excess  luminosity is uneclipsed.  Even at $\phi_o = 0$
it is almost unobscured, which accounts for the relative
shallowness  of mid-eclipse.  Subsequent occultation causes the
distortion and delay of mid-eclipse.

At quiescence, the bright spot is  immersed  in the region
$\phi_o$ = -0.019 to -0.002 (Smak 1979); i.e.  it is totally eclipsed
by $\phi_o$ = 0.  In contrast, the bright region  undergoing  eclipse
in light curve B is more extensive, not becoming fully eclipsed
until $\phi_o$ = 0.03.  However,  the emergences of the quiescent
bright  spot  and  the  bright  superhump  region  occur  at
approximately the same orbital phase:  i.e.  the  radius of the
following lune is similar at quiescence and supermaximum.

In  summary,  when  $\phi_s$ = 0 at $\phi_o$ = 0.87,  the  excess
brightness  which  gives  rise to the superhump  arises  in  an
extensive area of the  following  lune  of  the accretion disk,
extending from roughly the position of the quiescent  hot  spot
to  the  limb of the disk.  Even when $\phi_s$ = 0 at $\phi_o$ = 0.46, this
same region has  an  excess  of  brightness over the preceeding
lune,  but  its  effect is less dramatic  in  the  vicinity  of
mid-eclipse.


3.    MODELS OF THE SUPERHUMP PHENOMENON

3.1  General Remarks

The observational properties reviewed  in  Section  2 have
become known only gradually over the past  decade.  At various
stages,  with only partial information available, models of the
superhump pehomena  have  been  proposed:  some  authors  have
offered  more  than one model.  The available data are now very
extensive  and rich  in  information;  the  requirements  of  a
convincing  model  are  correspondingly  more  demanding.
Confronted with  this  mass  of  information none of the existing
models is satisfactory.

3.2  Critique of Existing Models

Vogt's  (1981) discussion of the various models eliminates
most from further consideration. The essence of his argument is
that any successful model must obviously be able to account for
the absorption  line $\gamma$-velocity variations seen in Z Cha during
supermaxima.  The inadequacy of previous models prompted him to
propose alternative structures:

(a)  The white  dwarf is magnetic and rotates with the observed
period $P_{beat}$.  Superhumps arise  from  the presence of a bright
accretion pole modulated in brightness by variations in rate of

mass transfer as the secondary revolves around the primary (i.e. a significant amount of material has to pass straight from the secondary to the pole without proceeding through the disk). The variation in $\gamma$-velocities is attributed to gas flow through the accretion column: special geometry is required to allow negative $\gamma$-velocities to be observed.

(b) An eccentric accretion ring or disk, whose line of apsides rotates in an inertial frame with period $P_{beat}$, is established immediately prior to a supermaximum. Superhumps arise from the resulting variation in kinetic energy acquired by the mass transfer stream as its length varies from the secondary to the bright spot during an orbital cycle. An eccentricity e $\sim$ 0.6 is required and periastron of the ring needs to be $\sim 1/2$ $r_d$. The $\gamma$-velocity variation results from the non-zero radial velocity component of gas in the ring seen projected against the bright central region of the inner accretion disk (or primary). Variations in strength of the absorption lines (S1) are attributed to supposed higher densities of ring material near periastron.

As we have seen in Section 2.5 that the extra luminosity in the superhump is located in the outer parts of the accretion disk, Vogt's model (a) is eliminated.

Vogt's model (b), until now the most promising empirical model (although difficult to accept on dynamical grounds) is also contradicted by the observations: our eclipses show no evidence for the bright spot occurring on an eccentric ring. On the contrary, during supermaximum the bright spot remains at almost the same radius vector that it occupies at quiescence. The most direct way of seeing the conflict of the eccentric ring model with observation is to consider eclipse A of Figure 6. At this time, according to Vogt's model (see the right hand diagram in his Figure 9: Vogt 1981) the major axis of the elliptical ring is approximately perpendicular to the direction of the observer. At eclipse, the initial stage of ingress would be that portion of the eccentric ring ner periastron, or the accretion disk lying within it. Yet the ingress phase of eclipse A starts at least as early as that of quiescence, showing that luminous gas in that direction extends to $\sim 80$ percent of the Roche radius of the primary. If this is required to be periastron, the eccentric ring would have to extend well outside the Roche limit and could not maintain its existence for the two weeks of supermaximum.

Two further objections to the eccentric ring hypothesis are the absence of observational evidence for the initial large pulse of mass transfer from the secondary, needed to establish the ring, and the absence of a naturally occurring explanation for the 180 deg. phase shift (P9) of the late superhumps.

As we have concluded that the superhump, at least when it is eclipsable, arises from an increase of luminosity in the region of the bright spot, it is necessary to reconsider the model proposed by Papaloizou & Pringle (1979). In this model the supermaximum arises from a greatly increased rate of mass transfer, the which is modulated as a result of a small orbital eccentricity. In order to understand the lack of orbital modulation of the superhump (P7) it is postulated that the entire outer rim of the accretion disk varies in brightness as a result of the modulation in mass transfer. The line of apsides of the orbit rotates with period $P_{beat}$.

The areas in which the model in its present form appears inadequate are as follows:

(i) If the absorption lines (S2) arise in the vicinity of the bright spot, their radial velocity would contain components of both the orbital velocity and the disk rotation. This does not provide a constant, displaced $\gamma$-velocity nor the normal orbital velocity amplitude.

(ii) Alternatively, if the lines are formed by gas at the nearside rim of the disk, seen in projection against the bright spot or a combination of the spot and centre of the disk, then at least in the period $0.4 \lesssim \phi_o \lesssim 0.6$ of Vogt's March 28/29, 1978, spectra, negative velocities should have occurred, whereas they were always positive.

(iii) If the entire rim of the accretion disk is modulated in luminosity, any lines produced by absorption against such a background should have $\gamma \sim 0$.

(iv) Photometric properties (P9) and (P10) and the $dP/dt$ behaviour in (P6) do not have a ready explanation.

Although the Papaloizou & Pringle (1979) model remains attractive and may increase its plausibility when the origin and behaviour of the narrow absorption lines are more fully explored, there is an alternative model which appears capable, or at least pregnant with the possibility, of explaining all of the photometric and spectroscopic properties of the superhumps.

## 4.   AN INTERMEDIATE POLAR MODEL FOR SUPERHUMPS

## 4.1 Magnetic Fields:  Polars and Intermediate Polars

Over the past six years it has become increasingly evident that the magnetic fields of the white dwarfs in some cataclysmic variables are strong enough at least partially to disrupt the inner parts of their accretion disks.  In such systems accretion onto the white dwarf is directed by the field lines onto one or both magnetic poles, producing an accretion column close to the stellar surface. The most extreme members of this class are the Polars, in which fields $\sim 10^7$ gauss cause synchronous rotation of the primary (i.e. $P_\tau = P_o$) through magnetic interaction with the secondary.  Systems with smaller fields could show various degress of asynchronism:  such objects have been identified and referred to as Intermediate Polars (Warner 1983a,b).  In these, rotation of the primary is made evident by periodic modulation of the optical and/or X-ray flux; this implies anisotropic emission of radiation from the primary. Models of the Intermediate Polars (Warner 1983a,b) require a beam of X-rays, presumably from the accretion column, which sweeps around and irradiates the disk.  Non-axisymmetric structures such as the bloated disk region in the vicinity of the bright spot, may intercept the beam and reprocess it to cause optical brightness variations with a synodic period $P_{syn} = P_o P_\tau / (P_o - P_\tau)$ .  Alternatively, any front-back asymmetry of the accretion disk, as viewed by us, leads to optical modulation with period $P_\tau$. Both $P_{syn}$ and $P_\tau$ are detectable in some systems. The observations appear to demand a beam whose angular width (as seen from the primary) is considerably less than that resulting simply from a bright spot on the surface of the primary (Warner 1983b).

The Intermediate Polars and the Polars are in a state of steady accretion; continuous quasi-radial infall of gas onto the white dwarf generates a detectable hard X-ray flux.  If in the dwarf novae accretion onto the primary is intermittent, any beam phenomena might be expected to be evident only during the periods of high mass inflow, i.e. during outbursts (and only from the moment when the inflowing material reaches the inner edge of the accretion disk - after the rise to maximum).

Table III shows the incidence of magnetic field related phenomena among cataclysmic variables with known orbital periods (Ritter 1983).  The increase towards short periods of the fraction of stars possessing strong magnetic fields is marked. Below the "period gap" one third of the systems have strong fields; only one star (EX Hya) is classified as an

Table 3

| Period Range (d) | Number of Stars | Polars | Intermediate Polars | SU UMa Stars |
|---|---|---|---|---|
| 0.056 - 0.088 | 27 | 9 | 1 | 13(+1?) |
| 0.118 - 0.204 | 27 | 2 | 5(+2?) | 1 |
| 0.206 - 2.00 | 27 | 0 | 2 | 0 |

Intermediate Polar. If there is a broad spread of field strengths, as there appears to be at longer orbital periods, then, unless we have yet to discover a new class of short period binaries, the short period Intermediate Polars must be contained within (or are) the SU UMa stars. EX Hya itself is an eruptive binary but with an amplitude only half of that of a typical SU UMa star. It also has a stronger X-ray flux than SU UMa stars of the same apparent magnitude. These properties place it intermediate in properties between the Polars and the SU UMa stars.

If the SU UMa stars represent a class of object with intermediate magnetic field strengths their X-ray behaviour during outbursts may differ significantly from that of ordinary dwarf novae. Below the period gap only three dwarf novae appear at present to be free of SU UMa characteristics - HT Cas, V2051 Oph and T Leo. Although further observations may show that they do possess supermaxima (see SU UMa itself, which was recently for three years in a state where supermaxima no longer appeared), a study of the X-ray, polarisation and radio emission properties of these three stars, as compared with the SU UMa systems, may demonstrate the presence of larger magnetic fields in the latter. In this connection it is significant that in a search for 4.75 GHz emission from dwarf novae (Benz et al. 1983) SU UMa during (normal) outbursts was the only object to be detected. Similar emission has been found in AM Her (Chanmugam & Dulk 1982).

The absence of optical polarisation in the SU UMa stars parallels the situation in the Intermediate Polars, where the fields are thought to be sufficient only to produce polarisation in the infrared (Lamb & Patterson 1983).

4.2 SU UMa stars as Intermediate Polars

In seeking an Intermediate Polar model for the superhump behaviour of SU UMa stars we must choose the rotation period $P_r$ of the primary, i.e. the period at which the illuminating beam sweeps around in an inertial frame. Of the two obvious possibilities, $P_r = P_s$ or $P_r = P_{beat}$ (Table 1), the first would imply nearly synchronised rotation of the primary, would produce a model like TV Col (Warner 1983a,b) and not too unlike EX Hya (Warner 1983a), and therefore at first sight would appear the most attractive proposition. Such a model resembles that proposed by Patterson (1979) but with the addition of beam illumination of the disk.

In the model with $P_r = P_s$ the beam illuminates the region of the bright spot only once per beat cycle. In Z Cha our observations (Section 2.5) show that such spot illumination must occur at the time when $\phi_s \sim 0$ at $\phi_o = 0$, i.e. in a situation similar to that of Vogt's March 28/29, 1978, spectroscopic observations (S2). This incidentally requires that superhump maxima always occur when the beam is illuminating some part of the following lune of the disk. The model then fails to explain Vogt's spectroscopic results (S2) for reasons similar to those given for the Papaloizou & Pringle model (Section 3.2, criticisms (i) and (ii)). Furthermore, as the production of superhumps in this model relies on a strong asymmetry as seen by the observer, their independence of inclination (P5) is not expected.

On the other hand, a model with $P_r = P_{beat}$ possesses features worthy of closer attention. In this model the beam from the slowly rotating primary is almost fixed in direction through one orbital revolution. During any orbital revolution asymmetries of the disk, e.g. the region of the bright spot, run into the beam and become more luminous by reprocessing. In the next orbit the beam has rotated progradely through a small angle so the illumination of the asymmetry occurs at a later orbital phase. With $P_r = P_{beat}$ we obviously obtain $P_s$ for the recurrence period of the reprocessing.

## 4.3 Spectroscopic Properties of the Model

At quiescence the narrow absorption lines in Z Cha (Section 2.2) are thought to arise in gas at the edge of the accretion disc absorbing light from the central radiating source which, from quiescent eclipse profiles (Cook & Warner 1983), provides about half of the light in the system. At supermaximum the radiating source consists of two components: the accretion disc as a whole and the region illuminated by the beam. The relative contributions of these is not known; from the height of the superhumps, early in the outburst beam reprocessing contributes not less than one third of the total light when the bright spot region is in the beam. However, beam illumination of the symmetrically distributed gas in the disc produces a constant luminosity so it is possible that there is always a significant fraction of the optical radiation arising from beam reprocessing.

The off-centre beam illuminated material provides an additional continuum source against which gas at the rim of the disk produces absorption lines whose strengths will vary in phase with the superhump profile (S1). As the beam rotates with period $P_{beat}$ so the systematic variations in radial velocity of the absorption lines will contain a component with period $P_{beat}$.

The model is best illustrated by considering the circumstances of Z Cha on the two nights when Vogt made his radial velocity measurements: Figure 7. The diagram shows the relative positions of the components at the times of superhump maxima as determined from the orbital elements and Vogt's superhump ephemeris. We do not know the direction of the beam on either occasion: only that it rotated 173 deg. between the two times. We also do not know what the beam illuminates: whether it is the bright spot and a raised rim downstream or perhaps a penetrating mass transfer stream. The beam positions are therefore tentative, but that for March 28/29 is chosen to be consistent with our deductions (Section 2.5) of the disk brightness asymmetry in similar circumstances.

As seen in Figure 7, on March 27/28 the absorption lines are produced by gas at the nearside edge of the disk as seen against the illuminated back rim, but on March 28/29 it is the front rim that is both illuminated and absorbing. In the former case the line of sight to the observer will intercept less absorbing gas than the latter: this explains the weaker absorption lines observed by Vogt on March 27/28 (he states that the line strengths on this night near $\phi_s = 0$ are only as great as at $\phi_s \sim 0.65$ on the following night).

During one orbital rotation, as the beam is almost fixed in direction, the section of disk rim that contributes the absorption lines remains fixed as seen by the observer. This produces the constant but deviant $\gamma$-velocity on which is superimposed the normal orbital variation. If the source of continuum radiation were only illumination by a narrow progradely rotating beam the observed radial velocity would be

$$V(t) = K_1 \sin\left(\frac{2\pi t}{P_o}\right) + V_d \sin\left[\frac{2\pi(t-t_o)}{P_{beat}}\right],$$

(2)

where $V_d$ is the Keplerian velocity at the outer edge of the disk, $t$ is measured from conjunction and $t_o$ is the time when the beam points directly at the observer.

For Z Cha, with $M_1 = M_\odot$ and $r_d = 1.5 \times 10^{10}$ cm (Vogt 1981, Cook & Warner 1983), we have $V_d = (GM_1/r_d)^{1/2} \sim 1000$ km s$^{-1}$. Therefore $\gamma$-velocity variations with the range measured in Z Cha are feasible. Because of the finite beam width, the $V_d$ to be used in (2) should be an average over the appropriate region of the rim; continuum provided by the symmetric portions of the disk should also be considered. The net effect will be asymmetric line profiles which, although in Z Cha simulate an approximately constant $\gamma$-velocity during one orbit, could, in other systems, generate a distorted radial velocity curve with non-constant and/or incorrect apparent $K_1$.

## 4.4 Photometric Properties of the Model

The profile of the superhump arises from convolution of the radiation pattern of the beam with the silhouette of the disk as seen from the primary. Any persistent structure in either of these will produce features recurring at period $P_s$ (P10). Systematic changes in either, through outburst, can account for the secular variation in hump amplitude and profile (P4).

As all supermaxima produce superhumps (P5) there must be two accreting poles in all systems. Through outburst the optically thick disk prevents detection of one of the beams. Near the end of outburst, however, at least in V436 Cen and VW Hyi (P9), the inner regions of the disk are sufficiently transparent to show the illuminating effects of the second beam (the outer regions must be optically thick because an orbital hump is also present at this time) which produces superhump features 180 deg. phase-shifted from the regular humps.

In highly inclined systems such as Z Cha the strong front-back asymmetry will modify the profile of the superhump according to the orbital phase at which they occur. For $\phi_s = 0$ near $\phi_o = 0$ we view mostly back illumination of the disk asymmetry. When $\phi_s = 0$ near $\phi_o = 0.5$ we see "reflected" light from the more visible far side of the disk. This could account for the changes in superhump profile around the beat cycle (Z2).

The distorted eclipse egresses (Z3, Section 2.5) imply that the region of the bright spot (even when not in the beam) is brighter than at quiescence by factors typically $\sim 2$. Possible explanations are that the rate of mass transfer is increased during supermaxima (perhaps due to heating of the atmosphere of the secondary by the outburst) or that the bright spot region processes not only a periodic component from the rotating beam but the surface of the primary. From the lack of

orbital   modulation   (Z3)   the latter explanation   appears   more
likely.


## 5.   THE SUPERHUMP PERIOD DRIFT

Finally we   consider   the   non-linear ephermerides for the
superhump maxima (P6) for which   no   model   yet   proposed has a
natural   explanation.   There   are   at   least   three   ways   of
interpreting the second order term $CE^2$ in equation (1):

(i)   Linear   decrease   in   superhump   period $P_s$.   In this case
$C = 1/2 \ PdP/dt$ give $dP/dt \sim -2 \times 10^{-7}$. By the final cycle $E_f$ of
the   superhump   period through outburst the   accumulated   phase
shift is $CE_f^2 / P_s \sim 1 \ 1/2$ cycles.

(ii)   Quadratic phase   shift   from   a constant superhump period
$P_s$.   If   the   underlying clock is   constant   but   there   is   a
displacement $\delta(E)$ in $T_{max}$ which is zero at the beginning (E=0)
and end ($E = E_f$) of   an   outburst   (see Figure 2 of Vogt 1983)
then   $\delta(E) = CE \ (E - E_f)$; the maximum   displacement   occurs   at
$E_m = 1/2 \ E_f$ and is $\delta(E_m) = 1/4 \ CE_f^2$ .   For the systems listed in
Table 2, $\delta(E_m)$ ranges up to $\sim 90$ deg. of orbital revolution.

This may   be   interpreted either as a variation in profile
of the superhumps (Patterson   1979)   of unknown origin (and not
caused by noticeable asymmetries in the superhump profile – see
Figure   9   of Haefner et al.   1979)   or,   in   the   Intermediate
Polar model proposed   in   this paper, as systematic movement of
the illuminated asymmetries during outburst.   For   example, an
initial   expansion of the disk, followed by   contraction   would
cause the   region   of the bright spot at first to move near the
line of centres of the stars and then to move away.

(iii)   In the Intermediate   Polar model a linear increase in $P_r$
would produce a linear decrease   in   superhump period $P_s$.   Then

$$\frac{dP_r}{dt} = - \frac{2CP_r}{P_0^2} \sim 2 \times 10^{-3}$$

for the   systems   listed   in   Table 2.   An increase in rotation
period $P_r$ could occur if the   magnetic   field   of the primary is
confined   to   its   outer   non-degenerate   envelope.   Then, as a
result of heating of the envelope by accretion during outburst,
expansion could lead to a slower   rotation.   Energetically this
appears feasible:   if energy $\Delta E$ is deposited into the envelope,
causing expansion expansion $\Delta R$ of the envelope   mass   $\Delta M$,   then
conservation of energy and angular momentum give

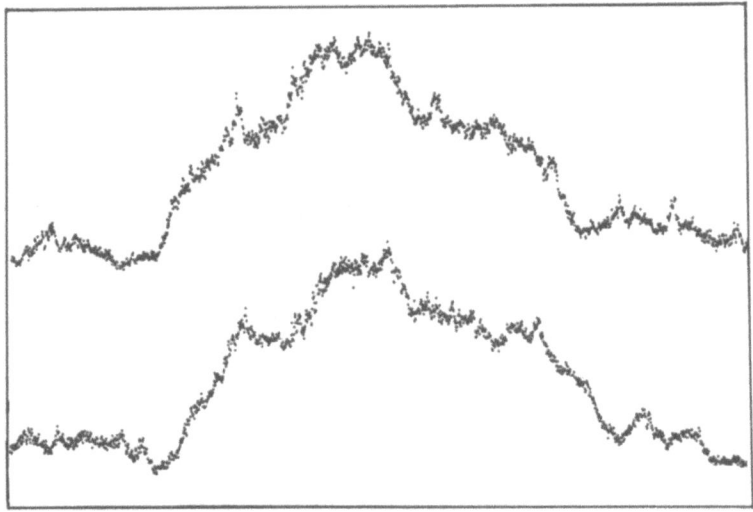

Figure 1.  Recurrent superhump structure in VW Hyi.  Adapted from
Figure 2 of Warner (1975).

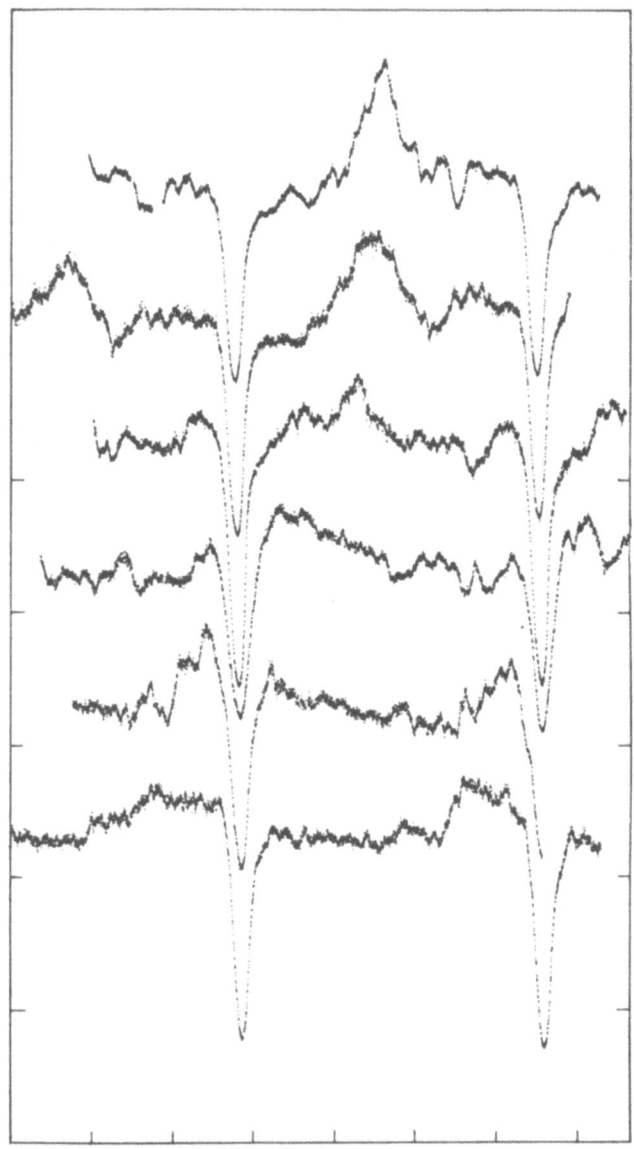

Figure 2. Superhumps in Z Cha. All light curves have been normal-
ised to approximately the same height in regions free of eclipse
or superhump. Ordinate carets are zero intensities for the various
light curves. Abscissa carets mark intervals of $0^d.0200$. From
the top down the dates of observation were 14 December 1982, 16
December 1982, 23 February 1980, 20 February 1980, 17 December
1982, 19 December 1982.

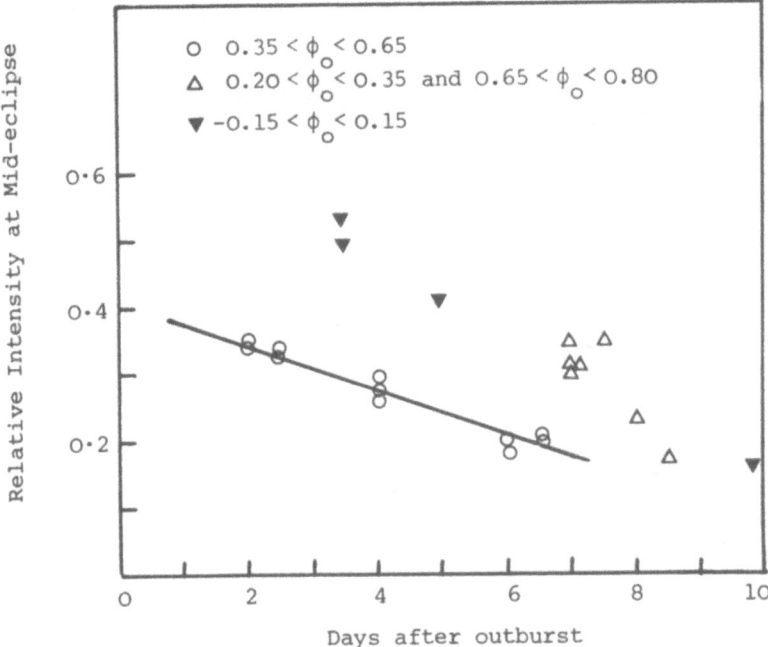

Figure 3.  Relative intensity of Z Cha at mid-eclipse versus days
from start of outburst, segregated according to orbital phase $\phi_o$
at which superhump maximum occurs.

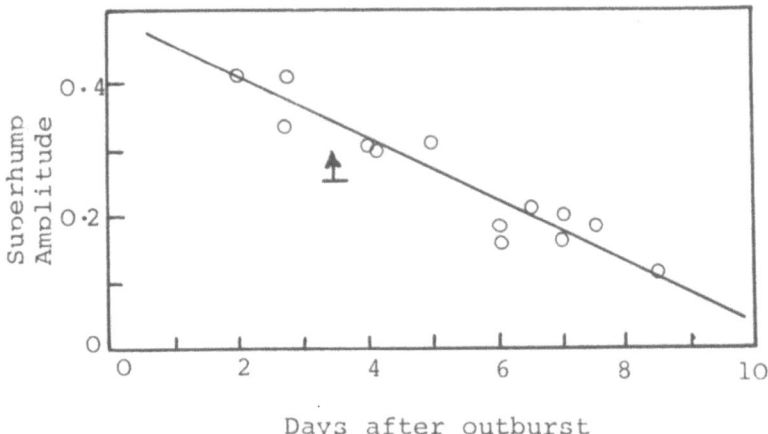

Figure 4.  Relative amplitude of superhump in Z Cha as a function
of days since start of outburst.

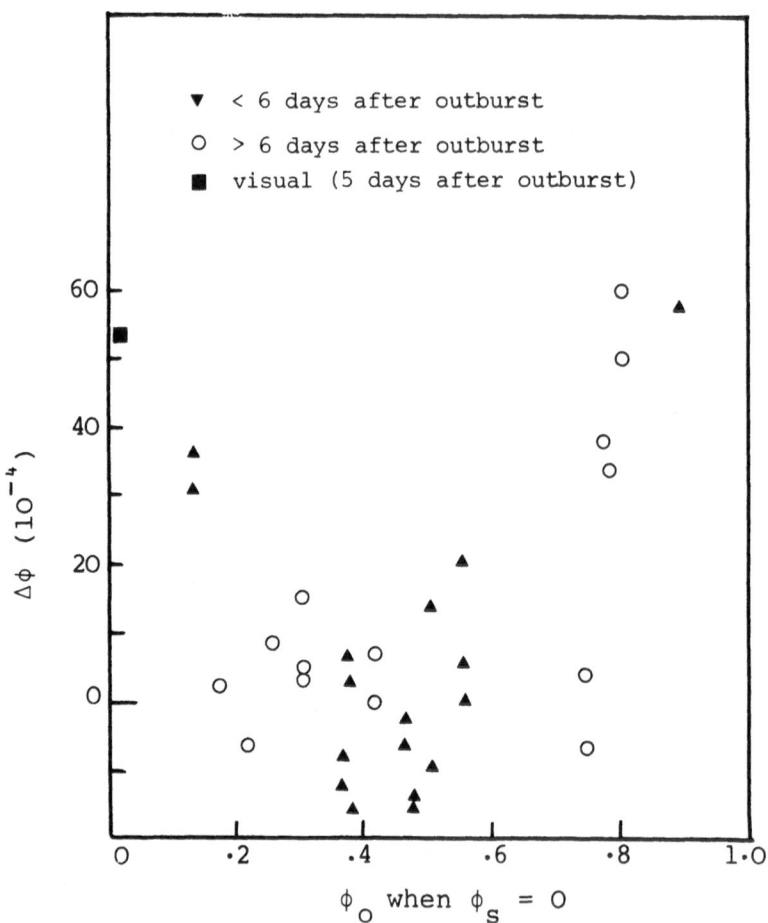

Figure 5.  Observed minus calculated phase of mid-eclipse in Z Cha
as a function of orbital phase at which maximum of superhump
occurs.

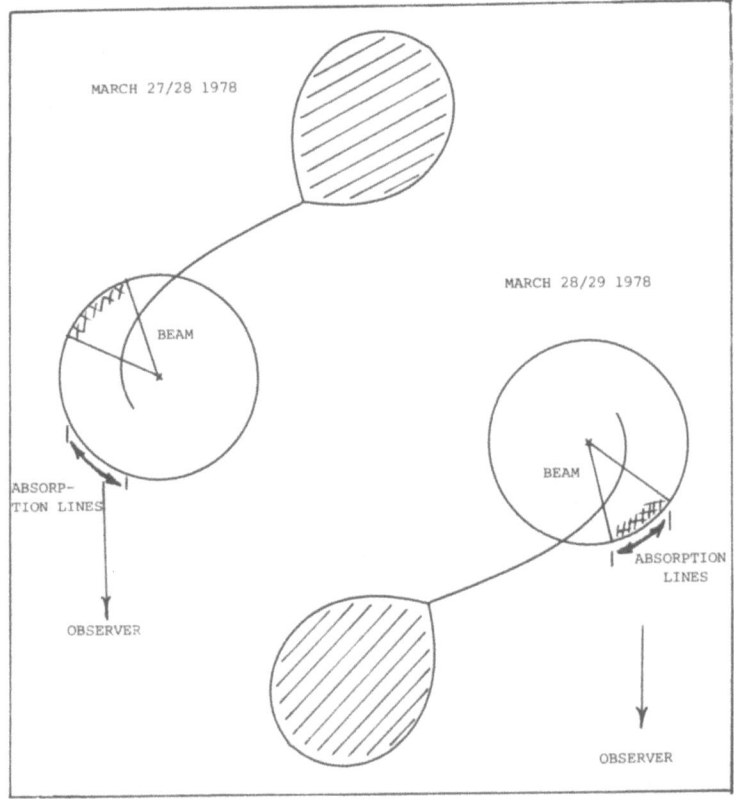

Figure 7. Formation of absorption lines according to the Intermediate Polar model of Z Cha.

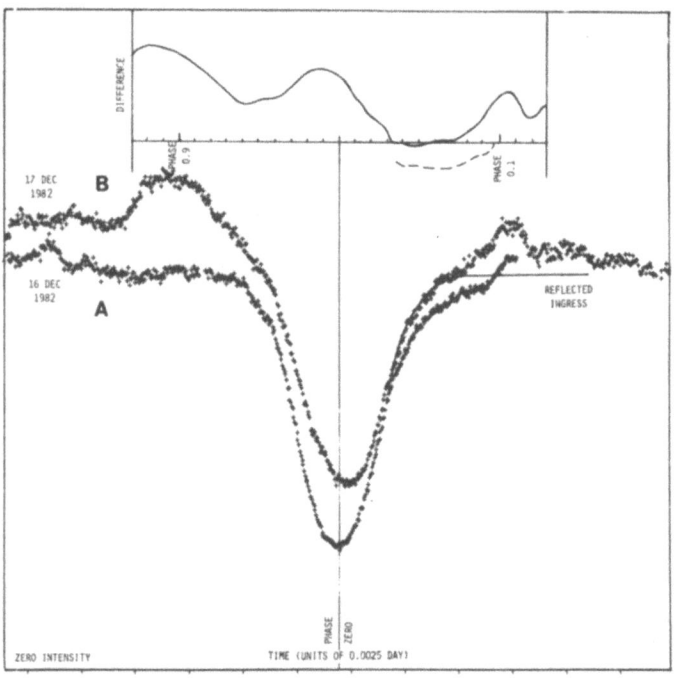

Figure 6.   Details of eclipses in Z Cha.

$$\Delta E \sim E \cdot \frac{\Delta M}{M} \cdot \frac{\Delta P_r}{P_r}$$

where E is the binding energy of the primary. By the end of an outburst

$$\frac{\Delta P_r}{P_r} \sim 2 \times 10^{-2}$$

so with $\Delta M/M \sim 10^{-7}$ we have $\Delta E \sim 10^{39}$ ergs which is only one percent of the total energy radiated during a supermaximum (Warner 1976). Therefore if a small fraction of the accretion energy of outburst goes into steadily expanding the outer envelope of the priamry ($\Delta R/R \sim 0.01$), the envelope and its magnetic field and associated accretion columns and irradiating beams may slip backwards relative to the core of the rotating primary, later to relax during quiescence.

In contradiction with this, however, is the fact that, if the outer envelope so easily slides over the core, then it should be rapidly spun-up by the accreted angular momentum.

## 6. CONCLUDING REMARKS

This review of published and personal information on the properties of SU UMa stars has concentrated on the structure underlying the superhump phenomenon. It may have raised more questions than it has answered. The subject is in need of rigorous speculation on a number of outstanding problems, viz.: What is the cause of the different types of outbursts? If the Intermediate Polar model is correct why does magnetically controlled accretion only appear during supermaxima? What is the evolutionary significance of very slow rotation of the primaries in the SU UMa stars and why are their rotation periods inversely correlated with orbital period (Table 1)?

## 7. ACKNOWLEDGEMENTS

The writer is indebted to Peter Eggleton and the Institute of Astronomy, Cambridge for support and hospitality.

REFERENCES

Bailey, J., 1979. Mon.Not.R.Astr.Soc., **188**, 681.
Bateson, F., 1979a. Pub.Var.Star Section Roy. Astr.Soc.
    New Zealand No. 7, p. 5.
Bateson, F., 1979b. Ibid. p. 29.
Bateson, F., 1981. Ibid, No. 9, p. 2.

Benz, A.O., Furst, E. & Kiplinger, A.L., 1983. Nature,
    **302**, 45.
Bond, H.E., Kemper, E. & Mattei, J., 1982. Astrophys.J.Lett.,
    **260**, L79.
Chanmugam, G. & Dulk, G.A., 1982. Astrophys.J.Lett.,
    **255**, L107.
Cook, M.C. & Warner, B., 1981. Mon.Not.R.Astr.Soc., **196**, 55P.
Cook, M.C. & Warner, B., 1984. Mon.Not.R.Astr.Soc., **207**, 705.
Haefner, R., Schoembs, R. & Vogt, N., 1979. Astron.Astrophys.,
    **77**, 7.
Lamb, D.Q. & Patterson, J., 1983. IAU Colloq. 72, "Cataclysmic
    Variables and Related Objects", ed. Livio & Shaviv. p. 229.
Papaloizou, J. & Pringle, J.E., 1979. Mon.Not.R.Astr.Soc.,
    **189**, 203.
Patterson, J., 1979. Astron.J., **84**, 804.
Patterson, J., 1981. Astrophys.J.Suppl., **45**, 517.
Patterson, J., McGraw, J.T., Coleman, L. & Africano, J.L.,
    1981. Astrophys.J., **248**, 1067.
Ritter, H., 1983. Catalogue of Cataclysmic Binaries, 2nd
    Edition. Max Planck Institute.
Robinson, E.L., 1976. Ann.Rev.Astr.Astrophys., **14**, 119.
Schoembs, R., 1979. ESO Messenger No. 16, p. 15.
Schoembs, R., 1982. Astron.Astrophys., **115**, 190.
Schoembs, R. & Vogt, N., 1980. Astron.Astrophys., **77**, 7.
Semeniuk, I., 1980. Astron.Astrophys.Suppl., **39**, 29.
Smak, J., 1979. Acta Astr. **29**, 309.
Stolz, B. & Schoembs, R., 1981. I.B.V.S. No. 1955.
Van Paradijs, J., 1983. Astron.Astrophys., **125**, L16.
Vogt, N., 1974. Astron.Astrophys., **36**, 369.
Vogt, N., 1981. Astrophys.J., **252**, 653.
Vogt, N., 1982. Mitt.Astr.Gesell., **57**, 79.
Vogt, N., 1983. Astron.Astrophys., **118**, 95.
Vogt, N., Schoembs, R., Krzeminski, W. & Pedersen, H., 1981.
    Astron.Astrophys., **94**, L29.
Warner, B., 1974. Mon.Not.R.Astr.Soc., **168**, 235.
Warner, B., 1974. Mon.Not.R.Astr.Soc., **170**, 219.
Warner, B., 1974. In IAU Symp. No. 73, "Structure and
    Evolution of Close Binary Systems", ed. P. Eggleton
    et al. Reidel. p. 85.
Warner, B., 1983a. In IAU Colloq. 72 "Cataclysmic Variables
    and Related Objects," ed. Livio and Shaviv, p. 155.
Warner, B., 1983b. In "Cataclysmic Variables and Low Mass
    X-Ray Binaries," Cambridge, Mass. In Press.
Warner, B., 1983c. I.B.V.S. No. 2397.
Whelan, J.A.J., Rayne, M.W. & Brunt, C.C., 1979. In IAU
    Colloq. 46, "Changing Trends in Variable Star Research,"
    ed. Bateson, et al., University of Waikato, p. 39.

# INTERACTING BINARIES - SUMMING UP

Virginia Trimble

| Astronomy Program | and | Department of Physics |
|---|---|---|
| University of Maryland | | University of Calif. |
| College Park, MD 20742 | | Irvine, CA 92717 |
| U.S.A. | | U.S.A. |

## 1. HISTORICAL INTRODUCTION

Our subject is, in one sense, a very old one (ancient and boring, many of our colleagues in extragalactic astronomy would say). John Michell (1767) really invented binary stars more than two hundred years ago, when he pointed out in Phil.Trans.Roy.Soc. Lond. that the number of close pairs of stars in the sky is so large that it makes sense only if most of the pairs are physically connected and not just chance superpositions. In another sense, the subject of interacting binaries is quite young. The first international meeting exclusively on them (IAU Colloquium No. 6) took place in 1969 in Elsinore, Denmark. It bothers me a little to see that none of the official participants in that meeting (Table 1) has been with us here these past two weeks (though Johannes Andersen, then a graduate student at Copenhagen, and I as a new postdoc were present as gate crashers). Evidently, 14 years is a generation of binary star astronomers, at least in their guise as conference participants. Table 1 summarizes some of the highlights of that meeting. Tables 2 and 3 do the same for the next two large, international meeting devoted to the full range of types of interacting binary systems, IAU Symposia 73, held here in Cambridge in 1975, and 88, held in Toronto in 1979. Many of the participants from those gatherings have been here, and, since binary star gatherings appear to occur about every four years, many of us can expect to be together again somewhere in 1987. Having now joined the august company of introducers and summarizers of these meetings, I look forward eagerly to being, in due course, the subject of one of John Faulkner's parodies.

*P. P. Eggleton and J. E. Pringle (eds.), Interacting Binaries, 393–410.*
© *1985 by D. Reidel Publishing Company.*

## 2.   STATISTICS AND OTHER FORMS OF UNSATISFACTORY DATA

J. Andersen (Copenhagen) told us at the beginning of the meeting that he doesn't believe in statistics, and we shouldn't either.    Now  the English word "believe" carries three related but distinct meanings.     You can say that you don't believe in tactical nuclear weapons, meaning that you don't think they are a good thing.  Or you can say that you don't believe in transubstantiation,  meaning that  you don't think it is true. Or you can say  that  you don't believe in ghosts, meaning that they don't exist.  Andersen undoubtedly had  in  mind the first of these three meanings, but I would like  to  show  you  that, sometimes,  we  shouldn't  believe  in  statistics in the third sense either......they don't even seem to exist.

Consider  two  cases, the distributions of  binary  system separations and of binary mass ratios and how they have evolved with  time.  Kuiper (1935) was the first to  try  to  find  the total distribution  function.   He  had  data  on  many  visual binaries  and  common  proper motion pairs, but only a very few spectroscopic systems, and there  was  no  overlap  at  all  in separations  between  the two groups.  Nevertheless, he bravely drew a dotted  line  between  the  two,  and suggested that the binaries were a single population of common origin.   Subsequent events  have  justified his dotted line (if not  his  cosmogonic conclusion).  A  later  sample,  vintage 1970 (data from Batten 1968 and Finsen and Worley  1970) shows many more spectroscopic systems, and the gap partially filled in.   The 108 systems that R.  Griffin (Cambridge) presented  here  fill  in  that  gap considerably  further.   The  rapid drop at  large  separations means  that Finsen and  Worley  (1970)  tabulated  only  visual binaries with  measured orbits, not common proper motion pairs. And, finally, a  nearly  complete  sample  of  binaries  for  a restricted  group  of  stars  (Abt & Levy 1976, for F-G main sequence stars near the sun) does indeed show  a single, fairly featureless  distribution.   This  statistic  at  least exists, though  you  may  not  want  to  believe  any  particular interpretation of it.

On  the  other hand consider the  corresponding  evolution with time of  the  deduced  distribution of binary mass ratios. Again the first word belongs to  Kuiper  (1935)  who observed a fraction  about  0.06  of  all  systems with mass ratio between 0.0-1.0, with the fraction in each  mass ratio bin of width 0.1 rising to about 0.18 in the 0-0.1  bin.  Jaschek (1972) finds a similar result but with an even higher proportion of systems at small mass ratio.  But suddenly something odd happens.  Samples collected through the mid 1970's show two comparable  peaks  at high  and  low  mass  ratio (Trimble 1974).  The most recent samples,  however,  for 0-type spectroscopic systems (Garmany <u>et</u>

al. 1980) and for very close binaries (Lucy & Ricco 1979) show only the peak close to mass ratio of unity. I am not at all sure that statistics on the distribution of mass ratios can currently be said to exist.

But even non-statistical data can be bad. J. Solheim (Tromso) drew our attention to the case of AM CVn, for which a seemingly-thorough analysis (Patterson et al. 1979) had given a rate of increase of the orbital period of 1.1 x $10^{-5}$ $yr^{-1}$, about a thousand times the theoretical prediction (Faulkner et al. 1972) based on mass transfer induced by gravitational radiation. But the enormous value now seems attributable to a miscounting of the number of orbit periods elapsed between widely separated data sets. That Herculean efforts of calculation failed utterly to "predict" the incorrect observed value goes some way to restore one's faith in theory! Solheim's current best-fit value, $\dot{P}/P = -9.4$ x $10^{-8}$ $yr^{-1}$, is heavily dependent upon the earliest data point and should perhaps be regarded as an upper limit. It still disagrees with the model value, in sign as well as amplitude; but, as R. Webbink (Illinois) remarked, all close binaries show inexplicable period changes!

And then there is SS Cyg, a prototype dwarf nova, and one of the few in whose spectrum two line sets are clearly visible, permitting unambiguous determination of the mass ratio. Except that very careful radial velocity measurements using conventional techniques yield a K for the absorption component of about 120 $\pm$ 5 km/sec (Cowley et al. 1980; Walker 1981), while equally careful application of a cross-correlation technique gives 153 $\pm$ 2 km/sec (Stover et al. 1980). E.L. Robinson (Austin) made the situation seem even more puzzling with new data that show K increasing to nearly 200 km/sec (via the cross-correlation technique) when the system is at maximum light. This can be explained by the light centroid shifting as the cool star is heated by the brightening of the disk, as well as by other more complex mechanisms, none of which really resolves the earlier discrepancy.

Finally, data can be perfectly good and still not solve the problem you had in mind. Andersen, for instance, pointed to the problem of measuring stellar composition (especially helium abundance) by matching observed luminosities and effective temperatures to calculated evolutionary tracks, and requiring both members of a binary system to have the same age and composition. KM Hya illustrates one sort of problem: the stars are equally well fit by Z = 0.04, X = 0.7 and Z = 0.02, X = 0.8 (and we already knew the truth must be somewhere in that ballpark). W Peg illustrates another: the cooler star can only be fit by X = 0.8 models, and the hotter

star only by X = 0.7 models (given the observed masses). This
is silly. Anderson is inclined to blame the mixing length
formalism, though similar problems appear for fairly early type
stars.

Part of the problem in this and other cases is the extreme
sensitivity of the parameter we want to determine to the
quantity available from observations. For double-line
spectroscopic binaries, masses scale as velocity amplitude
cubed. And, as C. Whyte (Sussex) reminded us, the mass
transfer rate derived for a cataclysmic variable in some models
scales as the fourth power of the disk temperature fitted to
the ultraviolet continuum. Caveat astronomer!


3.    WHICH OBJECTS BELONG TO US?

Several speakers began their talks by claiming that the
objects they were going to discuss were not interacting
binaries, or maybe not binaries at all, or at least it didn't
matter much. Just which classes of phenomena really do belong
to this conference, and how do we know? The cataclysmic
variables, discussed by R. Wade (Cambridge) are one of the
cleanest cases: there are clearly two stars, they clearly
interact via a gas stream, and the phenomena we observe would
clearly not happen without two stars and the interaction.
Barium stars are an odder case: virtually all are
spectroscopic binaries (McClure 1983) but this doesn't seem to
help us a bit in explaining their chemical peculiarities
(Dominy & Lambert 1983). And then there are the blue
stragglers, some of which are clearly spectroscopic binaries
(Peterson et al. 1983) and some equally clearly not
(Stryker & Hrivnak 1983), at least not with amplitudes greater
than 12 km/sec, though most of the models say they should all
be post-mass-transfer pairs. Some are perhaps merged pairs and
thus rapid rotators as observed (Deut 1966). Others may have
been left with velocity amplitudes too small to see, according
to A. Collier (Royal Greenwhich Observatory).

Table 4 is a highly prejudiced division of phenomena
sometimes blamed on close binary interactions into yes's
(always in binaries and it matters), maybe's (perhaps not
always in binaries; perhaps it matters), and more complicated
situations. I have been heavily influenced by P. Eggleton's
(Cambridge) remarks and the talks of R. Wade, M. Plavec (UCLA),
A. Willis (London), S. Kenyon (Illinois), B. Schaefer (MIT), E.
van den Heuvel (Amsterdam), and N. White (ESTEC). A few types
appear in more than one list. It cannot be helped.

A couple of points came up during the meeting that really don't belong to us at all, but are too interesting to pass by in silence. First, we now seem to have direct observational evidence for the three sorts of white dwarfs predicted by (single star) evolutionary models. R.E. Nather (Austin) remarked on the very high helium abundance in AM CVn, which implies that the mass-losing white dwarf is of the helium type. And J. Truran (Illinois) reported recent spectroscopic analyses of novae and their ejecta showing greatly enhanced abundances of CNO nuclides in some events and O, Ne, Mg, Al etc. in others. Apparently material dredged out of the interiors of carbon-oxygen and O-Ne-Mg white dwarfs is being blown out where we can see it. Second, C. Zwaan (Utrecht) pointed out that, during pre-main-sequence contraction, convective stars are likely to spin down, but radiative ones to spin up, thus enabling us to understand the very rapidly rotating K dwarfs ($v \sin i \gtrsim$ 100 km/sec) found in young clusters by J. Stauffer (Harvard).

## 4. FORMATION

Star formation in general is not very well understood, and this is doubly so for double stars. If anything, the situation has recently become more chaotic. It used to be believed that fission of a single protostar would make close pairs with mass ratios near one, while separate condensation or capture would make wide pairs with small mass ratios (Opik 1924 and many later authors). But Gingold and Monaghan (1983) have recently found that fragmentation of a differentially rotating cloud can make low mass ratio systems. And P. Artymowicz (Warsaw) reported here that accretion on to lumps of 0.2 $M_\odot$ (resulting from heirarchical fragmentation) eventually builds up systems of total mass 1-3 $M_\odot$ and mass ratios very close to one.

Several lines of evidence suggest that some close systems really are born with $M_2/M_1 <$ 1. W. Packet (Brussels) has modeled the early evolution of massive contact (SV Cen) systems, and finds he can match the objects we observe only if the initial mass ratio is 0.5. Second, as A. Willis (London) noted, about a third of known Wolf-Rayet stars are binaries with low mass companions. No plausible evolutionary scheme can turn one member of a binary into a degenerate dwarf while the other is still a ~10$M_\odot$ helium burner; if the companion is a neutron star, we should see (and don't) X-rays produced by accretion of the WR's strong wind; the remaining alternative is a 1-2 $M_\odot$ main sequence star, and thus low initial mass ratio. A similar argument applies to the barium stars with low mass companions. A neutron star is evolutionarily unlikely; a white dwarf should show up via an ultraviolet excess in IUE data, and doesn't in most cases (Lambert & Dominy 1983; unpublished work described at this meeting by H. Bond, Louisiana State), leaving a low mass main sequence companion the most likely alternative.

## 5.    INTERACTIONS OTHER THAN MASS TRANSFER

G. Savonije (Amsterdam) reported preliminary calculations of tidal interactions in close binaries, and C. Campbell (Cambridge) and G. Chanmugam (Louisiana State) discussed magnetic interactions, particularly in AM Her (polar) systems. Both processes tend to circularize orbits and synchronize rotation on time scales short enough to matter for the kinds of systems we see. In the tidal case, convective and radiative envelopes react differently, and we might expect the correlation between binary period and eccentricity to be different for the two, switching at about F3. This effect is not conspicuously present in the data. In the magnetic case, the torque competes with that due to accretion, and a likely outcome is near, but not exact, synchronization. This might show up when one magnetic pole of the white dwarf moves slowly around away from the mass-losing star and accretion switches to the other pole. Just possibly, this is now happening in VV Pup, according to Chanmugam. But probably neither set of calculations is yet ready for direct, detailed confrontation with observations!

## 6.    INTERACTIONS WITH MASS TRANSFER

### A. The Physics of Mass Loss and Accretion
When it comes to mass transfer in close binaries, it is clearly more blessed to give than to receive - or at least easier. The problem of how the secondary copes with everything dumped upon it has been with us since Benson (1970), in his thesis, found that the other half of the Roche lobe was already overflowing when only a few percent of a solar mass had been transferred, and promptly gave up astronomy. R. Webbink (Illinois) made the point that a convective envelope is inclined to shrink as its mass increases, rather than expanding like a radiative one. This goes some way to solve the problem, as well as reducing the discrepancy between results like Benson's and those of Shu & Lubow (1981) who conclude that the secondary can accommodate mass and angular momentum as fast as they arrive. But it is probably not the whole story.

There is, however, observational evidence that the secondary manages quite well, at least some of the time. C. Bailyn (Cambridge) evolved the triple system Lambda Tau

backwards in time for us, demonstrating that the larger orbit
could have remained stable over the life of the system only if
the closer pair had lost less than 50% of the angular momentum
transferred during the rapid transfer phase. And the system
could have evolved completely conservatively. Other systems,
further along in the slow transfer phase, do have combinations
of mass and separation that imply significant angular momentum
loss, at least 44% for AS Eri, and even more for some of the
Wolf Rayet systems, according to Willis.

Once rapid mass transfer ends, many systems appear as
Algol-type variables, with a gas stream flowing from the cool
star to the hot one. The impact of stream on star causes
considerable local heating. This shows up both
photometrically, in cycle-to-cycle changes in eclipse shape,
attributed by E. Olson (Illinois) to a bulge of hot gas around
the equator, and spectroscopically in emission lines coming
from what G. Peters (USC) called the high temperature accretion
region. Bailyn noted that the correlation between Algol mass
ratios and the amount of angular momentum lost from the systems
implies that most of the loss occurs during this slow transfer
phase.

   B. The Nature of Contact Systems
   It is possible, according to the accretion calculations
reported by P. Artymowicz (Warsaw), to form binaries that are
in contact from the very beginning. C. Whyte (Cambridge)
finds, however, that it is very difficult to evolve such
systems so that they continue to resemble observed ones for any
length of time. A major problem is the tendency for them to
spend about half their time in broken contact, detached but
still very close. But, as Whyte, S. Mochnacki (Dunlap
Observatory), and R. Webbink (Illinois) noted, we see very few
systems of this type, although they should be almost as
conspicuous eclipsers as the W UMa contact systems themselves.
V Pup and CN And were suggested as possible examples.

   Whyte favours evolution into contact controlled jointly by
gradual angular momentum loss and nuclear evolution of the
component stars. Such very slow evolution must occur at least
some of the time, since we see W UMa's in the very old galactic
cluster NGC 188, and, as Mochancki noted, these do not show
extreme mass ratios or any other evidence of having been in
contact much longer than their field analogues. Thus they have
probably evolved into a W UMa configuaration on a nuclear time
scale.

   C. Period Changes Attributable (perhaps) to Mass
Transfer
   The orbital periods of Algols (discussed by E. Olson,

Illinois) and of W Ursae Majoris stars (discussed by S.
Rucinski, Cambridge, and S. Mochnacki, David Dunlap
Observatory) both show large, erratic period changes, of both
signs, continuous and/or discrete, which are not properly
understood, though surely due one way or another to exhange of
angular momentum among the individual stars, the orbit, and the
outside world. Much the same can be said about the rotation
periods of neutron stars in X-ray binaries as described by N.
White (ESTEC) and S. Ilovaisky (Besançon). For these, however,
the sign of the change can be understood as reflecting the
geometry of the accretion process. For accretion from a disc,
the magnetic field of the neutron star can channel accretion
onto the poles, the star should spin down, and the X-rays
appear pulsed. For a higher accretion rate, the wind flow
dominates the field, accretion occurs spherically, the neutron
star spins up, and the pulses disappear. Roughly this
combination is observed for the source GX 301-2, according to
White.

## 7.  CATACLYSMIC VARIABLES

The origin of the term cataclysmic variable has been lost
in the mists of time (Payne-Gaposchkin 1977), though Kraft
(1962a,b) seems to have been the first to use it regularly in
print. He included the dwarf novae, classical novae, and
supernovae within the phrase. It now commonly means dwarf
novae, classical and recurrent novae, nova-like variables
(Robinson 1976), the new class of polars or AM Her stars
(magnetic CV's), and, sometimes, the symbiotic stars. My own
prejudice is that we ought to mean systems consisting of white
dwarf plus some other star transferring material to the WD in a
conspicuous fashion: thus we should include the "real"
nova-like variables (e.g. UX UMa), but not the oddities (P Cyg,
Eta Car, etc.) so labelled by McLaughlin (1960) and others, and
the "real" symbiotic stars like R Aqr and Z And, but not
anything best modeled by a single star or accreting on to a
main sequence companion (e.g. CI Cyg, according to S. Kenyon,
Illinois).

The "cataclysmic" part means that the system survives and
is likely to do whatever it has done again. The "variable"
part means that each member of the class will have a name of
the form XY Constellation (or V 1234 Constellation). This is
well and good, and helps us to find them when they are lost.
It also means that there is a great temptation to invent
subclasses called UR Con or MY Ast. This is a bad thing and
should be discouraged, but I do not know how.

R. Wade (Cambridge), athough his declared intention was to

undermine, or at least explore, the foundations of the subject, succeeded, I think, in convincing us instead that the basic model is in quite reasonable shape: there are two stars, they interact, and each of the pieces does more or less what it ought.

Among the subtypes, the novae are perhaps best understood. J. Truran's (Illinois) talk left the impression that nuclear (hydrogen) explosions on surfaces of CO and ONeMg white dwarfs will occur in the kinds of systems we see and can provide reasonable matches to the total energy output, composition of ejecta, etc., although some quick footwork is still needed to get the declining part of the light curve.

About dwarf novae I am not so certain. There was a good deal of cheerful-sounding coffee break chattter toward the end of the meeting rejoicing that a "consensus" had at least been achieved on the nature of the dwarf nova instability - located in the disc and attributable largely to ionization acting as an energy source and sink (but changes in viscosity and mass input rate might also contribute). I wonder, though, whether the seeming consensus may not have been attributable to the under-representation of dissident research groups. It is certainly true that some of the model light curves presented by D. Lin (Lick Observatory) and E. Meyer-Hofmeister (Munich) looked very much like real events. But other models, with only small changes in the input physics, did not resemble anything so far seen. One is left wondering how the real systems manage always to land in the right part of the parameter space.

The symbiotics are a mess (Kenyon).
Considerable progress has been made in identifying both progenitors and descendents of CV's. R. Taam (Illinois) presented results of two-dimensional calculations of binary systems enveloped by the expansion of a red giant. Mass outflow occurred at rates as large as $10^{26}$ g/sec, mostly in the equatorial plane. Complete expulsion of the envelope occurred most readily in wide systems (because the red giant was more extended and the envelope less tighly bound when the process started) and took only a few years. No wonder we don't catch many systems in the act!

H. Bond (Louisiana State) showed data for about a dozen systems in what looks like a post-common-envelope and pre-CV state. That is, they are close, but non-interacting, pairs with one main sequence component and one that is still ionizing a planetary nebulae or identifiable as a white dwarf or O subdwarf. "Close" means both that the stars must have spiraled together since the white dwarf was a red giant and that loss of angular momentum by gravitational radiation and/or winds will bring them into interaction within another Hubble time. These systems appear to be the short-period tail of a more extensive distribution. Bond also listed a somewhat larger number of WD + MS systems with periods greater than 1.5 days. These are too close to have evolved without extensive loss of angular momentum, but are unlikely to come into contact in the age of the universe.

R.E. Nather (Texas) discussed systems at the other end of the line: close pairs of interacting white dwarfs that are presumably post-CV systems. The first of these found, AM CVn, counted as a discovery, the second, GP Com, as a confirmation, and the recent identification of a third, PG 1346 + 082, reduces them to a well-known class of astronomical objects.

## 8.    X-RAY BINARIES

EXOSAT, at the time of the meeting, was just beginning to return its first results. J. Osbourne (Darmstadt) reported that Her X-1 had not turned back on at the predicted time in July. X-rays are still being produced, because the optical variations attributed to photon reprocessing continue, but they did not start reaching us on schedule. EXOSAT had also produced the first long, continuous stretch of data for Cyg X-3, covering seven of the 4.8 hour cycles. These reveal cycle-to-cycle changes in the light curve (especially on the rising branch), flickering, and variable absorption features. With axes unlabeled, the light curve might have been mistaken for that of a cataclysmic variable!

Other familiar sources also revealed new aspects. The outbursts of GX 301-2 and 0538-66 can now with some confidence be attributed to variable accretion produced by high orbital eccentricity, according to White and H. Henrichs (Amsterdam). S. Ilovaisky (Besançon) discussed the evidence for black holes as the accretors in LMC X-3, LMC X-1, and GX 339-4 (roughly in decreasing order of confidence). The first of these has a good orbit (Cowley et al. 1983) implying a mass of 7.14 $M_\odot$ for the compact object. The optical identification of the second is not quite certain, but the obvious star, R148, shows a mass function of 0.12 and a mass ratio of 0.5 (if the emission lines mark the centre of mass of the compact star), making the companion too massive for a neutron star for any orbital inclination greater than about 40 deg. The last has no radial velocity curve, but the optical and X-ray properties are consistent with those expected from a system containing a low mass normal star and a black hole.

Finally, T. Mazeh (Israel) presented an analaysis of the light curve of SS 433, based on the assumption that the primary minimum is due to the occultation of a precessing disc by a normal star which fills its Roche lobe. The result is a minimum mass for the compact component of 4.8 $M_\odot$, again too large for a neutron star. As most models for the object make use of a magnetized neutron star, the conclusion of this poster paper was "back to square one."

## 9. MISCELLANY

Chromospheres and coronae are very common among binary stars, occurring in W Ursae Majoris stars (S. Rucinski, Cambridge), in cataclysmic variables (K. Jensen, Goddard Space Flight Center), and in Algols (Plavec, Olson, Peters). At least in the case of the W UMa's, the phenomena are very much like those in single stars of the same masses and rotation rates, suggesting that the close companion is not really relevant. This is probably not the case for the Algols, where much of the heating comes from impact of a gas stream on the accreting star or for the CV's where the corona surrounds the accretion disc rather than either of the stars.

Instabilities of various kinds are also widespread. J. Pringle (Cambridge) discussed a newly-found one afflicting thick discs. It is global and occurs on the dynamical time scale. Neglected effects like viscosity cannot, therefore, be expected to supress it. G. Bath (Oxford) said very little about the instability expected when an extended star touches its Roche lobe, but he presented a graphic demonstration, using beakers, test tubes, rubberbands, and green water. This convinced most of the onlookers that mass transfer, once begun, could indeed be self-sustaining, though not all were persuaded that dwarf novae actually work this way.

Periodicities that appear during outbursts, die away with them, and are not understood very well are a feature of both CV's (where they are called superhumps, discussed by B. Warner, SAAO) and of some transient X-ray sources (where they are called pulsations, discussed by H. Henrichs, Amsterdam).

Non-spherical ejection of material from binaries occurs with both of the obvious assymetries. S. Tapia (Tucson) showed photographs of R Aqr that clearly reveal bipolar outflow. The streams or jets are curved into a gentle S-shape, and in the absence of scale for the picture, one might have taken it for a Very Large Array radio map of some active galactic nucleus. Equatorial ejection, on the other hand, is characteristic of mass outflow during the common envelope binary phase as modelled by R. Taam (Illinois).

Clever use of less than ideal observing facilities was revealed by the two-star photometer employed by H. Bond (Louisiana State) to identify eclipsing binaries among nuclei of planetary nebulae and by the polarimetry done from high northern latitudes by J. Brown (Glasgow). The amount and variation of polarization around a binary orbit period reveals the amount and location of ionized gas in streams, discs, and the rest; while the angle of polarization tells us the orientation of the system, otherwise quite unavailable for non-eclipsing binaries, and important for constraining the masses and evolutionary status of the components. The moral would seem to be that you can't keep a good astronomer down.

## 10. EXHORTATION

I should like to quote two of my predecessors in the role of concluding speaker. Jorge Sahade, at Elisnore, began his wrap-up with the statement (which does not appear in the published proceedings): "This conference has brought together people who have worked hard on these problems, people who have thought deeply about them, and a few who have done both." Let us all strive to belong to his third class!

And Bohdan Paczyński ended the previous Cambridge meeting by noting (and this is in the published proceedings) with pleasure that the future of binary star astronomy was in good hands, for the average age of the speakers at the meeting had been considerably less than his own. It is with distinctly modified rapture that I note the same effect here. I am considerably older than the median participant, or even speaker. Let us ensure that this will continue to be the case for future concluding speakers by urging our gifted younger colleagues to take an interest in the ancient but ever-fresh subject of interacting binaries!

## 11. APPRECIATION

As the last speaker, I have the happy duty of expressing our collective gratitude to those whose hard work and kindness has made the last two weeks possible. Thanks first to Prof. Donald Lynden-Bell for extending to us the hospitality of his beautiful Institute of Astronomy. Then to Peter Eggleton and Jim Pringle and their Scientific Organizing Committee for their brave efforts at keeping us fed and housed (or at least tented) for the fortnight. Special thanks to those who did most of the day-to-day work, Michael Ingham (Executive Secretary of the Institute) his wife Pauline, and the staff members Norah Tate, Jean Burris, Margaret Harding, and Alice Julier. And a very special thank you to Elena and Roberto

Terlevich for organizing the barbeque, and bon voyage to them as they leave Cambridge for Sussex, we trust not forever.


## 12. BENEDICTION

Binary star meetings seem to occur about every four years. Thus along about 1987 we can expect to find ourselves together again somewhere. And so until then, adios, do widzenia, au revoir, tot ziens, shalom, på återseende, arrivederci, aloha, and auf wiedersehen.

TABLE 1

IAU COLLOQUIUM NO. 6, 1969, Elsinore, Denmark

"Mass Loss and Evolution in Close Binaries"

Introductory Remarks:  Daniel M. Popper
Concluding Remarks:   Jorge Sahade

HOT TOPICS

> Activity in old novae
>
> Models for Beta Lyrae
>
> Observational evidence for mass loss from systems
> (including the first rocket UV data)
>
> Invention of RS CVn category (table of 22 objects
> presented by Popper; all still belong to the class)
>
> Data on masses of Algols
>
> Models for formation of Algols - Kippenhahn & Weigert
>                                 - Paczynski ·
>                                 - Plavec et al.

OFFICIAL PARTICIPANTS

| | |
|---|---|
| * A.H. Batten | D. Lauterborn |
| L. Binnendijk | * A.P. Linnell |
| K. Yu. Chen | * H. Mauder |
| * P. Conti | D.J.K. O'Connell |
| * M.G. Fracastoro | * B. Paczyński |
| * P. Harmaneç | * D.M. Popper |
| * T. Herczeg | S. Refsdal |
| * J. Horn | * J. Sahade |
| * M. Kitamura | * R.E. Wilson |
| * R.H. Koch | * F.B. Wood |
| * G. Larsson-Leander | |

* Still publishing in the field of binary star astronomy
  in 1982.

TABLE 2

IAU SYMPOSIUM No. 73, 1975, Cambridge, England

"Close Binary Systems"

Introductory Remarks:  Jorge Sahade
Concluding Remarks:  Mirek Plavec and Bogdan Paczyński

HOT TOPICS

Completion of conservative mass transfer program

Nature of dwarf nova instability

Solution (?!) of W UMa problem through thermal
relaxation oscillations in and out of contact

Common envelope binaries

Particle paths in accretion disks

X-ray binaries

Fast variations in cataclysmic binaries

Detection of hot spot in U Gem

TABLE 3

IAU SYMPOSIUM No. 88, 1979, Toronto, Canada

"Close Binary Systems:  Observations and Interpretation"

Introductory Remarks:  Mirek Plavec
Concluding Remarks:  Joe Smak, Roger Ulrich


HOT TOPICS

        Non-conservative evolution

        IUE Data

        SS 433

        Physics of mass transfer in Algols and contact systems,
        as probed by spectroscopy and polarimetry

        Double nuclei of planetary nebulae and other common
        envelope binaries

        Formation mechanisms

        RS CVn's and other manifestations of star spots and
        magnetic field effects

## TABLE 4

## DEFINING THE TERRITORY COVERED BY "INTERACTING BINARIES"

A. Yes (always in close binaries and the interaction matters)
   Cataclysmic variables
   Algols and Serpentids
   RS CVn stars
   W UrsaeMajoris stars
   X-ray sources (bright, point-like, galactic ones not
   associated with pulsars)
   Symbiotic stars (at least "real" ones)

B. Maybe (perhaps always in close binaries; perhaps it matters)
   Wolf-Rayet stars
   Blue Stragglers
   Barium Stars
   Gamma Ray Bursters

C. Inhibited in close binaries, through effects of mass
   transfer or loss
   OBC stars
   Extended supergiant envelopes needed for proper Type II
   Supernovae

D. Favoured in close binaries, through effects of mass
   transfer or loss
   Wolf-Rayet stars
   Of and OBN stars
   "millisecond" (short period, low magnetic field) pulsars

E. Favoured or prevented in close binaries through effect of
   synchronization of rotation BY Dra and UV Ceti flares,
   spots, etc.
   Chemical peculiarities in Am, Ap, Bp stars
   Be stars
   Delta Scuti variables
   Winds from OB stars, at least SMC X-1
   Coronae and chromospheres in general

REFERENCES

Abt, H.A. & Levy, S.G., 1976.  Astrophys.J.Sup. **30**, 273.
Batten, A.H., 1968.  Publ.Dom.Ap.Obs. 13, No.8.
Benson, R.S., 1970.  Thesis. U. Calif. Berkeley
Cowley, A.P., Crampton, D. & Hutchings, J.B., 1980.
    Astrophys.J. **241**, 269.
Deutsch, A.J., 1966.  Private communication.
Dominy. J.F. & Lambert, D.L., 1983.  Astrophys.J. **270**, 180.
Faulkner, J., Flannery, B.P. & Warner, B., 1972.  Astrophys.J.
    **175**, L79.
Finsen, W.C. & Worley, C.E., 1970.  Circ.Rep.Obs.Joh. **7**, 203.
Garmany, C.D., Conti, P.S. & Massey, P., 1980.  Astrophys.J.
    **242**, 1063.
Gingold, R.A. & Monaghan, J.T., 1983.  M.N.R.A.S., **204**, 715.
Jaschek, C., 1972.  P.A.S.P., **84**, 292.
Kraft, R.P., 1962a.  Astrophys.J., **135**, 408.
Kraft, R.P., 1962b.  Adv.Astron.Astrophys., **2**, 43.
Kuiper, G., 1935.  P.A.S.P., **47**, 15 & 121.
Lucy, L.B. & Ricco, E., 1979.  Astron.J., **84**, 401.
McClure, R.D., 1983.  Astrophys.J., **268**, 264.
McLaughlin, D.B., 1960.  In J.L. Greenstein (ed.) "Stellar
    Atmosheres, Vol. VI of Stars and Stellar Systems"
    (U. Chicago Press).
Michell, J., 1767.  Phil.Trans.Roy.Soc., p. 234 ff.
Opik, E., 1924.  Publ.Astron.Obs.Univ.Tartu XXV, No. 6.
Patterson, J., Nather, R.E. & Robinson, E.L., 1979.
    Atsrophys.J. **232**, 819.
Payne-Gaposchkin, C.H., 1977.  Astron.J., **82**, 665.
Peterson, R.C., Carney, B. & Latham, D., 1983.  Preprint,
    subm. to Astrophys.J.
Robinson, E.L., 1976.  Ann.Rev.Astron.Astrophys., **14**, 119.
Shu, F. & Lubow, S.H., 1981.  Ann.Rev.Astron.Astrophys. **19**, 299.
Stover, R.J., Robinson, E.L., Nather, R.E. & Montemayor, T.J.,
    1980 Astrophys.J., **240**, 597.
Stryker, L.L. & Hrivnak, B.J., 1983.  Preprint.
Trimble, V., 1974.  Astron.J., **79**, 967.
Walker, M.F., 1981.  Astrophys.J., **248**, 256.